Asymmetric Planetary Nebulae VII

Asymmetric Planetary Nebulae VII

Special Issue Editors

Quentin A. Parker
Noam Soker

MDPI • Basel • Beijing • Wuhan • Barcelona • Belgrade

MDPI

Special Issue Editors
Quentin A. Parker
University of Hong Kong
China

Noam Soker
Department of Physics, Technion—Israel Institute of Technology
Israel

Editorial Office
MDPI
St. Alban-Anlage 66
4052 Basel, Switzerland

This is a reprint of articles from the Special Issue published online in the open access journal *Galaxies* (ISSN 2075-4434) from 2018 to 2019 (available at: https://www.mdpi.com/journal/galaxies/special_issues/Planetary)

For citation purposes, cite each article independently as indicated on the article page online and as indicated below:

LastName, A.A.; LastName, B.B.; LastName, C.C. Article Title. *Journal Name* **Year**, *Article Number*, Page Range.

ISBN 978-3-03897-640-0 (Pbk)
ISBN 978-3-03897-641-7 (PDF)

Cover image courtesy of I. Bojicic, D.J. Frew and Q.A.Parker.

Contents

About the Special Issue Editors ix

Preface to "Asymmetric Planetary Nebulae VII" xi

Foteini Lykou, Albert A. Zijlstra, Jacques Kluska, Eric Lagadec, Peter G. Tuthill,
Adam Avison, Barnaby R. M. Norris and Quentin A. Parker
Infrared Observations of the Asymmetric Mass Loss of an AGB Star
Reprinted from: *Galaxies* **2018**, *6*, 108, doi:10.3390/10.3390/galaxies6040108 1

Raghvendra Sahai
Binary Interactions, High-Speed Outflows and Dusty Disks during the AGB-To-PN Transition
Reprinted from: *Galaxies* **2018**, *6*, 102, doi:10.3390/galaxies6040102 6

Robert G. Izzard and Adam S. Jermyn
Post-AGB Discs from Common-Envelope Evolution
Reprinted from: *Galaxies* **2018**, *6*, 97, doi:10.3390/galaxies6030097 16

Todd Hillwig
Surveying Planetary Nebulae Central Stars for Close Binaries: Constraining Evolution of
Central Stars Based on Binary Parameters
Reprinted from: *Galaxies* **2018**, *6*, 85, doi:10.3390/galaxies6030085 23

Griet C. Van de Steene, Bruce Hrivnak, Hans Van Winckel, J. Sperauskas, D. Bohlender
Spectroscopic and Photometric Variability of Three Oxygen Rich Post-AGB "Shell" Objects
Reprinted from: *Galaxies* **2018**, *6*, 131, doi:10.3390/galaxies6040131 31

Peter van Hoof, Stefan Kimeswenger, Griet Van de Steene, Adam Avison, Albert Zijlstra,
Lizette Gúzman-Ramirez, Falk Herwig and Marcin Hajduk
The Real-Time Evolution of V4334 Sgr
Reprinted from: *Galaxies* **2018**, *6*, 79, doi:10.3390/galaxies6030079 36

Nicole Reindl, Nicolle L. Finch, Veronika Schaffenroth, Martin A. Barstow,
Sarah L. Casewell, Stephan Geier, Marcelo M. Miller Bertolami and S. Taubenberger
Revealing the True Nature of Hen 2-428
Reprinted from: *Galaxies* **2018**, *6*, 88, doi:10.3390/galaxies6030088 42

Efrat Sabach
Jsolated Stars of Low Metallicity
Reprinted from: *Galaxies* **2018**, *6*, 89, doi:10.3390/galaxies6030089 49

Sagiv Shiber
The Morphology of the Outflow in the Grazing Envelope Evolution
Reprinted from: *Galaxies* **2018**, *6*, 96, doi:10.3390/galaxies6030096 56

Adam Frank, Zhuo Chen, Thomas Reichardt, Orsola De Marco, Eric Blackman and
Jason Nordhaus
Planetary Nebulae Shaped by Common Envelope Evolution
Reprinted from: *Galaxies* **2018**, *6*, 113, doi:10.3390/galaxies6040113 64

Natalia Ivanova and Jose L. A. Nandez
Planetary Nebulae Embryo after a Common Envelope Event
Reprinted from: *Galaxies* **2018**, *6*, 75, doi:10.3390/galaxies6030075 71

Amit Kashi
Simulations and Modeling of Intermediate Luminosity Optical Transients and Supernova Impostors
Reprinted from: *Galaxies* **2018**, *6*, 82, doi:10.3390/galaxies6030082 . . . . **79**

R. Wesson, D. Jones, J. García-Rojas, H.M.J. Boffin, R.L.M. Corradi
Close Binaries and the Abundance Discrepancy Problem in Planetary Nebulae
Reprinted from: *Galaxies* **2018**, *6*, 110, doi:10.3390/galaxies6040110 . . . . **90**

Daniela Barría and Stefan Kimeswenger
Analysis of Multiple Shell Planetary Nebulae Based on HST/WFPC2 Extended 2D Diagnostic Diagrams
Reprinted from: *Galaxies* **2018**, *6*, 84, doi:10.3390/galaxies6030084 . . . . **99**

Martin A Guerrero
X-ray Shaping of Planetary Nebulae
Reprinted from: *Galaxies* **2018**, *6*, 98, doi:10.3390/galaxies6030098 . . . . **105**

J.A. Toalá and S.J. Arthur
Simulations of the Formation and X-ray Emission from Hot Bubbles in Planetary Nebulae
Reprinted from: *Galaxies* **2018**, *6*, 80, doi:10.3390/galaxies6030080 . . . . **112**

Marcin Hajduk, Peter A. M. van Hoof, Karolina Sniadkowska, Andrzej Krankowski, Leszek Blaszkiewicz, Bartosz Dabrowski and Albert A. Zijlstra
Radio Continuum Spectra of Planetary Nebulae
Reprinted from: *Galaxies* **2019**, *7*, 6, doi:10.3390/galaxies7010006 . . . . **116**

Jan Cami, Jeronimo Bernard-Salas, Els Peeters, GregDoppmann and James de Buizer
The Formation of Fullerenes in Planetary Nebulae
Reprinted from: *Galaxies* **2018**, *6*, 101, doi:10.3390/galaxies6040101 . . . . **122**

SeyedAbdolreza Sadjadi and Quentin Andrew Parker
The Astrochemistry Implications of Quantum Chemical Normal Modes Vibrational Analysis
Reprinted from: *Galaxies* **2018**, *6*, 123, doi:10.3390/galaxies6040123 . . . . **128**

Katrina Exter
Spectroscopy of Planetary Nebulae with *Herschel*: A Beginners Guide
Reprinted from: *Galaxies* **2018**, *6*, 73, doi:10.3390/galaxies6030073 . . . . **136**

Carmen Sanchez Contreras, Javier Alcolea, Valentin Bujarrabal and Arancha Castro-Carrizo
ALMA's Acute View of pPNe: Through the Magnifying Glass... and What We Found There
Reprinted from: *Galaxies* **2018**, *6*, 94, doi:10.3390/galaxies6030094 . . . . **144**

Toshiya Ueta and Masaaki Otsuka
Understanding the Spatial Distributions ofthe Ionic/Atomic/Molecular/DustComponents in PNe
Reprinted from: *Galaxies* **2019**, *7*, 10, doi:10.3390/galaxies7010010 . . . . **153**

Sun Kwok
On the Origin of Morphological Structures of Planetary Nebulae
Reprinted from: *Galaxies* **2018**, *6*, 66, doi:10.3390/galaxies6030066 . . . . **159**

Eric Lagadec
AGBs, Post-AGBs and the Shaping of Planetary Nebulae
Reprinted from: *Galaxies* **2018**, *6*, 99, doi:10.3390/galaxies6030099 . . . . **164**

Xuan Fang, Martín A. Guerrero, Ana I. Gómezde Castro, Jesús A. Toalá, Bruce Balick and Angels Riera
UV Monochromatic Imaging of the Protoplanetary NebulaHen 3-1475 Using *HST* STIS
Reprinted from: *Galaxies* **2018**, *6*, 141, doi:10.3390/galaxies6040141 . . . . **172**

Lisa Löbling
Sliding along the Eddington Limit—Heavy-Weight Central Stars of Planetary Nebulae
Reprinted from: *Galaxies* **2018**, *6*, 65, doi:10.3390/galaxies6020065 . . . . **178**

Noam Soker
Planets, Planetary Nebulae, and Intermediate Luminosity Optical Transients (ILOTs)
Reprinted from: *Galaxies* **2018**, *6*, 58, doi:10.3390/galaxies6020058 . . . . **184**

About the Special Issue Editors

Quentin A. Parker, Professor, Associate Dean (Global) of the Faculty of Science at HKU and Director of the Laboratory for Space Research. An eminent researcher and observational astronomer whose interests include: phases of late stage stellar evolution especially planetary nebulae and supernova remnants; large scale wide-field surveys; astronomical instrumentation (fibre optics and narrow band filters), galactic archaeology, galaxy redshift surveys, classification systems, Chinese bronze and antiquities.

Noam Soker, Professor, Technion, Haifa, Israel. An eminent astrophysicist conducting theoretical research on a rich variety of objects: Heating hot gas in clusters of galaxies by jets launched from super-massive black holes; supernovae of exploding massive stars; the progenitors of supernovae ia (exploding white dwarfs); merger of white dwarfs; the shaping of clouds around dying stars including planetary nebulae; the influence of planets on stellar evolution; and violent mass transfer between stars.

Preface to "Asymmetric Planetary Nebulae VII"

It is with great pleasure that we present the Galaxies Special Issue publication of the refereed proceedings of the "Asymmetric Planetary Nebulae VII" international conference. This meeting took place in Hong Kong from 4–8 December 2017. This publication represents and encapsulates the best presentations both invited and contributed as the latest in the highly successful APN conference series. These well-regarded meetings cover the current up-to-date research, developments and insights into late stage stellar evolution. Particular emphasis is placed on the hypothesised physical shaping mechanisms that give rise to the many beautiful and mysterious forms of the resultant planetary nebulae as well as their connections to related objects.

Quentin A. Parker, Noam Soker

Special Issue Editors

galaxies

MDPI

Article

Infrared Observations of the Asymmetric Mass Loss of an AGB Star

Foteini Lykou [1,2,*], Albert A. Zijlstra [3], Jacques Kluska [4,5], Eric Lagadec [6], Peter G. Tuthill [7], Adam Avison [3], Barnaby R. M. Norris [7] and Quentin A. Parker [1,2]

[1] Department of Physics, The University of Hong Kong, Pokfulam Road, Hong Kong, China; quentinp@hku.hk

[2] Laboratory for Space Research, The University of Hong Kong, 100 Cyberport Road, Cyberport, Hong Kong, China

[3] Jodrell Bank Centre for Astrophysics, The University of Manchester, Oxford Road, Manchester M13 9PL, UK; albert.zijlstra@manchester.ac.uk (A.A.Z.); adam.avison@manchester.ac.uk (A.A.)

[4] Institute for Astronomy, KU Leuven, Celestijnenlaan 200D B2401, 3001 Leuven, Belgium; jacques.kluska@kuleuven.be

[5] School of Physics, University of Exeter, Stocker Road, Exeter EX4 4QL, UK

[6] Observatoire de la Côte d'Azur, Laboratoire Lagrange, Université Côte d'Azur, 06304 Nice, France; eric.lagadec@oca.eu

[7] Sydney Institute of Astronomy, School of Physics, The University of Sydney, Camperdown, NSW 2006, Australia; peter.tuthill@sydney.edu.au (P.G.T.); barnaby.norris@sydney.edu.au (B.R.M.N.)

* Correspondence: lykoufc@hku.hk

Received: 31 July 2018; Accepted: 9 October 2018; Published: 12 October 2018

Abstract: We report on the observations of the circumstellar envelope of the AGB star II Lup in the near- and mid-infrared with the use of direct imaging and interferometric techniques. Our findings indicate that the circumstellar envelope is not spherically symmetric and that the majority of the emission originates within 0.5 arcsec from the star.

Keywords: infrared interferometry; AGB stars; stellar evolution; observations; aperture masking

1. Introduction

The study of the asymmetries found in the majority of planetary nebulae have been the core subject of this conference series. However, as it has been stressed in the last two APNmeetings, these asymmetries ought to be generated by mechanisms that act as early as the AGB phase. It is currently believed that these mechanisms are the result of an interplay between two shaping mechanisms, binarity and magnetic fields, and both could be investigated in AGB stars.

Different spatial scales at different wavelength ranges help us to dissect different parts of the circumstellar envelope (CSE) of evolved stars. The outer and colder layers of a CSE are observed in the sub-mm and far-infrared wavelengths (e.g., ~ 1″ with ALMA and Herschel telescopes), while the warmer layers of the CSE can be observed in the mid-infrared (e.g., ~ 0.5″ with VISIR/Very Large Telescope (VLT)). The layers of the CSE closer to the central star, including the stellar photosphere and its hot, dusty atmosphere, can be explored in the near-infrared (e.g., ≤0.2″ with NACO/VLT).

Until now, there have been only a few imaging surveys that have explored this: two in the sub-mm wavelengths [1,2], one in the far-infrared [3] and one in the mid-infrared [4]. The initial target lists are very similar in all four surveys, and the majority of the objects were found to depart from spherical symmetry at large spatial scales (≥1″). However, many of the targets, especially the inner layers of the CSEs of the AGB stars, were unresolved at smaller spatial scales (≤0.4″). We therefore initiated a survey of 22 objects in the period 2009–2018 to look for any asymmetries in evolved stars and explored the possibility of binary interactions with the use of interferometry, to access the sub-arcsecond angular scales needed, and direct imaging in the infrared. The targets were selected from the initial list of [4]. Some of the most striking results of this survey have been presented in [5–7], and the analysis of the final sample is on-going.

One of the AGB stars in our sample is II Lup. It is a carbon-rich AGB star, and the typical masses for such stars range from 1–4 $M_\odot$. The star is placed near the tip of the AGB (m_{bol} = 3.73, [8]) in the evolutionary tracks of [9], and therefore, it is not yet hot enough to ionize its CSE. II Lup shows a peculiar variability in the near-infrared (*J–L* bands), where its light curve can be fitted by two periods: a short-term one at 575 days and a long-term one at ~19 years [10]. The latter was characterized as an obscuration event and is thought to be related to asymmetric[1] mass loss [10]. We reported the first-ever detection of asymmetries in II Lup's CSE in the near-infrared and sub-mm wavelengths in Lykou et al. [11]. Here, we present complementary (near- and mid-infrared) images to that work.

2. Results

The observations of II Lup in the near- and mid-infrared were carried with 8 m-class telescopes by Lykou et al. [11]. A brief description of the observing modes is given below, while the results can be easily compared to [11].

II Lup was observed with the VISIR mid-infrared instrument in March 2016 (JD = 2,457,468). VISIR is a spectrometer and imager on the Very Large Telescope (VLT) [12]. The observations were carried in burst mode, which can provide diffraction-limited images (e.g., θ_{res} = 0.25" at 8 µm). Observations of the science target and a calibrator were obtained with the PAH_1 and PAH_2 filters[2] (hereby, 8 µm and 11 µm for simplicity). The data were reduced and analysed using the method of [4]. The science data suffered from saturation from the central star; therefore, we present here a tentative analysis of this dataset.

Each science image was deconvolved following the Lucy–Richardson method, using the cropped images of the corresponding PSF calibrator (radius ~1 arcsec) and thus removing the noise of the otherwise empty field-of-view (10" × 10"). This significantly minimizes the computation time for the deconvolution. The process was stopped after 30 iterations. Each image was then smoothed (convolved) with a two-pixel Gaussian kernel. The deconvolved images (2.9" × 2.9") are shown in Figure 1.

As expected, the central star is unresolved, and any deviations from symmetry within the resolution elements in Figure 1 should be ignored. The morphology of the envelope at 8 µm is relatively similar to that at *L* and *M* (cf. Figure 7 in Lykou et al. [11]) with respect to its north-south orientation. At 11 µm, the CSE appears to be more round with a small displacement to the north; however, the brightness distribution is not entirely uniform. Although the data suffered from saturation, we can tentatively deduce that the CSE extends up to a radius of 0.47" and 0.6" at 8 µm and 11 µm, respectively (black contours in Figure 1). Therefore, the CSE appears to be a relatively compact object in the mid-infrared with respect to the size of the envelope in the far-infrared (e.g., 70 µm PACS/Herschel map where CSE size ~40"; see also [13]). However, the size of the photosphere must be less than 0.25".

II Lup was also observed in June 2010 (JD = 2,455,377) in the near-infrared (*K*, *L* and *M*) with the Sparse Aperture Masking mode on NACO/VLT [14–17]. This technique uses a nine-hole mask that converts the single-dish 8 m-class telescope into an interferometer with 36 baselines[3] to produce diffraction-limited images (e.g., θ_{res} = 72 mas in *M*). The data reduction used a custom-made pipeline, and the analysis and image reconstruction processes were performed as described in [6,11,17].

---

[1] The term "asymmetric" will hereby refer to any non-spherical symmetry.
[2] PAH_1: λ = 8.54 µm, $\Delta\lambda$ = 0.42 µm; PAH_2 : λ = 11.25 µm, $\Delta\lambda$ = 0.59 µm).
[3] The baseline range was 1.3–6.9 m for various azimuths.

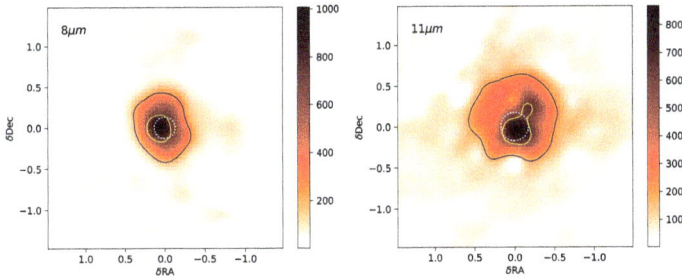

Figure 1. VISIR/Very Large Telescope (VLT) deconvolved images of II Lup at 8 μm (**left**) and 11 μm (**right**) in squared-root intensity scale. The resolution element is indicated by the dotted, white circles in the core of each image, while the black and yellow contours indicate the 5% and 50% levels of the peak intensity, respectively. North is up and east is to the left. The colour bars indicate a relative intensity scale.

Figure 2 shows the image reconstruction for the *M* data with the MiRA algorithm [18,19]. It is evident that the circumstellar envelope departs from spherical symmetry and extends up to 120 milliarcseconds (mas) north with another protrusion extending approximately 80 mas south-west. The photosphere of the AGB star is unresolved, and therefore, its size must be smaller than 32 mas, as shown by the *K*-band images of Lykou et al. [11]. The entire structure fits well inside the resolution element (white circle) of the VISIR 8 μm image (left panel, Figure 1); hence, we were able to resolve the CSE of this AGB star at the smallest spatial scale possible with this technique in the *M* band.

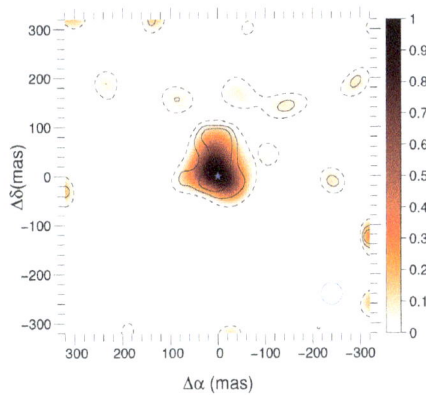

Figure 2. II Lup's *M* band image reconstructed with MiRA. The flux scale has been normalized to unity, and the contours represent the 5, 3 (solid) and 1 (dash) significance levels. The resolution element is indicated by a blue circle and the position of the central star by a blue asterisk.

The upper limits of the radii of CSE's detected layers are tabulated in Table 1. The angular sizes were converted to physical values for an adopted distance of 590 pc [13]. Using these sizes and assuming that the expansion velocity of II Lup ($v_{exp} = 23$ km s^{-1}; [20]) remains the same[4] throughout the CSE, the approximate time scales for the expansion of the layers ought to be shorter than 14.6, 57,

---

[4]   There are no high-resolution, infrared, spectroscopic measurements for this star in the current literature.

73 and 2430 years, respectively. This would suggest that the NACO observations detected a relatively recent mass-loss event.

Table 1. Upper-limits for the circumstellar envelope sizes for an adopted distance of 590 pc.

Band	Radius		Reference
	(arcsec)	(au)	
M	0.12	71	this work
8 μm	0.47	277	this work
11 μm	0.6	354	this work
70 μm	20	11,800	this work and [13]

3. Discussion

The near- and mid-infrared observations of II Lup reveal that the hot- and warm-dust layers of its circumstellar envelope are very compact (size $\leq 1.2''$) with respect to the larger and spherically-symmetric, cold-dust envelope (size $\sim 40''$; [3,13]). Although the mid-infrared images are not conclusive on the asymmetry of the circumstellar envelope, which was mainly due to the quality of the data and the imaging method used, the near-infrared data indicate an oblate envelope (Figure 2).

These images therefore indicate that the morphology of the dusty, circumstellar envelope of II Lup is not spherically symmetric, which confirms the hypothesis of [10] for this star, as well as the findings of [11]. The mechanism of these asymmetries could be the influence of a binary companion orbiting the AGB star, but no such star was found in spatial scales $\geq 0.2''$ (or else, orbital separations ≥ 118 AU at the adopted distance of 590 pc). However, if such a companion exists closer to the AGB star, this hypothesis can only be tested with new interferometric observations, preferably made with larger scale interferometers such as the VLTI. The analysis of the current images suggests that we have detected layers of the CSE that were recently formed (age ≤ 80 years). We expect that any planetary nebula created from this star in the future will be shaped by the same mechanism that created the asymmetries in the current circumstellar envelope.

Author Contributions: F.L., A.A.Z. and E.L. conceived of and designed the project. F.L., P.G.T. and B.R.M.N. performed the observations. F.L., E.L. and A.A. analysed the data. P.G.T., E.L. and J.K. contributed analysis tools. Q.A.P. contributed in the discussion. F.L. wrote the paper.

Funding: F.L. acknowledges support from the University of Hong Kong Postdoctoral Fellowships scheme and the Austrian Science Fund (AP23006 , PI: Josef Hron). J.K. acknowledges support from the Philip Leverhulme Prize (PLP-2013-110, PI: Stefan Kraus) and from the research council of the KU Leuven under Grant Number C14/17/082.

Conflicts of Interest: The authors declare no conflict of interest.

References

1. Sánchez Contreras, C.; Sahai, R. OPACOS: OVRO Post-AGB CO (1-0) Emission Survey. I. Data and Derived Nebular Parameters. *Astrophys. J. Suppl. Ser.* **2012**, *203*, 16. [CrossRef]
2. Castro-Carrizo, A.; Quintana-Lacaci, G.; Neri, R.; Bujarrabal, V.; Schöier, F.L.; Winters, J.M.; Olofsson, H.; Lindqvist, M.; Alcolea, J.; Lucas, R.; et al. Mapping the ^{12}CO J = 1 − 0 and J = 2 − 1 emission in AGB and early post-AGB circumstellar envelopes. I. The COSAS program, first sample. *Astron. Astrophys.* **2010**, *523*, A59. [CrossRef]
3. Cox, N.L.J.; Kerschbaum, F.; van Marle, A.J.; Decin, L.; Ladjal, D.; Mayer, A.; Groenewegen, M.A.T.; van Eck, S.; Royer, P.; Ottensamer, R.; et al. A far-infrared survey of bow shocks and detached shells around AGB stars and red supergiants. *Astron. Astrophys.* **2012**, *537*, A35. [CrossRef]
4. Lagadec, E.; Verhoelst, T.; Mékarnia, D.; Suárez, O.; Zijlstra, A.A.; Bendjoya, P.; Szczerba, R.; Chesneau, O.; Van Winckel, H.; Barlow, M.J.; et al. A mid-infrared imaging catalogue of post-asymptotic giant branch stars. *Mon. Not. R. Astron. Soc.* **2011**, *417*, 32–92. [CrossRef]

5. Lykou, F. Dusty Discs around Evolved Stars. Ph.D. Thesis, The University of Manchester, Manchester, UK, 2013.

6. Lykou, F.; Klotz, D.; Paladini, C.; Hron, J.; Zijlstra, A.A.; Kluska, J.; Norris, B.R.M.; Tuthill, P.G.; Ramstedt, S.; Lagadec, E.; et al. Dissecting the AGB star L2 Puppis: A torus in the making. *Astron. Astrophys.* **2015**, *576*, A46. [CrossRef]

7. Lykou, F.; Hron, J.; Zijlstra, A.A.; Tuthill, P.G.; Norris, B.R.M.; Kluska, J.; Paladini, C.; Lagadec, E.; Wittkowski, M.; Ramstedt, S.; et al. Unraveling disks in AGB stars. *EAS Publ. Ser.* **2015**, *71–72*, 217–222. [CrossRef]

8. Groenewegen, M.A.T.; Sevenster, M.; Spoon, H.W.W.; Perez, I. Millimetre observations of infrared carbon stars. II. Mass loss rates and expansion velocities. *Astron. Astrophys.* **2002**, *390*, 511. [CrossRef]

9. Miller Bertolami, M.M. New models for the evolution of post-asymptotic giant branch stars and central stars of planetary nebulae. *Astron. Astrophys.* **2016**, *588*, A25. [CrossRef]

10. Feast, M.W.; Whitelock, P.A.; Marang F. The case for asymmetric dust around a C-rich asymptotic giant branch star. *Mon. Not. R. Astron. Soc.* **2003**, *346*, 878–884. [CrossRef]

11. Lykou, F.; Zijlstra, A.A.; Kluska, J.; Lagadec, E.; Tuthill, P.G.; Avison, A.; Norris, B.R.; Parker, Q.A. The curious case of II Lup: A complex morphology revealed with SAM/NACO and ALMA. *Mon. Not. R. Astron. Soc.* **2018**, *480*, 1009. [CrossRef]

12. Lagage, P.O.; Pel, J.W.; Authier, M.; Belorgey, J.; Claret, A.; Doucet, C.; Dubreuil, D.; Durand, G.; Elswijk, E.; Girardot, P.; et al. Successful Commissioning of VISIR: The Mid-Infrared VLT Instrument. *Messenger* **2004**, *117*, 12–16.

13. Groenewegen, M.A.; Waelkens, C.; Barlow, M.J.; Kerschbaum, F.; Garcia-Lario, P.; Cernicharo, J.; Blommaert, J.A.D.L.; Bouwman, J.; Cohen, M.; Cox, N.; et al. MESS (Mass-loss of Evolved StarS), a Herschel key program. *Astron. Astrophys.* **2011**, *526*, A162. [CrossRef]

14. Rousset, G.; Lacombe, F.; Puget, P.; Hubin, N.N.; Gendron, E.; Fusco, T.; Arsenault, R.; Charton, J.; Feautrier, P.; Gigan, P.; et al. NAOS, the first AO system of the VLT: On-sky performance. *Proc. SPIE* **2003**, *4839*, 140–149. [CrossRef]

15. Lenzen, R.; Hartung, M.; Brandner, W.; Finger, G.; Hubin, N.N.; Lacombe, F.; Lagrange, A.M.; Lehnert, M.D.; Moorwood, A.F.; Mouillet, D. NAOS-CONICA first on sky results in a variety of observing modes. *Proc. SPIE* **2003**, *4841*, 944–952. [CrossRef]

16. Tuthill, P.G.; Monnier, J.D.; Danchi, W.C.; Wishnow, E.H.; Haniff, C.A. Michelson Interferometry with the Keck I Telescope. *Publ. Astron. Soc. Pac.* **2000**, *112*, 555–565. [CrossRef]

17. Tuthill, P.; Lacour, S.; Amico, P.; Ireland, M.; Norris, B.; Stewart, P.; Evans, T.; Kraus, A.; Lidman, C.; Pompei, E.; et al. Sparse aperture masking (SAM) at NAOS/CONICA on the VLT. *Proc. SPIE* **2010**, *7735*, 77351O. [CrossRef]

18. Thiébaut, E. MIRA: An effective imaging algorithm for optical interferometry. *Proc. SPIE* **2008**, *7013*, 70131. [CrossRef]

19. Kluska, J.; Malbet, F.; Berger, J.P.; Baron, F.; Lazareff, B.; Le Bouquin, J.B.; Monnier, J.D.; Soulez, F.; Thiébaut, E. SPARCO: A semi-parametric approach for image reconstruction of chromatic objects. Application to young stellar objects. *Astron. Astrophys.* **2014**, *564*, A80. [CrossRef]

20. De Beck, E.; Decin, L.; de Koter, A.; Justtanont, K.; Verhoelst, T.; Kemper, F.; Menten, K.M. Probing the mass-loss history of AGB and red supergiant stars from CO rotational line profiles. II. CO line survey of evolved stars: Derivation of mass-loss rate formulae. *Astron. Astrophys.* **2010**, *523*, A18. [CrossRef]

galaxies

MDPI

Article

Binary Interactions, High-Speed Outflows and Dusty Disks during the AGB-To-PN Transition

Raghvendra Sahai

Jet Propulsion Laboratory, California Institute of Technology, 4800 Oak Grove Drive, Pasadena, CA 91109, USA; sahai@jpl.nasa.gov; Tel.: +1-818-354-0452

Received: 9 July 2018; Accepted: 13 September 2018; Published: 25 September 2018

Abstract: It is widely believed that the dramatic transformation of the spherical outflows of AGB stars into the extreme aspherical geometries seen during the planetary nebula (PN) phase is linked to binarity and driven by the associated production of fast jets and central disks/torii. The key to understanding the engines that produce these jets and the jet-shaping mechanisms lies in the study of objects in transition between the AGB and PN phases. I discuss the results of our recent studies with high-angular-resolution (with ALMA and HST) and at high-energies (with GALEX, XMM-Newton and Chandra) of several such objects, which reveal new details of close binary interactions and high-speed outflows. These include two PPNe (the Boomerang Nebula and IRAS 16342-3814), and the late carbon star, V Hya. The Boomerang Nebula is notable for a massive, high-speed outflow that has cooled below the microwave background temperature, making it the coldest object in the Universe. IRAS 16342-3814 is the prime example of the class of water-fountain pre-planetary nebulae or PPNe (very young PPNe with high-velocity H_2O masers) and shows the signature of a precessing jet. V Hya ejects high-speed bullets every 8.5 years associated with the periastron passage of a companion in an eccentric orbit. I discuss our work on AGB stars with strongly-variable high-energy (FUV, X-ray) emission, suggesting that these objects are in the early stages of binary interactions that result in the formation of accretion disks and jets.

Keywords: planetary nebulae; AGB and post-AGB stars; binarity; accretion disks; jets; mass-loss; circumstellar matter; (sub)millimeter interferometry; ultraviolet radiation, X-rays

1. Introduction

The fundamental question that has motivated the Planetary Nebulae conference series is: How do the slowly expanding (5–15 km s^{-1}), largely spherical, circumstellar envelopes (CSEs) of AGB stars transform themselves into highly aspherical Planetary Nebulae (PNe), with collimated lobes and fast outflows ($\gtrsim few \times 100$ km s^{-1}) along one or more axes? The importance of collimated jets in forming ansae in PNe was recognized in [1]. Based on the wide variety of multipolar and point-symmetric morphologies seen in unbiased surveys of young PNe with HST, reference [2] proposed that collimated fast winds or jets (hereafter, CFWs), operating during the pre-planetary nebula (PPN) or very late-AGB phase, are the primary agent for producing asymmetric shapes in PNe. The CFWs are likely to be episodic, and either change their directionality (i.e., axis wobbles or precesses) or have multiple components operating in different directions (quasi)simultaneously. These CFWs sculpt the AGB CSE from the inside-out, producing elongated bubbles or lobes within the CSEs. Later, additional action of the fast radiative wind from the central star may further modify these lobes, and ionization may lead to loss of some structure [3]. If a dense equatorial torus is present, it may add additional confinement for the CFWs, as well as for the spherical radiative wind from the hot central star at a later stage of evolution.

Binary star interactions are believed to underlie the formation of the overwhelming majority of PNe, which represent the bright end-stage of most stars in the Universe. Close binary interactions also dominate a substantial fraction of stellar phenomenology, e.g., cataclysmic variables, type Ia supernovae progenitors, and low and high-mass X-ray binaries. Understanding the formation of aspherical PNe can help in addressing one of the biggest challenges for 21st century stellar astronomy—a comprehensive understanding of the impact of binary interactions on stellar evolution.

In this paper, I describe our observational techniques for searching for binarity (and signatures of associated active accretion) in AGB stars, as well as observational results from our recent studies of three key transition objects that have likely undergone recent (or are currently undergoing) close binary interactions. These objects show large and sudden mass-ejections prior to the formation of a planetary nebula, as well as disks, torii and (episodic) high-speed, collimated jets. Thus, they are Rosetta Stones for understanding aspherical PN formation. The paper is based on an invited talk that I gave at the Asymmetrical Planetary Nebulae (APN) VII meeting (Hong Kong, December 2017).

2. Binarity in AGB Stars

Observational evidence of binarity in AGB stars is difficult to come by because AGB stars are very luminous and variable, thus standard techniques for binary detection such as radial-velocity and photometric variations due to a companion star are not applicable. However, one can exploit the favorable secondary-to-primary photospheric flux contrast ratios reached in the UV for companions of spectral type hotter than about G0 (T_{eff} = 6000 K) and luminosity, L $\gtrsim$ 1 $L_\odot$. Reference [4] (hereafter, Setal08) first used this technique, employing GALEX [5] to find emission from 9/21 objects in the FUV (1344–1786 Å) and NUV (1771–2831 Å) bands. Since these objects (hereafter, fuvAGB stars) also showed significant UV variability, setal08uv concluded that the UV source was unlikely to be solely a companion's photosphere, and was dominated by emission from variable accretion activity.

Accretion activity is likely to produce X-ray emission as well, as observed in young stellar objects [6]. A survey of archival XMM and ROSAT data found two AGB stars and the symbiotic star, Mira, with X-ray emission [7]. A pilot survey for X-ray emission from a fuvAGB stars using XMM-Newton and Chandra by [8] (hereafter, Setal15) detected X-ray emission in 3/6 fuvAGB stars observed. The X-ray fluxes were found to vary in a stochastic or quasi-periodic manner on roughly hour-long times-scales. These data, together with previous and more recent studies (Figure 1) and [9], show that X-ray emission is found only in fuvAGB stars, with an FUV/NUV ratio $\gtrsim$ 0.17 (e.g., Table 1). There are two exceptions: V Hya and V Eri. The non-detection of V Hya, which has a high FUV/NUV ratio, is likely related to the fact that the companion is in an eccentric orbit, and the accretion rate, which is highly variable, was probably low when the X-ray observations were done (2.5 years after periastron passage) [10]; (hereafter Setal16). For V Eri, we can only speculate that the non-detection may be because V Eri was in a relatively low-accretion phase when the X-ray data were taken.

From modeling the X-ray spectra, Setal15 found that the observed X-ray luminosities are (0.002–0.11) $L_\odot$, and the X-ray emitting plasma temperatures are $\sim$(35–160)$\times 10^6$ K. These high X-ray temperatures argue against the emission arising directly in an accretion shock, unless it occurs on a white dwarf (WD) companion. However, none of the detected objects is a known WD-symbiotic star, suggesting that, if WD companions are present, they are relatively cool (<20,000 K). The high X-ray luminosities argue against emission originating in the coronae of MS companions. A likely origin of the X-ray emission is that it arises in hot plasma confined by magnetic fields associated with a disk around a binary companion. The plasma may be generated by an accretion shock on the disk that gives rise to the FUV emission in these objects. Based on the time-scale ($\sim$1.3 h) of the quasi-periodic variations in Y Gem—similar to the period of material orbiting close to the inner radius of an accretion disk around a sub-solar mass companion, i.e., with $M_c \lesssim 0.35\,M_\odot$ (implying a semi-major axis a $\lesssim$3 $\times 10^{10}$ cm)—Setal15 argued that the most likely model for the X-ray emission from fuvAGB stars is that it arises at or near the magnetospheric radius in a truncated disk, or the boundary layer between the disk and star.

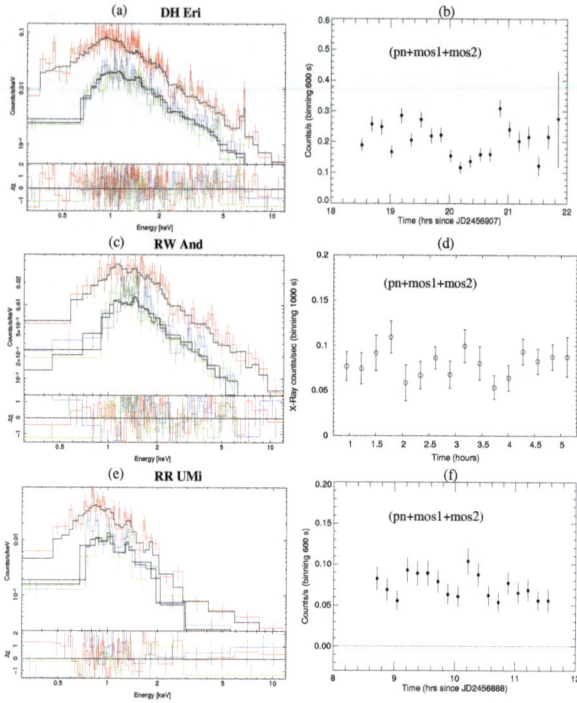

Figure 1. X-ray spectra (colored curves: XMM-Newton/EPIC pn, red; MOS1, green; and MOS2, blue) and model fits (black curves) of the fuvAGB stars DH Eri, RW And and RR UMi (panels **a**, **c**, **e**), and X-ray light curves (black symbols) (panels **b**, **d**, **f**). As for most fuvAGB stars, DH Eri and RW And are each fitted by a single hard component (Tx $\sim 5 \times 10^7$ K) (panels a & C, respectively), but in RR UMi (panel **e**) the dominant emission component has Tx that is significantly lower (Tx(1) $\sim 6.6 \times 10^6$ K); a weaker harder component (Tx(2) $\sim 10^8$ K) is also present (*adapted from Sahai, Sánchez Contreras and Sanz-Forcada 2018, in prep.*).

We note that, in RR UMi, which has a relatively low FUV/NUV ratio amongst the X-ray emitting fuvAGB stars, the X-ray spectrum has a hard and a soft component, with the softer one being more dominant (Figure 1e). We infer that, whereas the hard component is related to weak accretion activity, the soft one likely represents the expected coronal emission from the main-sequence (MS) companion. This hypothesis is supported by the fact that, in contrast to the other X-ray detected objects, RR UMi also has very low FUV variability (as expected when accretion activity is relatively weak).

We searched the Mikulski Archive for Space Telescopes (MAST) for GALEX data on fuvAGB stars, using a comprehensive input catalog of $\sim$4000 AGB stars (includes O-rich, C-rich and S-stars) collected from various published sources and searches for counterparts in the GALEX archive within a user-specified search radius. We have generated a photometric database that provides: (i) relevant parameters such as FUV and NUV fluxes, errors, and exposure times, for detected objects for each "visit", (ii) the background "sky-noise" for stars that were observed but not detected; and (iii) variability properties for each of the detected objects. We found about $\sim$100 fuvAGB stars in the FUV band at a $\gtrsim 5\,\sigma$ level. Even for the hottest sources in our catalog (spec. type M4), the detected FUV fluxes ($\gtrsim$20 µJy) are significantly in excess above photospheric emission. We found that many fuvAGB stars show strong UV variability as well (e.g., Y Gem: [11]), with time intervals where the FUV and NUV variations are correlated, suggesting changes in the emission measure and/or the obscuring column, and where they are anti-correlated, suggesting changes in the temperature of the emitting material.

We tabulate below, for AGB stars that have been observed at two or more epochs and have X-ray observations with adequate sensitivity (i.e., integration times greater than $\sim$5 ksec with XMM-Newton or Chandra), FUV variability measures (standard deviation/median or *var*1 and (maximum-minimum)/median or *var*2 and the FUV/NUV ratio, versus the presence/absence of X-ray emission. We found that the FUV variability measures also show a correlation (Table 1) with the absence/presence of X-ray emission, but somewhat weaker than that shown by the FUV/NUV ratio: excluding the special case of V Hya, 5/7 stars with X-ray emission have high FUV variability (*var*1 > 0.23, *var*2 > 0.58), whereas all four stars without X-ray emission have low FUV variability (*var*1 < 0.19, *var*2 < 0.37). We hypothesize that the variable X-ray and FUV emission is directly or indirectly related to variable accretion onto the accretion disk and/or the companion; however, our sample is small, and X-ray data on a larger number of objects with FUV variability is needed to test our hypothesis more robustly.

Table 1. UV Variability and X-Ray Emission of fuvAGB Stars.

Name	FUV *var*1 [1]	FUV *var*2 [1]	FUV/NUV	X-Ray?
Y Gem	1.9	5.13	1.3	Y
EY Hya	0.39	0.93	0.83	Y
CI Hyi	0.67	1.33	0.54	Y
RW And	0.23	0.58	2.3	Y
R UMa	0.15	0.36	1.0	Y
RR UMi	0.033	0.066	0.3	Y
UY Leo	0.30	0.73	0.24	Y?
V Hya [2]	0.27	0.54	1.7	N
V Eri	0.19	0.37	0.96	N
del01 Aps	0.012	0.024	0.11	N
NU Pav	0.13	0.26	0.11	N
EU Del	0.10	0.19	0.10	N

[1] *var*1 = σ/median, *var*2 = (max-min)/median. [2] eccentric orbit, X-ray data taken far from periastron passage.

A detailed UV spectroscopic study of Y Gem with HST/STIS by [12] (hereafter Setal18) reveals strong flickering in its UV continuum on time-scales of $\lesssim$20 s, a characteristic signature of an active accretion disk. The *TESS* mission, which can provide 1σ noise sensitivity in 2 min of 690 ppm for fuvAGB stars as faint as $\sim$8.9 mag in the *TESS* photometric band, can thus be used to detect optical flickering with amplitudes that are a factor $\sim$200 lower than that in the FUV. We expect that such data will reveal some combination of the $\sim$hour-long quasi-periodic variations as well as stochastic (flickering) variations, based on current UV and X-ray data for a few fuvAGB stars. In addition, new kinds of variability may become apparent, since 2-min cadence data of the very high photometric sensitivity that *TESS* can achieve has never been reported for AGB stars.

Reference [13] compared the UV properties of their sample with those of 12 AGB stars with known binary companions [14], and supported Setal08's inference that the detection of FUV emission in AGB stars is a strong indicator of binarity.

In contrast, reference [15] claimed that GALEX-detected UV emission is an inherent characteristic of AGB stars (likely combination of photoshperic and chromospheric emission). Their conclusion is based on a sample of 468 AGB stars, in which they detected 179 in both bands and 38 in the FUV band. They stated that there is evidence that the NUV emission appears to vary in phase with visible light curves in a few AGB stars, and then concluded that the UV emission is an inherent characteristic of AGB stars, and not likely to be indicative of a binary. However, we note that an inspection of their Figure 1 (which shows the American Association of Variable Star Observers (AAVSO) visual light curves and the X-ray data for nine stars) reveals only one star (R Cet) for which the correlation between optical and X-ray variability is clearly seen. Furthermore, the GALEX FUV/NUV ratio for chromospheric emission is likely to be lower than 0.17, as, e.g., revealed by spectroscopic observations of chromospheric emission in two objects: ∼0.05–0.1 in TW Hor, ∼0.1 in the red supergiant α Ori.

Another mechanism for producing high-energy radiation from single AGB stars may be long-duration flares due to magnetic reconnection events, suggested to explain an X-ray flare from the only AGB star that has shown one—the primary of the symbiotic star, Mira [16]. A high-sensitivity search in two stars with inferred strong magnetic fields did not find X-ray emission [17]. Thus, the (admittedly scant) observations so far do not show a relationship between the presence of magnetic fields and high-energy emission from single AGB stars.

3. The Effects of Binarity

3.1. Large Episodic Mass-Ejections that End the AGB/RGB Phase

The well-studied carbon star, V Hya, shows evidence for high-speed, collimated outflows and dense equatorial structures (discussed below), and is a key object in understanding the early transition of AGB stars into aspherical PNe. Setal16 found that this star is ejecting massive high-speed compact clumps (hereafter, bullets) periodically, leading to a model in which the bullet ejection is associated with the periastron passage of a close binary companion in an eccentric orbit around V Hya with an orbital period of ∼8.5 years. The detailed physical properties of this ejection suggests that the companion approaches the primary very close to the latter's stellar envelope at every periastron passage, suggesting that V Hya is a good candidate where the binary interaction will result in a CE configuration (see also [18]).

IRAS 16342-3814 (hereafter, IRAS 16342) is the best studied and nearest (∼2 kpc) example of "water-fountain" PPN—a class of young PPNe with unusually fast radial H_2O outflows with $V_{exp} \gtrsim 50 \, \text{km s}^{-1}$ [19] showing that jet activity is extremely recent ($\lesssim 100$ years, reference [20]), and indicating that these objects have become PPNe fairly recently. Its morphology is well-resolved with optical (HST) and near-infrared (Keck Adaptive Optics) imaging [21,22]. Radio interferometry (VLA, VLBA) shows water masers spread over a range of radial velocities encompassing 270 km s^{-1} [23]. From a study in which emission from ^{12}CO J=3-2 and other molecular lines with ∼0.35″ resolution was mapped using ALMA, reference [24] (hereafter, Setal17) found that ∼455 years ago, the progenitor AGB star of IRAS 16342 underwent a sudden, very large increase (by a factor > 500) in its mass-loss rate; the average value over this period is $> 3.5 \times 10^{-4} \, M_\odot \, \text{yr}^{-1}$ The Boomerang Nebula, long understood as an "extreme" bipolar pre-planetary Nebula (PPN), is the coldest known object in the Universe, with a massive high-speed outflow that has cooled significantly below the cosmic background temperature [25]. ALMA observations confirmed this finding, and revealed unseen distant regions of this ultra-cold outflow (UCO), out to $\gtrsim 120,000$ AU [26]. The very large mass-loss rate (∼0.001 $M_\odot$ yr^{-1}) characterizing the UCO and the central star's very low-luminosity (300 $L_\odot$) are unprecedented, making it a key object for testing theoretical models for mass-loss during post-main sequence evolution and for producing the dazzling variety of bipolar and multipolar morphologies seen in PNe [27]. In the UCO, the mass-loss rate ($\dot{M}$) increases with radius, similar to its expansion velocity (V)—taking $V \propto r$, $\dot{M} \propto r^{0.9-2.2}$. The mass in the UCO is $\gtrsim 3.3 \, M_\odot$, and the Boomerang's MS progenitor mass is $\gtrsim 4 \, M_\odot$. The UCO's kinetic energy is very high, $KE_{UCO} > 4.8 \times 10^{47}$ erg, and the most likely source of this energy is the gravitational energy released via binary interaction in a common envelope event (CEE). The Boomerang's primary was an RGB or early-AGB star when it entered the CE phase; the companion finally merged into the primary's core, ejecting the primary's envelope that now forms the UCO.

Although numerical simulations of CEE are becoming increasingly sophisticated, they are still very uncertain and require strong observational constraints (e.g., [28] and references therein). The Boomerang is the youngest and least-evolved known example of such an interaction, and it already raises a potential difficulty for such simulations in that the UCO morphology does not appear to be concentrated towards the equatorial plane, as numerical simulations usually indicate. A detailed inspection of such simulations by Iaconi et al. [28] shows that the ejecta's mass distribution tends to become increasingly isotropic with time. Since the UCO is observed at an age that is a factor ~ 650 larger than the 2000-day timespan in these simulations, it is not implausible that the lower-density polar regions seen on small scales in the simulation get filled in with time (e.g., small perturbations in the velocity vectors away from radial would allow material to move towards the axis). New simulations of CEE or grazing envelope ejection (GEE: reference [29], and references therein) that can reproduce the relatively well-defined properties of the Boomerang Nebula, will be very useful in improving our understanding of this important channel for binary star evolution.

3.2. Central Disks and Torii

Like most bipolar or multipolar PPN, all three of the objects described above (Section 3.1) directly show the presence of compact, dense equatorial structures in the form of disks and/or torii. In V Hya, reference [30] found, using HST STIS observations, a hot, central disk-like structure of diameter 0.6″ (240 AU at V Hya's distance of 400 pc) expanding at a speed of 10–15 km s^{-1}.

In IRAS 16342, the central region had (literally) remained in "shadow" because of its high optical depth at the longest wavelengths that it had been imaged (12 μm, reference [31]). Setal17's ALMA study provided an unprecedented close view of this region, revealing a compact source in ^{29}SiO J=8-7 (v=0) emission and dust thermal emission at 0.89 mm. In addition, a high-density ($> 3.3 \times 10^6$ cm^{-3}) tilted torus is revealed in H^{13}CN J=4-3 emission. The torus has a size of 1300 AU, and its inclination is 43°, consistent with the axis of its bipolar lobes [21] and the high-velocity H$_2$O outflow axis [23]. The deprojected torus expansion velocity is $V_{tor} = 20$ km s^{-1}, and its expansion time-scale is 160 years.

In the Boomerang, Setal17 found a dense central waist of size (FWHM) $\sim$1740 AU$\times$275 AU. The ^{12}CO J = 3-2 line profile from the central waist is relatively narrow and may include components due to rotation and expansion. Assuming its outer regions to be expanding, it has an age of $\lesssim$1925 years. The waist has a compact core seen in thermal dust emission at 0.87–3.3 mm, which harbors 4–7 $\times 10^{-4} M_\odot$ of very large ($\sim$mm-to-cm sized), cold ($\sim$20–30 K) grains.

The sizes of the above equatorial structures are much larger than the observed/expected sizes of accretion or circumbinary disks. The formation process is not understood. Reference [32] suggested that such waists can form via the interaction of a CFW with its progenitor AGB wind. The waist mass provides an important constraint on theoretical models—e.g., CE evolution would likely cause expulsion of most of the stellar envelope ([33], leading to much larger values of the waist mass ($\sim$0.1$M_\odot$) than wind-accretion modes. Quantitative predictions of the expected size and mass of the waist would be very helpful in testing these theoretical models.

3.3. Collimated Jet-Like Outflows and Binary Accretion Modes

All of the objects discussed above show evidence for episodic, collimated jet-like outflows, but with significant differences. If the physical properties of the fast outflows in PPNe are accurately determined, these can be used to estimate the jet momentum, $M_j V_j$, and the accretion time-scale, t_{acc}, which in turn can constrain the class of binary interaction and associated accretion modes (e.g., Bondi–Hoyle–Lyttleton wind-accretion and wind Roche lobe overflow (wRLOF), via an innovative analytical modelling approach described by [34] (hereafter, BL14). In BL14's approach, the intrinsic jet momentum is estimated from the observed fast outflow's momentum, assuming that the interaction between the intrinsic jet outflow and the ambient circumstellar envelope is momentum-conserving.

In Y Gem, Setal18 found UV lines with P-Cygni-type profiles from species such as Si IV and C IV with emission and absorption features that are red- and blue-shifted by velocities of $\sim$500 km s^{-1} from the systemic velocity. Setal18 concluded, from these (and previous) observations, that material from the primary star is gravitationally captured by a low-mass MS companion, producing a hot accretion disk around the latter. The disk powers a fast outflow that produces blue-shifted features due to absorption of UV continuum emitted by the disk, whereas the red-shifted emission features arise in heated infalling material from the primary. The accretion luminosity implies a mass-accretion rate $> 5 \times 10^{-7} M_\odot$ yr^{-1}. Setal18 concluded that Roche lobe overflow is the most likely binary accretion mode since Y Gem does not show the presence of a wind.

In V Hya, the collimated ejection of material is in the form of bullets, and the ejection axis flip-flops around an average orientation, in a regular manner. These data support a model in which the bullets are a result of collimated ejection from an accretion disk (produced by gravitational capture of material from the primary) that is warped and precessing, and/or that has a magnetic field that is misaligned with that of the companion or the primary star (Setal16). The average momentum rate of the bullet ejections is ($\sim 8.2 \times 10^{25}$ g cm s^{-2}), implying a minimum required accretion rate of $\dot{M}_a \sim 3.3 \times 10^{-8} M_\odot$ yr^{-1}. Since the secondary must get very close to the primary's stellar envelope at periastron passage (Setal16), the binary separation is comparable to the stellar radius (at 400 pc, $R_* \sim 2$ AU, see [35]), i.e., $a_{or} \sim 2$ AU, so the Roche-lobe overflow mode (RLOF) is the appropriate accretion mode, which can easily supply the required accretion rate.

IRAS 16342 shows two very high-speed, knotty, jet-like molecular outflows, whose axes are not colinear. The Extreme High Velocity Outflow (EHVO) and the High Velocity Outflow (HVO) have (deprojected) expansion speeds of 360–540 and 250 km s^{-1} and ages of 130–305 and $\lesssim$110 years. The spiral structure seen in the position-velocity (PV) plot of the HVO most likely indicates emission from a precessing high-velocity bipolar outflow, as inferred previously from near-IR imaging [22], that entrains material in the near and far bipolar lobe walls. The measured expansion ages of the above structural components imply that the torus (age~160 years) and the younger high-velocity outflow (age~110 years) were formed soon after the sharp increase in the AGB mass-loss rate. The relatively high momentum rate for the dominant jet-outflow in IRAS 16342 ($> 5 \times 10^{28}$ g cm s^{-2}) implies a correspondingly high minimum accretion rate of $\dot{M}_a = 1.9 \times 10^{-5} M_\odot$ yr^{-1}. Setal17 compared this rate with the expected mass-accretion rates derived for different accretion models shown in BL14's Figure 1, and concluded that standard Bondi–Hoyle–Lyttleton (BHL) wind-accretion and wind-RLOF models with WD or MS companions were unlikely; enhanced RLOF from the primary or accretion modes operating within common envelope evolution were needed. We revisit this conclusion because the BHL rate shown in BL14's Figure 1b is derived using the primary's AGB mass-loss properties $\dot{M}_w = 10^{-5} M_\odot$ yr^{-1} (and $V_w = 10$ km s^{-1}, together with assumed orbital separation $a_{or} = 10$ AU, companion mass $M_c = 0.6 M_\odot$, primary mass $M_p = 1.0 M_\odot$) and not the much higher value of $\dot{M} \sim 1.3 \times 10^{-4} M_\odot$ yr^{-1} (comparable to the mass-loss rate in IRAS16342), mentioned in BL14's Section 3.1 and referenced in the figure caption (E. Blackman, *priv. comm.*), so the expected BHL accretion rate could be quite high in IRAS16342. The BHL rate estimate is valid only if the orbital separation is much larger than the Roche-lobe radius, i.e., $a_{or} >> R_{roche}$. For IRAS 16342, we found that $R_{roche}/a_{or} = 0.38 + log(M_p/M_c) = 0.5 - 0.54$, assuming $M_c = 0.6 - 1 M_\odot$ and $M_p \sim 4 M_\odot$ (the primary was likely relatively massive, $\sim 4.5 M_\odot$, in order to have experienced HBB, needed to produce the very enhanced $^{13}C/^{12}C$ ratio observed in this object). Since the progenitor AGB star of IRAS 16342 must have had a radius $\gtrsim$1 AU (e.g., Dijkstra et al. [36] modeled its SED using an M9III star with $T_{\text{eff}} = 2670$ K and R = 372 $R_\odot$), the binary separation must be $\gtrsim$5 AU in order for the BHL accretion rate to be valid. Hence, we considered $a_{or} > 5 - 10$ AU, and found $\dot{M}_{BHL} \lesssim (0.3 - 1) \times 10^{-5} M_\odot$ yr^{-1}, taking $\dot{M}_w = 3.5 \times 10^{-4} M_\odot$ yr^{-1} and $V_w = 23$ km s^{-1} for IRAS16342 from Setal17. We conclude that the HVO in IRAS 16342 is not driven by accretion via BHL, but requires wind-RLOF or modes that provide higher accretion rates.

The Boomerang's central region has an overall bipolar structure; in detail, this structure is comprised of multiple, highly collimated lobes on each side of the central disk, both in scattered light and in ^{13}CO J=3-2 emission. The velocity of the molecular material in the dense walls of the collimated lobes is not particularly high, and we expect it is likely to be substantially lower than the velocity of the unshocked jet-outflow that has carved out these lobes, since the jet outflow has interacted with a very massive envelope (the UCO). Optical spectroscopy, indicating that the pristine fast outflow may have a speed of about 100 km s^{-1} [37], support this expectation. Assuming momentum-conservation to derive the fast outflow momentum (as in BL14) is likely to provide a severe underestimate of the intrinsic jet momentum—numerical simulations are needed. However, in this object, the UCO's extreme properties(see Section 3.1) directly imply CE evolution.

4. Concluding Remarks and Future Prospects

We discuss observational results that address several key aspects in our quest to understand binary interaction as the underlying cause for the formation of aspherical planetary nebulae from AGB stars. These include the use of UV and X-ray observations to establish the presence of binarity and associated active accretion in AGB stars, as well as high-angular resolution mm-wave and optical observations to study the properties of jet-like outflows in objects in transition from the AGB to the PN phase. Further progress now requires: (i) high-angular resolution mm-wave observations (e.g., with ALMA) of a large sample of PPNe (to derive the jet properties and the AGB mass-loss properties immediately preceding the transition to the post-AGB phase); (ii) studies of accretion activity in statistical samples of fuvAGB stars using UV spectroscopy and high-sensitivity, high-time-cadence photometry to detect flickering, e.g., with the *TESS* mission (to understand the prevalence of binarity and assoicated accretion activity in AGB stars as a class); and (iii) X-ray surveys of AGB stars, including those with known strong B-fields, e.g., to test the relationship of primary's B-field and X-ray emission, and those with low FUV/NUV flux ratios (to understand the contribution and properties of X-ray emission from the companion's corona).

Funding: This research was funded by Space Telescope Science Intsitute: GO 14713.001, and NASA: 12-ADAP12-0283, 17-ADAP17-0206.

Acknowledgments: I thank Eric Blackman for a helpful discussion related to binary acretion modes. The author's research described here was carried out at the Jet Propulsion Laboratory, California Institute of Technology, under a contract with NASA, and funded in part by NASA via an ADAP award, and an HST GO award (GO 14713.001) from the Space Telescope Science Institute.

Conflicts of Interest: The author declares no conflict of interest.

References

1. Soker, N. On the formation of ansae in planetary nebulae. *Astron. J.* **1990**, *99*, 1869–1882. [CrossRef]
2. Sahai, R.; Trauger, J.T. Multipolar Bubbles and Jets in Low-Excitation Planetary Nebulae: Toward a New Understanding of the Formation and Shaping of Planetary Nebulae. *Astron. J.* **1998**, *116*, 1357–1366. [CrossRef]
3. Sahai, R.; Morris, M.R.; Villar, G.G. Young Planetary Nebulae: Hubble Space Telescope Imaging and a New Morphological Classification System. *Astron. J.* **2011**, *141*, 134. [CrossRef]
4. Sahai, R.; Findeisen, K.; Gil de Paz, A.; Sánchez Contreras, C. Binarity in Cool Asymptotic Giant Branch Stars: A GALEX Search for Ultraviolet Excesses. *Astron. J.* **2008**, *689*, 1274–1278. [CrossRef]
5. Morrissey, P.; Conrow, T.; Barlow, T.A.; Small, T.; Seibert, M.; Wyder, T.K.; Budavári, T.; Arnouts, S.; Friedman, P.G.; Forster, K.; et al. The Calibration and Data Products of GALEX. *Astrophys. J. Suppl. Ser.* **2007**, *173*, 682–697. [CrossRef]
6. Favata, F. Accretion, fluorescent X-ray emission and flaring magnetic structures in YSOs. *Mon. Not. R. Astron. Soc.* **2005**, *76*, 337.
7. Ramstedt, S.; Montez, R.; Kastner, J.; Vlemmings, W.H.T. Searching for X-ray emission from AGB stars. *Astron. Astrophys.* **2012**, *543*, A147. [CrossRef]
8. Sahai, R.; Sanz-Forcada, J.; Sánchez Contreras, C.; Stute, M. A Pilot Deep Survey for X-Ray Emission from fuvAGB Stars. *Astrophys. J.* **2015**, *810*, 77. [CrossRef]
9. Sahai, R.; Sanz-Forcada, J.; Sánchez Contreras, C. Variable X-Ray and UV emission from AGB stars: Accretion activity associated with binarity. *J. Phys. Conf. Ser.* **2016**, *728*, 042003. [CrossRef]
10. Sahai, R.; Scibelli, S.; Morris, M.R. High-speed Bullet Ejections during the AGB-to-Planetary Nebula Transition: HST Observations of the Carbon Star, V Hydrae. *Astrophys. J.* **2016**, *827*, 92. [CrossRef]
11. Sahai, R.; Neill, J.D.; Gil de Paz, A.; Sánchez Contreras, C. Strong Variable Ultraviolet Emission from Y Gem: Accretion Activity in an Asymptotic Giant Branch Star with a Binary Companion? *Astrophys. J.* **2011**, *740*, L39. [CrossRef]

12. Sahai, R.; Sánchez Contreras, C.; Mangan, A.S.; Sanz-Forcada, J.; Muthumariappan, C.; Claussen, M.J. Binarity and Accretion in AGB Stars: HST/STIS Observations of UV Flickering in Y Gem. *Astrophys. J.* **2018**, *860*, 105. [CrossRef] [PubMed]

13. Ortiz, R.; Guerrero, M.A. Ultraviolet emission from main-sequence companions of AGB stars. *Mon. Not. R. Astron. Soc.* **2016**, *461*, 3036–3046. [CrossRef]

14. Famaey, B.; Pourbaix, D.; Frankowski, A.; van Eck, S.; Mayor, M.; Udry, S.; Jorissen, A. Spectroscopic binaries among Hipparcos M giants,. I. Data, orbits, and intrinsic variations. *Astron. Astrophys.* **2009**, *498*, 627–640. [CrossRef]

15. Montez, R., Jr.; Ramstedt, S.; Kastner, J.H.; Vlemmings, W.; Sanchez, E. A Catalog of GALEX Ultraviolet Emission from Asymptotic Giant Branch Stars. *Astrophys. J.* **2017**, *841*, 33. [CrossRef]

16. Soker, N.; Kastner, J.H. Magnetic Flares on Asymptotic Giant Branch Stars. *Astrophys. J.* **2003**, *592*, 498–503. [CrossRef]

17. Kastner, J.H.; Soker, N. Constraining the X-Ray Luminosities of Asymptotic Giant Branch Stars: TX Camelopardalis and T Cassiopeia. *Astrophys. J.* **2004**, *608*, 978–982. [CrossRef]

18. Barnbaum, C.; Morris, M.; Kahane, C. Evidence for Rapid Rotation of the Carbon Star V Hydrae. *Astrophys. J.* **1995**, *450*, 862. [CrossRef]

19. Likkel, L.; Morris, M.; Maddalena, R.J. Evolved stars with high velocity H_2O maser features—Bipolar outflows with velocity symmetry. *Astron. Astrophys.* **1992**, *256*, 581–594.

20. Imai, H. Stellar molecular jets traced by maser emission. In *Astrophysical Masers and Their Environments*; Chapman, J.M., Baan, W.A., Eds.; IAU Symposium: Paris, France, 2007; Volume 242, pp. 279–286.

21. Sahai, R.; te Lintel Hekkert, P.; Morris, M.; Zijlstra, A.; Likkel, L. The "Water-Fountain Nebula" IRAS 16342-3814: Hubble Space Telescope/Very Large Array Study of a Bipolar Protoplanetary Nebula. *Astrophys. J. Lett.* **1999**, *514*, L115–L119. [CrossRef]

22. Sahai, R.; Le Mignant, D.; Sánchez Contreras, C.; Campbell, R.D.; Chaffee, F.H. Sculpting a Pre-planetary Nebula with a Precessing Jet: IRAS 16342-3814. *Astrophys. J. Lett.* **2005**, *622*, L53–L56. [CrossRef]

23. Claussen, M.J.; Sahai, R.; Morris, M.R. The Motion of Water Masers in the Pre-Planetary Nebula IRAS 16342-3814. *Astrophys. J.* **2009**, *691*, 219–227. [CrossRef]

24. Sahai, R.; Vlemmings, W.H.T.; Gledhill, T.; Sánchez Contreras, C.; Lagadec, E.; Nyman, L.Å.; Quintana-Lacaci, G. ALMA Observations of the Water Fountain Pre-planetary Nebula IRAS 16342-3814: High-velocity Bipolar Jets and an Expanding Torus. *Astrophys. J. Lett.* **2017**, *835*, L13. [CrossRef] [PubMed]

25. Sahai, R.; Nyman, L.Å. The Boomerang Nebula: The Coldest Region of the Universe? *Astrophys. J. Lett.* **1997**, *487*, L155–L159. [CrossRef]

26. Sahai, R.; Vlemmings, W.H.T.; Huggins, P.J.; Nyman, L.Å.; Gonidakis, I. ALMA Observations of the Coldest Place in the Universe: The Boomerang Nebula. *Astrophys. J.* **2013**, *777*, 92. [CrossRef]

27. Balick, B.; Frank, A. Shapes and Shaping of Planetary Nebulae. *Ann. Rev. Astron. Astrophys.* **2002**, *40*, 439–486. [CrossRef]

28. Iaconi, R.; Reichardt, T.; Staff, J.; De Marco, O.; Passy, J.C.; Price, D.; Wurster, J.; Herwig, F. The effect of a wider initial separation on common envelope binary interaction simulations. *Mon. Not. R. Astron. Soc.* **2017**, *464*, 4028–4044. [CrossRef]

29. Shiber, S.; Soker, N. Simulating a binary system that experiences the grazing envelope evolution. *Mon. Not. R. Astron. Soc.* **2018**, *477*, 2584–2598. [CrossRef]

30. Sahai, R.; Morris, M.; Knapp, G.R.; Young, K.; Barnbaum, C. A collimated, high-speed outflow from the dying star V Hydrae. *Nature* **2003**, *426*, 261–264. [CrossRef] [PubMed]

31. Verhoelst, T.; Waters, L.B.F.M.; Verhoeff, A.; Dijkstra, C.; van Winckel, H.; Pel, J.W.; Peletier, R.F. A dam around the Water Fountain Nebula? The dust shell of IRAS16342-3814 spatially resolved with VISIR/VLT. *Astron. Astrophys.* **2009**, *503*, 837–841. [CrossRef]

32. Soker, N.; Rappaport, S. The Formation of Very Narrow Waist Bipolar Planetary Nebulae. *Astrophys. J.* **2000**, *538*, 241–259. [CrossRef]

33. Nordhaus, J.; Blackman, E.G. Low-mass binary-induced outflows from asymptotic giant branch stars. *Mon. Not. R. Astron. Soc.* **2006**, *370*, 2004–2012. [CrossRef]

34. Blackman, E.G.; Lucchini, S. Using kinematic properties of pre-planetary nebulae to constrain engine paradigms. *Mon. Not. R. Astron. Soc.* **2014**, *440*, L16–L20. [CrossRef]

35. Knapp, G.R.; Jorissen, A.; Young, K. A 200 km/s molecular wind in the peculiar carbon star V Hya. *Astron. Astrophys.* **1997**, *326*, 318–328.
36. Dijkstra, C.; Waters, L.B.F.M.; Kemper, F.; Min, M.; Matsuura, M.; Zijlstra, A.; de Koter, A.; Dominik, C. The mineralogy, geometry and mass-loss history of IRAS 16342-3814. *Astron. Astrophys.* **2003**, *399*, 1037–1046. [CrossRef]
37. Neckel, T.; Staude, H.J.; Sarcander, M.; Birkle, K. Herbig-Haro emission in two bipolar reflection nebulae. *Astron. Astrophys.* **1987**, *175*, 231–237.

galaxies

MDPI

Article

Post-AGB Discs from Common-Envelope Evolution

Robert G. Izzard [1,2,*] and Adam S. Jermyn [2]

[1] Department of Physics, University of Surrey, Guildford, Surrey GU2 7XH, UK
[2] Institute of Astronomy, University of Cambridge, Madingley Road, Cambridge CB3 0HA, UK;
     adamjermyn@gmail.com
*   Correspondence: r.izzard@surrey.ac.uk

Received: 26 July 2018; Accepted: 4 September 2018; Published: 11 September 2018

Abstract: Post-asymptotic giant branch (post-AGB) stars with discs are all binaries. Many of these binaries have orbital periods between 100 and 1000 days so cannot have avoided mass transfer between the AGB star and its companion, likely through a common-envelope type interaction. We report on preliminary results of our project to model circumbinary discs around post-AGB stars using our binary population synthesis code *binary_c*. We combine a simple analytic thin-disc model with binary stellar evolution to estimate the impact of the disc on the binary, and vice versa, fast enough that we can model stellar population and hence explore the rather uncertain parameter space involved with disc formation. We find that, provided the discs form with sufficient mass and angular momentum, and have an inner edge that is relatively close to the binary, they can both prolong the life of their parent post-AGB star and pump the eccentricity of orbits of their inner binaries.

Keywords: binary stars; post-AGB; discs

---

1. Introduction

About half the stars exceeding the mass of the Sun are binaries, and of these many interact during their lifetime [1]. Many key astrophysical phenomena occur in binary systems, such as thermonuclear novae, X-ray bursts, type Ia supernovae and merging compact objects as detected by gravitational waves. Despite their importance to fundamental astrophysics, many aspects of binary evolution remain poorly understood.

The evolution of binary stars differs from that of single stars mostly because of mass transfer. In binaries wide enough that one star becomes a giant, yet short enough that the giant cannot fit inside the binary, mass transfer begins when the radius of the giant star exceeds its Roche radius. Because the giant is convective, and likely significantly more massive than its companion, mass transfer is unstable and accelerates. This leads to a common envelope forming around the stars. Drag between the stars and the envelope cause the orbit to decay and, if enough energy is transferred to the envelope, the envelope is ejected [2]. Stars on the giant branch or asymptotic giant branch (GB or AGB stars) are stripped almost down to their cores, leaving a post-(A)GB binary.

During the common envelope process, tides are expected to be highly efficient and hence the binary that emerges is expected to have little, if any, eccentricity ($e \approx 0$). Yet, post-AGB binaries are often highly eccentric, up to $e = 0.6$. The source of this eccentricity is unknown but is probably related to a similar phenomena observed in the barium stars which are thought to have involved mass transfer from an AGB star [3,4].

Many post-AGB stars have discs, and all those with discs are in binary systems [5]. Investigations into the link between these discs and their stellar systems' peculiar eccentricity, based on eccentricity pumping by Lindblad resonances, suggest that the discs can cause the post-AGB systems' eccentricity if they are sufficiently massive and live for long enough [6]. Recent ALMA observations, e.g., of IRAS 08544-4431, show that post-AGB discs are mostly Keplerian, have masses of about $10^{-2}\,M_\odot$, outer diameters of 10^{16} cm $\approx 10^5$ $R_\odot$, angular momenta similar to their parent binary systems (around 10^{52} g cm^2 s^{-1}), and both slow mass loss from the outer part of the disc and inflow at its inner edge at a rate of about 10^{-7} $M_\odot$ year^{-1} [7].

In this work we combine a fast, analytic model of circumbinary discs with a synthetic binary stellar-evolution code to estimate the number of post-(A)GB discs and their properties. We include mass loss from the disc caused by illumination from the central star, ram-stripping by the interstellar medium, and include a viscous-timescale flow onto the central binary. The disc extracts angular momentum from its central binary star system. Resonances excited in the disc pump the central binary's eccentricity. While our results are preliminary, they show that eccentricity pumping is efficient in some systems, and the discs may live a considerable time.

2. Circumbinary Disc Model

We assume that circular, Keplerian discs form when a common envelope is ejected and some small fraction of the envelope mass, $f_M \ll 1$, is left behind as a disc containing a fraction, $f_J \ll 1$, of the envelope's angular momentum. The disc thus has mass M_{disc} and angular momentum J_{disc}. We assume our discs are thin such that $H/R < 1$, where $H(R)$ is the scale height at a radius R, and have viscous timescales that are short compared to their lifetimes such that they spread instantaneously [8]. Given that observed discs are mostly cool and neutral (i.e., $T \lesssim 1000$ K in most of the disc), we fix the opacity to $\kappa = 0.01$ m^2 g^{-1} and assume an α-viscosity model with $\alpha = 10^{-3}$.

Given the above simplifications, we write the heat-balance equation in the disc as,

$$\sigma T^4 = \mathcal{A} + \mathcal{B}\left(1 + \mathcal{C}\right), \tag{1}$$

where $T = T(R)$ is the temperature in the disc mid-plane (cf. [9] where these terms are derived in detail). The σT^4 term is the heating by the post-(A)GB star, which is balanced in equilibrium by viscous heating ($\mathcal{A}$) and re-radiation ($\mathcal{B}$ and $\mathcal{C}$). We neglect the impact of mass changes on the heat balance because observed discs have low mass-loss rates ($\lesssim 10^{-7}$ $M_\odot$ year^{-1}). The terms $\mathcal{A}$, $\mathcal{B}$ and $\mathcal{C}$ are functions of radius, such that we can rewrite Equation (1) as,

$$T^4 = a\Sigma^2 T R^{-3/2} + bT^{1/2}R^{-3/2} + cR^{-3}, \tag{2}$$

where $\Sigma = \Sigma(R)$ is the surface density at radius R. At a given radius R we then choose the largest term in the right hand side of Equation (2) and set the other terms to zero, allowing a simple solution for $T(R)$ everywhere in the disc (cf. [10]). Errors in T^4 are up to a factor of 3, hence errors in T are at most $3^{1/4} \approx 30\%$ and typically much less, which is good enough for our purposes.

To verify that we can neglect mass changes, note that the temperature, T_{acc}, associated with accretion at a rate $\dot{M}$ at radius R and binary mass M_{binary} [9],

$$\sigma T^4_{acc} = \frac{GM_{binary}\dot{M}}{2\pi R^3}, \tag{3}$$

where the gravitational mass is dominated by the binary. Evaluating this yields,

$$T_{acc}(R) = 290 \, \mathrm{K} \left(\frac{\dot{M}}{10^{-7} \, M_\odot \, \mathrm{year}^{-1}}\right)^{1/4} \left(\frac{M_{binary}}{M_\odot}\right)^{1/4} \left(\frac{R}{10^2 \, R_\odot}\right)^{-3/4}. \tag{4}$$

At the inner edges of our discs $R \approx 10^2$ $R_\odot$ so T_{acc} is a factor of several smaller than the typical 1000 K temperatures we find. The outer edges, at $R \approx 10^5$ $R_\odot$, have $T_{acc} \approx 2$ K which both cooler than the disc and the Cosmic Microwave background so this term can be safely ignored.

We next scale the density and outer radius so that the integrals of mass and angular momentum throughout the disc match M_{disc} and J_{disc} respectively, while the inner radius R_{in} is fixed by the torque on the disc caused by the inner binary which we take from [11] with a multiplier of 10^{-3}. Thus, given three constraints—M_{disc}, J_{disc} and the inner binary torque—we know $T(R)$ throughout the disc. From this we construct any other required physical quantities, and can integrate these to calculate, for example, the luminosity of the disc. With our chosen torque prescription, our discs' inner radii are typically around twice the orbital separation, as assumed in other works (e.g., [12]).

We treat mass loss from the circumbinary disc as a slow phenomenon. At the inner edge of the disc we include mass inflow onto the central binary at the local viscous rate. At the outer edge we strip material when its pressure is less than that of the interstellar medium, assumed to be 3000 K/k_B where k_B is the Boltzmann constant. Irradiation by the post-(A)GB star, particularly in X-rays, causes mass loss and we model this with the prescription of [13]. During most of the disc's lifetime, the viscous inflow is most important, but X-ray losses dominate when the star transitions to become a hot, young white dwarf and these losses quickly evaporate the disc. Our neglect of mass loss in Equation (1) is incorrect in this brief phase but because the disc is terminated very rapidly, on timescales of years, such systems will be rarely observed.

The evolution of the binary stars is calculated using *binary_c* [14–17]. Common envelopes are ejected with the formalism of [18] using an efficiency $\alpha_{\text{CE}} = 1$ to match observed post-AGB systems with periods between 100 and about 1000 days which are observed to have circumbinary discs. The envelope binding energy parameter, λ_{CE}, is fitted to the models of [19], and 10% of the envelope's recombination energy is used to aid ejection. Because typically $\lambda_{\text{CE}} \gg 1$ during the AGB, such envelopes are nearly unbound and common envelope ejection is efficient with only modest orbital shrinkage.First giant branch stars have $\lambda \lesssim 1$ because they are more tightly bound, so their orbits shrink significantly. We also assume that stars exit the common envelope with a small eccentricity, $e = 10^{-5}$, to which we apply Lindblad resonance pumping [6]. If α_{CE} is less than 1.0, as suggested by e.g., [20–22], then more recombination energy can be included to prevent orbital shrinkage. We do not pretend to better understand common envelope evolution with our simple model.

At the end of the common envelope phase our treatment differs from [18] in that we keep a thin envelope on the (A)GB star such that it just fills its Roche lobe when the envelope is ejected. Typically this is a 10^{-2}–10^{-3} $M_\odot$ hydrogen-rich envelope which keeps the star cool ($\sim$5000 K) during the post-(A)GB phase relative to the white dwarf ($\gtrsim 10^4$ K) it will become. The star continues its nuclear burning which reduces the mass of the envelope. Accretion from the circumbinary disc replenishes the shell and extends the lifetime of the post-(A)GB phase, as we show in the following section.

3. Example System

As an example binary star system we choose an initial primary mass $M_1 = 1.5\,M_\odot$, initial secondary mass $M_2 = 0.9\,M_\odot$, initial separation $a = 800\,R_\odot$ and metallicity $Z = 0.02$. The separation is chosen such that Roche-lobe overflow is initiated just after the primary starts thermally pulsing on the AGB. Common envelope evolution follows with $\alpha_{\text{CE}}\lambda_{\text{CE}} = 2.1$ so the orbit shrinks to 195 $R_\odot$. The primary is then a 0.577 $M_\odot$ post-AGB star, with an envelope mass of $7 \times 10^{-3}\,M_\odot$ and a 0.95 $M_\odot$ main-sequence dwarf companion. The 0.05 $M_\odot$ accreted on to the secondary is from the wind of the AGB star prior to common envelope evolution. The orbital period is then about 255 days, typical of post-AGB binaries with circumbinary discs. We model circular orbits but our model is equally applicable to initially mildly eccentric binaries. Tides are expected to be efficient as the primary ascends the AGB and will quickly circularize the system. In the following discussion, and the figures, times are measured from the moment the common envelope is ejected.

A circumbinary disc is formed with $f_M = 0.02$ and $f_J = 0.107$, giving $M_{\text{disc}} = 0.012\,M_\odot$ and $J_{\text{disc}} = 10^{52}$ g cm^2 s^{-1}. Both f_M and f_J are chosen to give us a disc with mass and angular momentum similar that of IRAS 08544-4431 [7] and other post-AGB systems with circumbinary discs (e.g., the Red Rectangle). As the disc evolves, it feeds off the angular momentum of the inner binary, but the total angular momentum gained during its lifetime is small. Mass flows through the inner edge onto the binary at between 10^{-7} (initially) and 10^{-8} $M_\odot$ year^{-1}, although this does not significantly alter the evolution of the disc. Rather, X-ray driven mass loss, caused by the post-AGB star increasing in temperature at approximately constant luminosity, leads to sudden termination of the disc at 6.5×10^4 years, as shown in Figure 1.

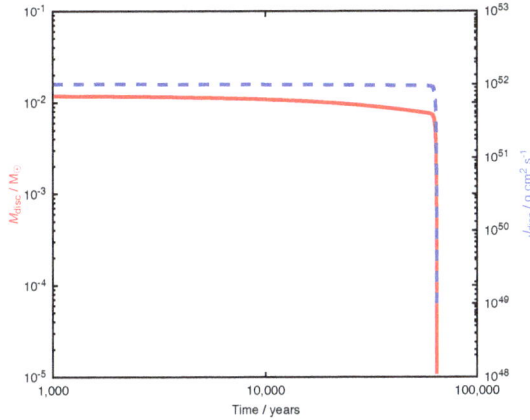

Figure 1. Mass (left axis, red solid line) and angular momentum (right axis, blue dashed line) evolution in our example circumbinary disc system. During most of the evolution of the disc, its mass changes slowly ($\sim 10^{-8} \, M_\odot \, year^{-1}$) because of viscous flow through its inner edge onto the inner binary system. As the post-AGB star in the binary heats up, its X-ray flux increases until it drives sufficient wind that the disc is quickly evaporated after about 6.5×10^4 years where a time of 0 years is when the common envelope is ejected.

The inner edge of the disc is at $317 \, R_\odot$, well outside the inner binary orbit, while the outer edge is at $4.7 \times 10^4 \, R_\odot$. The former is set by the applied torque, while the latter is set by the disc angular momentum. Until the disc is evaporated, neither the inner nor outer radius changes significantly. Figure 2 shows the evolution of said radii.

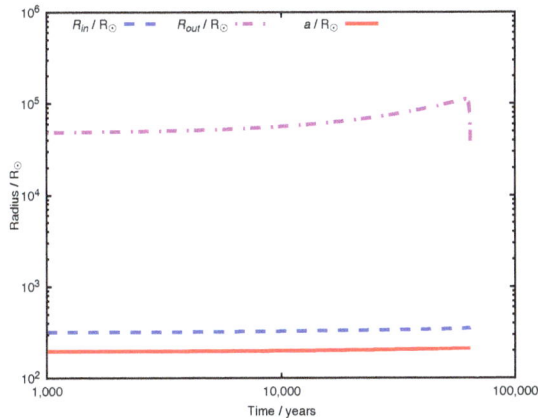

Figure 2. Evolution of the inner and outer radii, R_{in} and R_{out} respectively, and the binary orbital separation a, in our example circumbinary disc system. Time is measured from the moment of common envelope ejection.

The eccentricity of the inner binary system is pumped to about 0.25 by the time the disc is evaporated. This is similar to the eccentricity observed in post-AGB systems, and it is certainly non-zero.

Mass accreted onto the post-AGB star from the inner edge of the disc replenishes its hydrogen-rich envelope, thus cooling the star and prolonging its lifetime. Because our post-AGB star has a core mass of only 0.57 $M_\odot$, its nuclear burning rate is similar to the disc's viscous accretion rate. The stellar wind mass loss rate is less than $10^{-11} M_\odot$ year^{-1} because we apply the rate of [23] in our ignorance of the mechanism of post-AGB wind loss. To test how long accretion extends the post-AGB, we evolved an identical example system but with the inner-edge viscous inflow disabled. Figure 3 shows that the post-AGB star in the system with accretion lives for an extra 3×10^4 years, an approximate doubling of the its post-AGB lifetime. The extra lifetime of such systems may explain why they do not show residual nebulosity from envelope ejection. Observed planetary nebulae, which may be ejected common envelopes, have dynamical timescales of about 10^4 years. By the time the post-AGB star is hot enough to ionize such envelopes they are likely too diffuse to be observed as planetary nebulae [24].

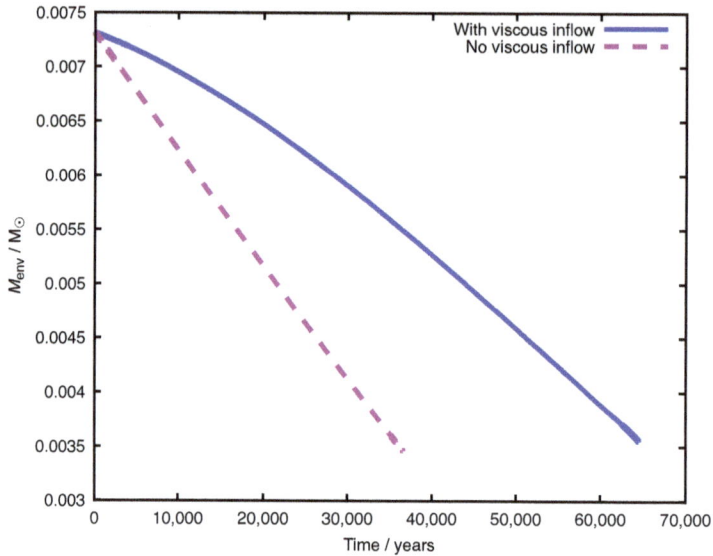

Figure 3. Post-AGB envelope mass vs. time since common envelope ejection. The blue, solid line is our example system with mass inflow from the inner edge onto the inner binary, while the magenta, dashed line is without. Mass flowing onto the post-AGB star replenishes its hydrogen envelope thus, for a while, keeps it relatively cool, limits its X-ray flux and prevents it from evaporating the disc. In this case, its lifetime is extended from about 35,000 years to 65,000 years. In our example system, post-AGB wind mass loss is negligible.

4. Stellar Populations and Improved Physics

Our model is simple yet it contains the essential physics of circumbinary discs around post-(A)GB stars. It is also fast enough that we can evolve a typical binary system containing a circumbinary disc in just a few seconds of CPU time. Speed is a an essential requirement of binary population synthesis studies because the parameter space is large. We can explore the consequence of the initial mass and angular momentum of our discs through the input parameters f_M and f_J, and we can estimate the effect of changing uncertain input physics, e.g., a stronger or weaker X-ray wind or binary torque.

We can also model post-first giant branch (helium core) systems and post-early-AGB systems. From an observational point of view, these differ in that their stellar evolution is truncated at an earlier stage than in post-thermally-pulsing-AGB, hence they are dimmer. These systems also overflow their Roche lobes at shorter periods and have more bound envelopes, so their orbits and resultant discs are more compact. Our assumption of constant opacity likely breaks at this point, although our assumed instantaneous viscous spreading of the disc is certainly valid. We are working on improving the model to take this into account.

Our discs are low in mass relative to their stars and we never put more than 10% of the common envelope mass in the disc. The example system we report in Section 3 has a Toomre Q parameter of about 10^8 [25] so is not gravitationally unstable. However, this does not preclude the formation of rocks in the disc, after all we know the discs contain small grains which emit in the infra red. In our models we expect a small number of discs to form in systems that exit the red giant branch just before helium ignition. These systems will contain sdB stars rather than white dwarfs, thus stay relatively cool. The relatively low X-ray flux from sdB stars means their discs are not evaporated quickly, at least not by means modelled here, so long-lived discs and the formation of so-called debris, i.e., rocks, in them discs seems quite likely.

The formation of second-generation planets in our discs seems not to be favoured. Our discs live for less than 10^5 years, too short for planet formation in the canonical sense, and our discs are quite warm, hotter than 1000 K, near the inner edge where they are densest. That said, circumbinary discs have material concentrated in their orbital plane of the system for some time, so if even minor planets survive the common-envelope phase, they could accrete material from the disc. The consequences for the disc may be that it does not survive but this is currently beyond the scope of our model.

Our discs do succeed in pumping the eccentricity of their inner binary, in the case of our example system up to 0.25. In part this is because our discs are, by design, quite massive (about $0.01\,M_\odot$), but this is to match observed discs such as IRAS 08544-4431, so is reasonable. Circumbinary discs seem to be good candidates to explain the eccentricities of at least some post-mass-transfer objects such as barium stars. We have not yet tested wind mass transfer as a mechanism for disc formation, but this is an obvious extension to our work and likely contributes to the observed population of post-(A)GB binaries.

Author Contributions: Conceptualization, data curation, funding acquisition, project administration, resources, software, supervision and writing—original draft; R.G.I. Formal analysis, investigation, methodology, validation, visualization and writing—review and editing; R.G.I. and A.S.J.

Funding: This research was funded by the Science and Technology Facilities Council under grant number ST/L003910/1.

Acknowledgments: RGI thanks the STFC for funding his Rutherford fellowship under grant ST/L003910/1, Churchill College, Cambridge for his fellowship and access to their library. Hello to Jason Isaacs. ASJ thanks the UK Marshall Commission for financial support.

Conflicts of Interest: The authors declare no conflict of interest.

References

1. De Marco, O.; Izzard, R.G. Dawes Review 6: The Impact of Companions on Stellar Evolution. *Publ. Astron. Soc. Aust.* **2017**, *34*, e001. [CrossRef]

2. Ivanova, N.; Justham, S.; Chen, X.; De Marco, O.; Fryer, C.L.; Gaburov, E.; Ge, H.; Glebbeek, E.; Han, Z.; Li, X.D.; et al. Common envelope evolution: Where we stand and how we can move forward. *Astron. Astrophys. Rev.* **2013**, *21*, 59. [CrossRef]

3. Jorissen, A.; Van Eck, S.; Mayor, M.; Udry, S. Insights into the formation of barium and Tc-poor S stars from an extended sample of orbital elements. *Astron. Astrophys.* **1998**, *332*, 877–903.

4. Karakas, A.I.; Tout, C.A.; Lattanzio, J.C. The eccentricities of the barium stars. *Mon. Not. R. Astron. Soc.* **2000**, *316*, 689–698. [CrossRef]

5. Bujarrabal, V.; Alcolea, J.; Van Winckel, H.; Santander-García, M.; Castro-Carrizo, A. Extended rotating disks around post-AGB stars. *Astron. Astrophys.* **2013**, *557*, A104. [CrossRef]

6. Dermine, T.; Izzard, R.G.; Jorissen, A.; Van Winckel, H. Eccentricity-pumping in post-AGB stars with circumbinary discs. *Astron. Astrophys.* **2013**, *551*, A50. [CrossRef]

7. Bujarrabal, V.; Castro-Carrizo, A.; Winckel, H.V.; Alcolea, J.; Contreras, C.S.; Santander-García, M.; Hillen, M. High-resolution observations of IRAS 08544-4431. Detection of a disk orbiting a post-AGB star and of a slow disk wind. *Astron. Astrophys.* **2018**, *614*, A58. [CrossRef] [PubMed]

8. Lynden-Bell, D.; Pringle, J.E. The evolution of viscous discs and the origin of the nebular variables. *Mon. Not. R. Astron. Soc.* **1974**, *168*, 603–637. [CrossRef]

9. Perets, H.B.; Kenyon, S.J. Wind-accretion Disks in Wide Binaries, Second-generation Protoplanetary Disks, and Accretion onto White Dwarfs. *Astrophys. J.* **2013**, *764*, 169. [CrossRef]

10. Haiman, Z.; Kocsis, B.; Menou, K. The Population of Viscosity- and Gravitational Wave-driven Supermassive Black Hole Binaries Among Luminous Active Galactic Nuclei. *Astrophys. J.* **2009**, *700*, 1952–1969. [CrossRef]

11. Armitage, P.J.; Natarajan, P. Accretion during the Merger of Supermassive Black Holes. *Astrophys. J. Lett.* **2002**, *567*, L9–L12. [CrossRef]

12. Rafikov, R.R. On the Eccentricity Excitation in Post-main-sequence Binaries. *Astrophys. J.* **2016**, *830*, 8. [CrossRef]

13. Owen, J.E.; Clarke, C.J.; Ercolano, B. On the theory of disc photoevaporation. *Mon. Not. R. Astron. Soc.* **2012**, *422*, 1880–1901. [CrossRef]

14. Izzard, R.G.; Tout, C.A.; Karakas, A.I.; Pols, O.R. A new synthetic model for asymptotic giant branch stars. *Mon. Not. R. Astron. Soc.* **2004**, *350*, 407–426. [CrossRef]

15. Izzard, R.G.; Dray, L.M.; Karakas, A.I.; Lugaro, M.; Tout, C.A. Population nucleosynthesis in single and binary stars. I. Model. *Astron. Astrophys.* **2006**, *460*, 565–572. [CrossRef]

16. Izzard, R.G.; Glebbeek, E.; Stancliffe, R.J.; Pols, O.R. Population synthesis of binary carbon-enhanced metal-poor stars. *Astron. Astrophys.* **2009**, *508*, 1359–1374. [CrossRef]

17. Izzard, R.G.; Preece, H.; Jofre, P.; Halabi, G.M.; Masseron, T.; Tout, C.A. Binary stars in the Galactic thick disc. *Mon. Not. R. Astron. Soc.* **2018**, *473*, 2984–2999. [CrossRef]

18. Hurley, J.R.; Tout, C.A.; Pols, O.R. Evolution of binary stars and the effect of tides on binary populations. *Mon. Not. R. Astron. Soc.* **2002**, *329*, 897–928. [CrossRef]

19. Tauris, T.M.; Dewi, J.D.M. Research Note On the binding energy parameter of common envelope evolution. Dependency on the definition of the stellar core boundary during spiral-in. *Astron. Astrophys.* **2001**, *369*, 170–173. [CrossRef]

20. Zorotovic, M.; Schreiber, M.R.; Gänsicke, B.T.; Nebot Gómez-Morán, A. Post-common-envelope binaries from SDSS. IX: Constraining the common-envelope efficiency. *Astron. Astrophys.* **2010**, *520*, A86. [CrossRef]

21. De Marco, O.; Passy, J.C.; Moe, M.; Herwig, F.; Mac Low, M.M.; Paxton, B. On the α formalism for the common envelope interaction. *Mon. Not. R. Astron. Soc.* **2011**, *411*, 2277–2292. [CrossRef]

22. Davis, P.J.; Kolb, U.; Knigge, C. Is the common envelope ejection efficiency a function of the binary parameters? *Mon. Not. R. Astron. Soc.* **2012**, *419*, 287–303, [CrossRef]

23. Vassiliadis, E.; Wood, P.R. Evolution of low- and intermediate-mass stars to the end of the asymptotic giant branch with mass loss. *Astrophys. J.* **1993**, *413*, 641–657. [CrossRef]

24. Keller, D.; Izzard, R.G.; Stanghellini, L.; Lau, H.B. Planetary nebulae: Binarity, composition and morphology. Unpublished work, 2018.

25. Toomre, A. On the gravitational stability of a disk of stars. *Astrophys. J.* **1964**, *139*, 1217–1238. [CrossRef]

galaxies

MDPI

Article

Surveying Planetary Nebulae Central Stars for Close Binaries: Constraining Evolution of Central Stars Based on Binary Parameters

Todd Hillwig

Department of Physics & Astronomy, Valparaiso University, Valparaiso, IN 46383, USA;
Todd.Hillwig@valpo.edu

Received: 7 July 2018; Accepted: 3 August 2018; Published: 6 August 2018

Abstract: The increase in discovered close binary central stars of planetary nebulae is leading to a sufficiently large sample to begin to make broader conclusions about the effect of close binary stars on common envelope evolution and planetary nebula formation. Herein I review some of the recent results and conclusions specifically relating close binary central stars to nebular shaping, common envelope evolution off the red giant branch, and the total binary fraction and double degenerate fraction of central stars. Finally, I use parameters of known binary central stars to explore the relationship between the proto-planetary nebula and planetary nebula stages, demonstrating that the known proto-planetary nebulae are not the precursors of planetary nebulae with close binary central stars.

Keywords: planetary nebulae; stars: binaries; central stars of planetary nebulae; proto-planetary nebulae

1. Introduction

The study of binary stars in planetary nebulae (PNe) has the potential to provide information about the formation of PNe and the evolution of the central stars (CSs). A number of different surveys, using different methods, have identified binary central stars of planetary nebulae (CSPNe). Photometric surveys looking for variability due to a companion tend to find close binary systems with orbital periods of about a week or less (e.g., [1–3]). Such photometric surveys are sensitive to both cool companions and double degenerate systems [4,5]. Infrared surveys are designed to be sensitive to cool companions at all orbital periods [6], and radial velocity surveys are sensitive to stellar mass companions in orbital periods of up to several years [7,8].

Once binary CSPNe are identified, follow-up work can determine system parameters. A number of such systems have recently been studied, with various sets of orbital and stellar parameters published (see the updated list of known close binary CSPNe maintained by David Jones at http://drdjones.net/?q=bCSPN). Because it is more difficult to determine parameters for binaries with long periods, the majority of binary CSPNe with known physical parameters are close binaries. Of the close binary CSPNe, the large majority have been discovered using photometry.

In addition to providing discovery data for binaries, the light curves of those binaries can also tell us a great deal about the nature of the binary system. For example, close binary CSPNe with a main sequence companion have light curves dominated by an irradiation effect in which the inner hemisphere of the cool star is irradiated and heated by the hot CS. This behavior results in a nearly sinusoidal light curve with one maximum and one minimum per orbit. However, if the companion to the hot CS is a compact object (e.g., a white dwarf, WD), then any detected variability will likely be through eclipses or, more likely, ellipsoidal variability that is due to one of the stars nearly filling its Roche lobe and thus being elongated by the mutual gravity of the two stars. Typically, it will be the CS that nearly fills its Roche lobe, since these objects can still be large, not having contracted yet to the

WD cooling track. As the CS contracts, the ellipsoidal variability decreases in amplitude, eventually becoming unobservable. A light curve dominated by ellipsoidal variability has two maxima and two minima per orbit due to the different projected surface areas (brighter when seen edge-on, fainter when the two stars are aligned along the line of sight). To date, all well-studied close binary CSPNe with light curves dominated by ellipsoidal variability have been shown to be double degenerate (DD) systems. Technically, the CS is typically not yet fully degenerate, but we use this terminology here to provide a better comparison with true DD systems, which these will become. They are also related to core degenerate (CD) systems [9], though in this case the common envelope (CE) has detached and left a close binary system.

With the growing number of discovered binary CSPNe, we are beginning to reach a point at which statistically relevant statements can be made about how binary companions may influence both the ejection of the nebula and the evolution of the central star. Below I review some conclusions that have already been discussed in the literature, along with several connections that are currently being explored but still need confirmation. I then discuss what our current knowledge of close binary CSPNe can tell us about the evolution of PNe with close binary nuclei, especially in the context of proto-planetary nebulae (PPNe).

2. Relationships between Close Binary CSs and Their PNe

2.1. The Connection between Central Binaries and PN Shaping

One of the earliest predictions of close binary evolution via common envelope evolution was by Paczynski [10]. Shortly afterward, discussions began about how important CE evolution might be to the ejection of stellar envelopes and the production of PNe [11]. These discussions led to exploring the relationship between binary interactions and the non-spherical structures of most PNe (e.g., [1,12]). A more recently described option is grazing envelope evolution (GEE), in which the binary orbit shrinks as mass is lost from a giant star via Roche lobe overflow. However, jet launching from the accreting star strips away enough of the overflowing mass to prevent the CE [13].

Even though it seems intuitive that CE or GEE evolution could, and perhaps would, affect the shape of the ejected envelope, the details of how this happens are not fully understood. Our knowledge of CE evolution has seen a dramatic increase over the past five years or so, but as a computational problem we still do not understand the full process from spiral-in, through ejection, and on to a visible PN [14]. GEE is a more recent theory and does not have the volume of work performed at this time, and it is not clear that it can result in short-period systems with separations of a few $R_\odot$ or less [15]. For this reason, the remainder of this work will focus on CE evolution, but the reader should take into account that GEE could also play a role.

However, enough systems with known binary inclination and nebular inclination now exist to demonstrate a direct relationship between PN shapes and close central binaries [16]. The clear observational connection between binary CSs and their PNe combined with computational work (e.g., [17]) allows us to conclude that the close binary CSs either strongly affect, or completely determine, the shape of the PN.

Binary CSs also appear to be connected to PNe with high abundance discrepancy factors (ADFs). ADFs are the result of very different abundance values calculated from collisional versus recombination line strengths. Recent work [18,19] shows a strong correlation between high ADF and binary CSs in PNe. In addition, there is evidence in at least one case that the ADF is caused by two clearly distinct spatial structures, such that the abundance differences result from different physical parameters in the two spatially distinct regions [19,20]. These regions may well be another example of PN shaping resulting from binary interactions.

2.2. The CE Phase and Evolution of the CS

CE evolution in this context begins when one of the two stars in a binary system moves off the main sequence and expands to the point of coming into contact with its Roche lobe with a deep convective envelope. Actual spiral-in occurs later in the process, but the CE phase is triggered by the Roche lobe of the more massive star and its expansion off the main sequence. For binaries that start off relatively close, the CE interaction may take place before the expanding star reaches the AGB, resulting in post-RGB objects. If is not known if CE evolution on the RGB can result in a visible PN. Typically, the slower evolution of post-RGB stars would mean that the envelope disperses before the core becomes hot enough to ionize the surrounding gas. However, Hall et al. [21] discuss some circumstances under which a post-RGB core may be able to ionize a visible PN.

Until recently, no CSPNe had been identified that were clearly consistent with post-RGB evolution. Hillwig et al. [22] provide a list of five CSs that are potential post-RGB objects, four of which are in known close binary systems (it is possible that the fifth, HaWe 13 [23], is a close binary, but no direct evidence currently exists). Of the five, the most secure candidate for a post-RGB CS is that of ESO 330-9. Modeling of the CS spectrum and binary modeling are both consistent with a low-mass ($M_{CS} = 0.38 - 0.45 \, M_\odot$) post-RGB object [22].

While it is not clear that any of these PNe harbor post-RGB CSs, confirming the nature of these stars, and other potential post-RGB stars [22] will have important implications for our understanding of CE evolution and PN production.

2.3. Companions to Binary CSPNe and the Close Binary Fraction

As noted above, the light curves of close binary CSPNe can give us information about the most-likely companion to the CS. A light curve dominated by ellipsoidal variability in all well-studied cases so far have been shown to be double degenerate (DD) systems, where the companion to the hot CS is a compact object—most likely a WD. Likewise, light curves dominated by an irradiation effect have been shown to have cool main sequence stars as companions. This makes physical sense due to the nature of the system. Because the CS is very hot, any main sequence companion will exhibit a strong irradiation effect. If the cool companion is large enough to nearly fill its Roche lobe, and thus exhibit ellipsoidal variability, the resulting irradiation effect would be far stronger than the ellipsoidal variability. Thus, for a cool companion we expect the light curve to be dominated by irradiation. The combination of temperatures and luminosities mean the irradiation effect should be visible for the entire PN lifetime.

Of the currently known close binary CSPNe in the literature (for a current list, see the online table maintained by David Jones at http://drdjones.net/?q=bCSPN), 48 show photometric variability and are firmly identified as the CS of a PN. Of those, 13 have light curves dominated by ellipsoidal variability. If we take all of those to be DD systems, then we find that roughly one-quarter of all detected close binary CSPNe have a WD companion.

The caveat is that these are the detected systems. Ellipsoidal variability will only be observable while one or both stars fill a significant fraction of their Roche lobe. In each of the well-studied cases it appears that the deformed star is the CS. So as the CS contracts toward the WD track, it will detach farther from its Roche lobe, meaning the ellipsoidal variability will decrease relatively rapidly. It is unclear how long the ellipsoidal variability for a typical binary CSPN will be observable. The author and collaborators are working on determining a realistic result for the duration of the observable ellipsoidal variability in these systems. A back-of-the-envelope approach results in a fractional observable lifetime (relative to the length of time the PN is visible) of about one-third to one-quarter. This is based on the observable lifetime of PNe relative to the length of time the CS will be large enough to nearly fill its Roche lobe and display ellipsoidal variability. So, if ellipsoidal variability is only visible for about one-third to one-quarter of the lifetime of a PN, then we can assume that we only observe about one-third to one quarter of all DD systems to have photometric variability.

With these pieces of information, we would expect there to be roughly as many DD close binary CSPNe as there are those with a main sequence companion. If we use close binary fractions from the literature for photometric studies [2,24] of about 10–20%, or the value from the author's recent survey (paper in preparation) of ~12%, and assume that we have only observed roughly one-quarter to one-third of the close DD systems, we arrive at a corrected *close* binary fraction of ~20–30%. With an improved observable lifetime of ellipsoidal variability, and with a more precise value of the close binary fraction, the uncertainty will shrink, with a goal of identifying the close binary percentage to within a 5% range in the near future.

3. The Relationship between PPNe and PNe

The discovery of objects in the intermediate evolutionary step between the end of the AGB and the PN phase began the study of PPNe [25–27]. Many PPNe show very strongly aspherical structures [28–30], so discussions of PN shaping naturally involved PPNe, and it seemed clear that the mechanism that shapes PNe must act very early, such that bipolar structures are apparent even in the early PPN stages.

Various photometric, spectroscopic, and radial velocity studies of PPNe to search for close binaries that may be responsible for shaping have all ended with the absence of any evidence for close binary cores [31,32], or in fact for any binary systems with orbital periods less than years.

The high luminosities and low temperatures of the CSs of PPNe mean that they must be relatively large—much larger than the orbital separation of any of the known close binary CSPNe. The distances to PPNe are not well known, though Gaia data will hopefully provide answers for some PPNe. Luminosities are difficult to determine without well-known distances, though typical luminosities are expected to be in the several thousands of $L_\odot$. With typical temperatures determined from spectra of 5000–10,000 K, we can estimate radii of the PPN CSs to be tens of $R_\odot$ or more. With typical orbital separations for the known close binary CSPNe of a few $R_\odot$ or less, for these PPNe to be the progenitors of the close binaries, they would have to still be in the CE phase. However, current studies show that the CE phase lasts on the order of weeks, or at most years, depending on when we begin to consider the system as entering the CE [33]. These PPN CSs are clearly not evolving on that timescale.

Indeed, Figure 1 shows evolutionary tracks from Miller Bertolami [34] along with the *beginning* positions of the CSs of a number of well-studied close binary CSPNe. These beginning positions, the locations of the points in Figure 1, are defined as the location along the CS evolutionary track where the CS would just fill its Roche lobe, the idea being that if the CS were any larger, the system would still be in the CE phase (or at least would be undergoing significant mass transfer).

The placement of each data point in Figure 1 was determined by choosing a mass value for the two stars in each binary from the literature (often from a range of values, or on occasion choosing between two or more different values from different studies). As such, the mass values in Table 1 *should not* be reproduced as a list of known masses—they are representative values.

The luminosity of each CS was then assigned based on its mass, using the models of Miller Bertolami [34]. Each CS was assigned a luminosity approximately equal to that for a CS of that mass, at a post-AGB age of zero years.

The temperature of the CS was determined by first calculating the volume radius of the Roche lobe of the CS using the method of Eggleton [35] where the separation, a, can be found from P_{orb} and the masses using Kepler's third law. Once the radius is calculated, the temperature is approximated from the luminosity and radius using the Stefan–Boltzmann equation.

Table 1 shows the physical parameters from the literature and the calculated values for each CSPN.

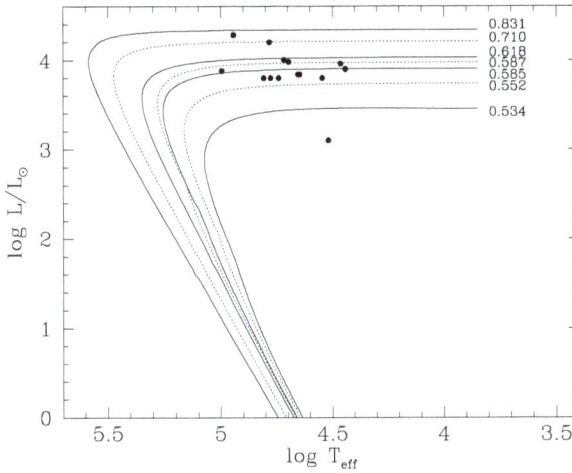

Figure 1. Evolutionary tracks from Table 3 in Miller Bertolami [34] for central stars of planetary nebulae (CSPNe) with $Z = 0.01$ and for final central star (CS) masses as labeled on the upper-right next to each track. Line styles alternate between sold and dashed to make them easier to follow on the plot. Also shown are the locations of the hot CS of a number of close binary CSPNe at their maximum possible radius (see text for description).

Table 1. Literature data for binary CSPNe along with calculated maximum radii and effective temperatures for the CSs at a luminosity appropriate for the CS mass.

PN Name	P_{orb} (days)	M_{CS} ($M_\odot$)	M_2 ($M_\odot$)	R_{max} ($R_\odot$)	$\log L/L_\odot$	$\log T_{eff}$	Ref.
V458 Vul	0.06812255	0.58	1	0.27	3.88	4.99	[36]
TS 01	0.163508	0.80	0.54	0.57	4.28	4.94	[37]
NGC 6337	0.1736133	0.56	0.35	0.53	3.80	4.81	[16]
Abell 41	0.226453066	0.56	0.56	0.61	3.80	4.78	[38,39]
HaTr 7	0.3221246	0.56	0.14	0.88	3.80	4.74	[22]
DS 1	0.35711296	0.70	0.3	0.96	4.20	4.78	[40,41]
Abell 63	0.46506921	0.63	0.29	1.10	4.00	4.72	[42]
Abell 46	0.47172909	0.51	0.14	1.09	3.10	4.51	[42]
Lo 16	0.48626	0.6	0.4	1.08	3.98	4.69	in prep
NGC 6026	0.528086	0.57	0.57	1.09	3.84	4.65	[43]
HFG 1	0.5816475	0.57	1.09	1.13	3.84	4.65	[44]
Abell 65	1.0037577	0.56	0.22	1.79	3.80	4.54	[45]
LTNF 1	2.2911667	0.59	0.25	3.14	3.96	4.46	[46]
Sp 1	2.90611	0.58	0.48	3.45	3.90	4.44	[16]

Figure 1 demonstrates that these "close" binaries are in fact close enough that the core of the evolving star could not have started its post-AGB evolution at the zero-age point (the top right of the plot, near the mass values) on the evolutionary tracks without over-filling its Roche lobe. The CSs of PPNe are observed to be single objects and stable over at least several decades through direct observation (though many show pulsations). They do not show evidence of binarity, significant mass transfer, or large-scale changes, and the circumstellar envelope has detached from the CS [32]. If these objects entered a CE in the past, they are now at a stage after the CE has been ejected. This has several implications. First, the particular PNe with close binary CSs shown in Figure 1 would *not* have gone through the PPN phase as we know it. Observed PPNe have CSs that are too cool and luminous (thus too large) to exist in a close binary system unless they are undergoing mass transfer or are in the CE phase. We also now know that the CE phase evolves rapidly to ejection of the envelope (or merger of the cores)—faster than the directly observed lifetimes of these PPNe. Indeed, for realistic stellar masses, none of the known close binary systems (with $P_{orb} \lesssim 7$ days) could begin their post-CE evolution at a stage (the zero-age post-AGB point) that would result in an object similar to observed PPN CSs.

One caveat to this is the possibility of a self-regulated CE phase, which may result in a longer-lived CE [47]. This phase is not well understood in terms of the rate of occurrence or observable properties, but it seems unlikely that it would result in the well-behaved stellar-like properties of the observed PPN CSs. This is especially true given the consistent evolution of the studied PPNe, such that they would either all be undergoing self-regulated CE evolution, or it becomes even less likely that any of them are.

So it seems clear that the known PPNe are not progenitors of PNe with close binary CSs, despite exhibiting strong nebular shaping. Note, however, that there could still be a wider binary companion that could engulf the current CS in a CE phase later in its evolution, producing a DD system. However, none of the known PPNe will proceed directly to the PN phase with a close binary CS. At present there are several alternative possibilities for the observed morphology of PPNe: the shaping is due to a companion (likely of low mass) that has been disrupted or has merged with the CS of the PPN, the shaping is caused by a wider companion, or the shaping is unrelated to a companion.

The second implication of Figure 1 is that the binary companion clearly affects the evolution of the CS. If the CSs of the binary systems in Table 1 had evolved as single stars, we would expect them to begin their evolution as a CSPN roughly at the zero-age point on the evolutionary tracks in Figure 1. However, due to their binary companions, they either need to start their evolution at a more evolved stage (higher temperature) on those evolutionary tracks, or they would over-fill their Roche lobe and transfer mass onto their companion. Either results in a change relative to single star evolution, or even CE evolution that would result in a wider final separation. Determining the result of the change in evolution is beyond the scope of the present work, but understanding that process may provide a key element in the energetics of CE evolution. For example, the core evolution may be accelerated as the companion spirals in, the final mass of the CS may differ from that of an equivalent single star, the core may undergo significant non-thermal equilibrium evolution, or some combination of these may occur. However, it seems that these interactions are likely to affect the overall energy budget of the CE evolution (though whether it is a significant effect is also a question).

4. Discussion

I have summarized some of the recent conclusions about the relationship between close binary CSs and their PNe. These conclusions are possible due to the increase in the number of discovered close binary CSPNe and by the increase in the number of systems with full binary modeling. I also demonstrated that the PPN-to-PN connection is not as clear as it may seem. The observed sample of PPNe are *not* progenitors of PNe with close binary stars. In addition, close binary CSPNe seem to alter the post-AGB evolution of the core of the evolving star, with possible implications for CE evolution in addition to PN shaping and close binary evolution.

Funding: This material is based upon work supported by the National Science Foundation under Grant No. AST-1109683. Any opinions, findings, and conclusions or recommendations expressed in this material are those of the author(s) and do not necessarily reflect the views of the National Science Foundation.

Conflicts of Interest: The author declares no conflict of interest.

References

1. Bond, H.E.; Livio, M. Morphologies of Planetary Nebulae Ejected by Close-Binary Nuclei. *Astrophys. J.* **1990**, *355*, 568–576. [CrossRef]
2. Miszalski, B.; Acker, A.; Moffat, A.F.J.; Parker, Q.A.; Udalski, A. Binary planetary nebulae nuclei towards the Galactic bulge. I. Sample discovery, period distribution, and binary fraction. *Astron. Astrophys.* **2009**, *496*, 813–825. [CrossRef]

3. De Marco, O.; Long, J.; Jacoby, G.H.; Hillwig, T.; Kronberger, M.; Howell, S.B.; Reindl, N.; Margheim, S. Identifying close binary central stars of PN with Kepler. *Mon. Not. R. Astron. Soc.* **2015**, *448*, 3587–3602. [CrossRef]
4. Hillwig, T.C. The physical characteristics of binary central stars of planetary nebulae. In Proceedings of the Asymmetric Planetary Nebulae 5 Conference, Bowness-on-Windermere, UK, 20–25 June 2010; p. 275.
5. Santander-Garcia, M.; Rodríguez-Gil, P.; Jones, D.; Corradi, R.L.M.; Miszalski, B.; Pyrzas, S.; Rubio-Díez, M.M. The binary central stars of PNe with the shortest orbital period. In Proceedings of the Asymmetric Planetary Nebulae 5 Conference, Bowness-on-Windermere, UK, 20–25 June 2010; p. 259.
6. Douchin, D.; De Marco, O.; Frew, D.J.; Jacoby, G.H.; Jasniewicz, G.; Fitzgerald, M.; Passy, J.C.; Harmer, D.; Hillwig, T.; Moe, M. The binary fraction of planetary nebula central stars-II. A larger sample and improved technique for the infrared excess search. *Mon. Not. R. Astron. Soc.* **2015**, *448*, 3132–3155. [CrossRef]
7. De Marco, O.; Bond, H.E.; Harmer, D.; Fleming, A.J. Indications of a Large Fraction of Spectroscopic Binaries among Nuclei of Planetary Nebulae. *Astrophys. J.* **2004**, *602*, L93. [CrossRef]
8. Jones, D.; Van Winckel, H.; Aller, A.; Exter, K.; De Marco, O. The long-period binary central stars of the planetary nebulae NGC 1514 and LoTr 5. *Astron. Astrophys.* **2017**, *600*, L9. [CrossRef]
9. Kashi, A.; Soker, N. A circumbinary disc in the final stages of common envelope and the core-degenerate scenario for Type Ia supernovae. *Mon. Not. R. Astron. Soc.* **2011**, *417*, 1466–1479. [CrossRef]
10. Paczynski, B. Common Envelope Binaries. *Struct. Evol. Close Bin. Syst.* **1976**, *73*, 75.
11. Bond, H.E.; Liller, W.; Mannery, E.J. UU Sagittae: Eclipsing nucleus of the planetary nebula Abell 63. *Astrophys. J.* **1978**, *223*, 252. [CrossRef]
12. Morris, M. Models for the structure and origin of bipolar nebulae. *Astrophys. J.* **1981**, *249*, 572–585. [CrossRef]
13. Soker, N. Close Stellar Binary Systems by Grazing Envelope Evolution. *Astrophys. J.* **2015**, *800*, 114. [CrossRef]
14. Ivanova, N.; Nandez, J. Planetary Nebulae Embryo after a Common Envelope Event. *Galaxies* **2018**, *6*, 75. [CrossRef]
15. Abu-Backer, A.; Gilkis, A.; Soker, N. Orbital Radius during the Grazing Envelope Evolution. *Astrophys. J.* **2018**, *861*, 136. [CrossRef]
16. Hillwig, T.C.; Jones, D.; De Marco, O.; Bond, H.E.; Margheim, S. Frew, D. Observational Confirmation of a Link between Common Envelope Binary Interaction and Planetary Nebula Shaping. *Astrophys. J.* **2016**, *832*, 125. [CrossRef]
17. Nordhaus, J. Common Envelope Evolution: Implications for Post-AGB Stars and Planetary Nebulae. In *Planetary Nebulae: Multi-Wavelength Probes of Stellar and Galactic Evolution*; IAU Symposium; Cambridge University Press: Cambridge, UK, 2017; Volume 323, p. 207.
18. Corradi, R.L.M.; García-Rojas, J.; Jones, D.; Rodríguez-Gil, P. Binarity and the Abundance Discrepancy Problem in Planetary Nebulae. *Astrophys. J.* **2015**, *803*, 99. [CrossRef]
19. Jones, D.; Wesson, R.; García-Rojas, J.; Corradi, R.L.M.; Boffin, H.M.J. NGC 6778: Strengthening the link between extreme abundance discrepancy factors and central star binarity in planetary nebulae. *Mon. Not. R. Astron. Soc.* **2016**, *455*, 3263–3272. [CrossRef]
20. García-Rojas, J.; Corradi, R.L.M.; Monteiro, H.; Jones, D.; Rodríguez-Gil, P. Cabrera-Lavers, A. Imaging the Elusive H-poor Gas in the High adf Planetary Nebula NGC 6778. *Astrophys. J.* **2016**, *824*, L27. [CrossRef]
21. Hall, P.D.; Tout, C.A.; Izzard, R.G.; Keller, D. Planetary nebulae after common-envelope phases initiated by low-mass red giants. *Mon. Not. R. Astron. Soc.* **2013**, *435*, 2048–2059. [CrossRef]
22. Hillwig, T.C.; Frew, D.J.; Reindl, N.; Rotter, H.; Webb, A. Margheim, S. Binary Central Stars of Planetary Nebulae Discovered through Photometric Variability. V. The Central Stars of HaTr 7 and ESO 330-9. *Astrophys. J.* **2017**, *153*, 24. [CrossRef]
23. Frew, D.J.; Parker, Q.A.; Bojičić, I.S. The Hα surface brightness-radius relation: A robust statistical distance indicator for planetary nebulae. *Mon. Not. R. Astron. Soc.* **2016**, *455*, 1459–1488. [CrossRef]
24. Bond, H.E. Binarity of Central Stars of Planetary Nebulae. *arXiv* **1999**, arXiv:astro-ph/9909516.
25. Parthasarathy, M.; Pottasch, S.R. The far-infrared (IRAS) excess in HD 161796 and related stars. *Astron. Astrophys.* **1986**, *154*, L16.
26. Hrivnak, B.J.; Kwok, S.; Volk, K.M. The High-Latitude F Supergiant IRAS 18095+2704: A Proto-Planetary Nebula. *Astrophys. J.* **1988**, *331*, 832–837. [CrossRef]
27. Van der Veen, W.E.C.J.; Habing, H.J.; Geballe, T.R. Objects in transition from the AGB to the planetary nebula stage: New visual and infrared observations. *Astron. Astrophys.* **1989**, *226*, 108.

28. Hrivnak, B.J.; Langill, P.P.; Su, K.Y.L.; Kwok, S. Subarcsecond Optical Imaging of Proto-Planetary Nebulae. *Astrophys. J.* **1999**, *513*, 421. [CrossRef]

29. Ueta, T.; Meixner, M.; Bobrowsky, M. A Hubble Space Telescope Snapshot Survey of Proto-Planetary Nebula Candidates: Two Types of Axisymmetric Reflection Nebulosities. *Astrophys. J.* **2000**, *528*, 861. [CrossRef]

30. Sahai, R. Multi-polar Structures in Young Planetary and Protoplanetary Nebulae. In *Symposium-International Astronomical Union*; Cambridge University Press: Cambridge, UK, 2003; Volume 209, p. 471.

31. Hrivnak, B.J. A Search for Binaries in Proto-Planetary Nebulae. In Proceedings of the Asymmetrical Planetary Nebulae IV, La Palma, Spain, 18–22 June 2017.

32. Hrivnak, B.J.; Van de Steene, G.; Van Winckel, H.; Sperauskas, J.; Bohlender, D. Lu, W. Where are the Binaries? Results of a Long-term Search for Radial Velocity Binaries in Proto-planetary Nebulae. *Astrophys. J.* **2017**, *846*, 96. [CrossRef]

33. Galaviz, P.; De Marco, O.; Passy, J.-C.; Staff, J.E.; Iaconi, R. Common Envelope Light Curves. I. Grid-code Module Calibration. *Astrophys. J. Suppl. Ser.* **2017**, *229*, 36. [CrossRef]

34. Miller Bertolami, M.M. New Models for the Evolution of Post-Asymptotic Giant Branch Stars and Central Stars of Planetary Nebulae. *Astron. Astrophys.* **2016**, *588*, A25. [CrossRef]

35. Eggleton, P.P. Aproximations to the radii of Roche lobes. *Astrophys. J.* **1983**, *268*, 368. [CrossRef]

36. Rodríguez-Gil, P.; Santander-García, M.; Knigge, C.; Corradi, R.L.M.; Gänsicke, B.T.; Barlow, M.J.; Drake, J.J.; Drew, J.; Miszalski, B.; Napiwotzki, R.; et al. The orbital period of V458 Vulpeculae, a post-double common-envelope nova. *Mon. Not. R. Astron. Soc.* **2010**, *407*, L21. [CrossRef]

37. Tovmassian, G.; Yungelson, L.; Rauch, T.; Suleimanov, V.; Napiwotzki, R.; Stasińska, G.; Tomsick, J.; Wilms, J.; Morisset, C.; Pena, M.; et al. The Double-degenerate Nucleus of the Planetary Nebula TS 01: A Close Binary Evolution Showcase. *Astrophys. J.* **2010**, *714*, 178. [CrossRef]

38. Bruch, A.; Vaz, L.P.R.; Diaz, M.P. An analysis of the light curve of the post common envelope binary MT Serpentis. *Astron. Astrophys.* **2001**, *377*, 898–910. [CrossRef]

39. Shimanskii, V.V.; Borisov, N.V.; Sakhibullin, N.A.; Sheveleva, D.V. MT Ser, a binary blue subdwarf. *Astron. Rep.* **2008**, *52*, 479–486. [CrossRef]

40. Drilling, J.S. LSS 2018: A double-lined spectroscopic binary central star with an extremely large reflection effect. *Astrophys. J.* **1985**, *294*, L107. [CrossRef]

41. Hilditch, R.W.; Harries, T.J.; Hill, G. On the reflection effect in three sdOB binary stars. *Mon. Not. R. Astron. Soc.* **1996**, *279*, 1380–1392. [CrossRef]

42. Afşar, M.; Ibanoglu, C. Two-colour photometry of the binary planetary nebula nuclei UU Sagitte and V477 Lyrae: Oversized secondaries in post-common-envelope binaries. *Mon. Not. R. Astron. Soc.* **2008**, *391*, 802–814. [CrossRef]

43. Hillwig, T.C.; Bond, H.E.; Afşar, M.; De Marco, O. Binary Central Stars of Planetary Nebulae Discovered through Photometric Variability. II. Modeling the Central Stars of NGC 6026 and NGC 6337. *Astron. J.* **2010**, *140*, 319. [CrossRef]

44. Shimanskii, V.V.; Borisov, N.V.; Sakhibullin, N.A.; Surkov, A.E. The Nature of the Unique Precataclysmic Variable V664 Cas with Two-Peaked Balmer Lines in Its Spectrum. *Astron. Rep.* **2004**, *48*, 563–576. [CrossRef]

45. Hillwig, T.C.; Bond, H.E.; Frew, D.J.; Schaub, S.C.; Bodman, E.H.L. Binary Central Stars of Planetary Nebulae Discovered through Photometric Variability. IV. The Central Stars of HaTr 4 and Hf 2-2. *Astron. J.* **2016**, *152*, 34. [CrossRef]

46. Shimanskii, V.V.; Borisov, N.V.; Pozdnyakova, S.A.; Bikmaev, I.F.; Vlasyuk, V.V.; Sakhibullin, N.A.; Spiridonova, O.I. Fundamental parameters of BE UMa revised. *Astron. Rep.* **2008**, *52*, 558–575. [CrossRef]

47. Ivanova, N.; Justham, S.; Chen, X.; De Marco, O.; Fryer, C.L.; Gaburov, E.; Ge, H.; Glebbeek, E.; Han, Z.; Li, X.D.; et al. Common envelope evolution: Where we stand and how we can move forward. *Astron. Astrophys. Rev.* **2013**, *21*, 59. [CrossRef]

MDPI

Article

Spectroscopic and Photometric Variability of Three Oxygen Rich Post-AGB "Shell" Objects

Griet C. Van de Steene [1,*], **Bruce J. Hrivnak** [2], **Hans Van Winckel** [3], **Julius Sperauskas** [4] and **David Bohlender** [5]

[1]   Royal Observatory of Belgium, Astronomy and Astrophysics, Ringlaan 3, 1180 Brussels, Belgium
[2]   Department of Physics and Astronomy, Valparaiso University, Valparaiso, IN 46383, USA; Bruce.Hrivnak@valpo.edu
[3]   Instituut voor Sterrenkunde, K.U. Leuven University, Celestijnenlaan 200 D, B-3001 Leuven, Belgium; h.vanwinckel@kuleuven.ac.be
[4]   Institute of Theoretical Physics and Astronomy, Vilnius University Observatory, Ciurlionio 29, 2009 Vilnius, Lithuania; julius.sperauskas@ff.vu.lt
[5]   National Research Council of Canada, Herzberg Astronomy and Astrophysics, 5071 West Saanich Road, Victoria, BC V9E 2E7, Canada; david.bohlender@nrc-cnrc.gc.ca
*   Correspondence: g.vandesteene@oma.be

Received: 28 June 2018; Accepted: 29 November 2018; Published: 3 December 2018

Abstract: Light, color, and radial velocity data (2007–2015) for HD 161796, V887 Her, and HD 331319, three oxygen-rich post-AGB stars, have thus far not provided direct support for the binary hypothesis to explain the shapes of planetary nebulae and severely constrain the properties of any such undetected companions. The light and velocity curves are complex, showing similar periods and variable amplitudes. Nevertheless, over limited time intervals, we compared the phasing of each. The color curves appear to peak with or slightly after the light curves, while the radial velocity curves peak about a quarter of a cycle before the light curves. Thus it appears that these post-AGB stars are brightest when smallest and hottest. The spectra of these objects are highly variable. The Hα line has multiple, variable emission and absorption components. In these oxygen-rich post-AGB stars atmospheric lines, such as near-infrared Ca II triplet and low-excitation atomic lines, also have multiple components and sometimes show line doubling, indicative of shocks induced by pulsation.

Keywords: late-stage stellar evolution; planetary nebulae; multi-wavelength photometry; radial velocity; stellar evolution; pulsation; shock wave

1. Introduction

Stars at the proto-planetary nebula (PPN) phase evolve from the Asymptotic Giant Branch towards the PN phase becoming hotter at almost constant luminosity. These low mass stars are surrounded by a circumstellar envelope ejected during the AGB phase. The details of the physical processes and stellar winds that shape these envelopes remain uncertain. The outstanding question in PN research is the role and importance of binarity in the shaping of PNe.

We selected seven bright (V < 10 mag) post-AGB stars observable with 1-m class telescopes from the northern hemisphere (Hrivnak et al. [1]). They all have F- and G spectral types. Of these 7 metal-poor PPNe, 3 are O-rich and 4 are C-rich. In this contribution, we will consider the 3 O-rich objects: HD 161796 (IRAS 17436+5003), HD 331319 (IRAS 19475+3119), and V887 Her (IRAS 18095+2704). The objects have spectral energy distributions typical for "shell" objects: they are double peaked, with a peak in the visible arising from the (reddened) photosphere and a second peak in the mid-infrared arising from re-radiation from cool (T ≤ 200 K) dust. However, they have no near-infrared excess, the presence of which is an indication of the presence of a disk. The disk-type post-AGB stars have been shown to be binaries (Manick et al. [2], Van Winckel et al. [3]).

HD 161796 is highly aspherical, and shows a low density polar outflow and a high density equatorial region (Min et al. [4]). HD 331319 is a quadrupolar nebula with two collimated bipolar lobes (Sahai et al. [5]). V887 Her is an extended source with a significant circumstellar shell (Ueta et al. [6]) and OH masers (Lewis et al. [7]).

To prove or exclude the presence of binaries in these shell objects is crucial to our understanding of the way(s) that bipolar PPNe and PNe form. If it turns out that these PPNe are not binaries, then it raises the possibility that there may be more than one way to produce bipolar PNe.

2. Observations and Analysis

High-resolution spectra were obtained at the Dominion Astrophysical Observatory 1.2 m telescope since 2007 and with the HERMES Spectrograph on the 1.2 m Mercator telescope (Raskin et al. [8], Van Winckel et al. [9]) since 2009. Radial velocities of the HERMES spectra were determined via the cross-correlation technique using a mask in the range 477–655 nm for F0-type stars and fitting the peak of the obtained correlation function with a gaussian (Hrivnak et al. [1]). Radial velocities of DAO spectra were also obtained via cross correlation with a set of bright IAU RV standards observed with the same instrumentation. In this case velocities were determined by fitting a parabola to the upper half of the cross correlation profiles (Hrivnak et al. [1]).

Photometric observations of these objects were obtained from 2007 to 2015 at the Valparaiso University Observatory with a 0.4 m telescope equipped with a CCD camera and filters standardized to the Johnson B and V and Cousins R systems. Differential photometry was employed, using an aperture of 11″ diameter (Hrivnak et al. [10]). For V887 Her ASAS-N data were available and used (Kochanek et al. [11]).

3. Results and Discussion

3.1. Radial Velocity and Photometry

The RV amplitudes are small (<4.0 km/s, Hrivnak et al. [12]) and the periods found in the RV data are in the range 35 to 103 days, similar to the pulsation periods of the objects. So far, no clear evidence for long-period RV variations due to binary companions has been found in any of these objects. This sets significant constraints on the properties of any undetected binary companion: they must be of low mass, ≤ 0.2 $M_\odot$, or long period, longer than 30 years. Thus the present observations do not provide direct support for the binary hypothesis to explain the shapes of thes PPNe.

The light and velocity curves are complex with multiple periods and variable amplitudes. However, over limited time intervals during 2007–2015, we were able to identify dominant periods in the light, color, and velocity curves and compare the phasing of each. See figures in Hrivnak et al. [12] which shows the phased light, color, and velocity curves of the objects. The color curves appear to peak with or slightly after the light curves while the RV curves peak about a quarter of a cycle before the light curves. For all three objects plus two previously studies PPNe (Hrivnak et al. [13]), (a) the light and color curves are approximately in phase, with the suggestion that the color curve perhaps peaks slightly ($\sim$0.05 P) after the light. (b) The RV curve is approximately -0.25 P out of phase with the light curve. Thus it appears that these PPNe are brightest when smallest and hottest. These results differ from those found for Cepheid variables, where the light and velocity differ by half a cycle, and are hottest at about average size and expanding. However, they do appear to have similar phasing to the larger amplitude pulsations seen in RV Tauri variables. Attempts at modeling pulsation in these post-AGB stars are few (Fokin et al. [14]) and the observed periods are longer than models indicate.

3.2. Spectroscopic Variability

The average Hα profiles show broad wings with central emission and absorption components (Figure 1). The individual spectra show that these Hα components vary in depth, in width, and position (Figure 2). The strength of the blue and red emission are changing alternately. The red emission and absorption component is generally weaker than the blue. In the inverse P-Cygni profile phase, the blue emission peak can be very strong.

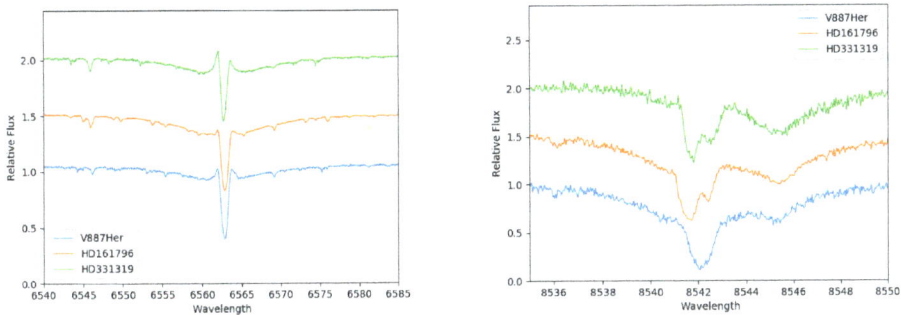

Figure 1. (**left**) The averaged Hα profiles of all spectra corrected for RV and normalized. (**right**) The averaged Ca II 854.2 nm profiles of all spectra corrected for RV and normalized.

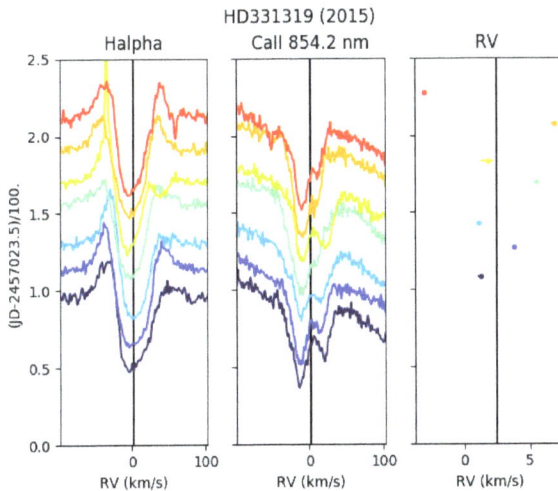

Figure 2. HD 331319 line profiles observed in 2015. At the left Hα, in the middle the Ca II line at λ 854.2 nm, and at the right the RV's measured in the spectrum. The y-axes are Julian Dates, modified as indicated.

The near-infrared Ca II triplet and some low-excitation metal lines such as Ba II and Fe II also show strong variability (Figure 2). In these lines, line splitting in two can be observed sometimes. This splitting was observed in C-rich post-AGB stars (Klochkova [15]) with a complex circumstellar envelope, but here we confirm that it is also observed in such O-rich post-AGB stars. Further analysis will help clarify where these components originate in these stars. The observed splitting and short-term variability in low excitation atomic lines provide evidence that shock waves are present in the outer layers of the atmosphere or circumstellar envelope (Fokin et al. [14], Klochkova [15], Zacs et al. [16]). The lack of line splitting phenomena in the high excitation potential lines is most likely because these lines are formed deeper in the atmosphere.

The complex spectral variability and lack of stable periods in the light and velocity curves makes the temporal changes hard to phase. Provisionally, in Figure 3 we phased V887 Her to the period of 102.3 days and the same epoch as Hrivnak et al. [12]. We searched for evidence for a jet as seen in BD+46 442 (Bollen et al. [17]), in which a broad absorption with a velocity up to ∼−400 km/s is observed. In the phased spectra of V887 Her there is an extended blue absorption at phase 0.35. However the velocity at the blue edge of the absorption is only −50 km/s, no more than 2–3 times the typical expansion velocity of PNe and AGB stars, and lasts less than 0.10 phase cycle, while in BD+46 442 it is at least 0.25 phase cycle. These are not indications for a strong jet. These "Shell" objects have Hα profiles similar to "disc" sources, but without the presence of strong jets.

The analysis of the high-resolution spectra of these post-AGB stars continues, in order to better investigate and understand the shocks, pulsation, and kinematics in these objects.

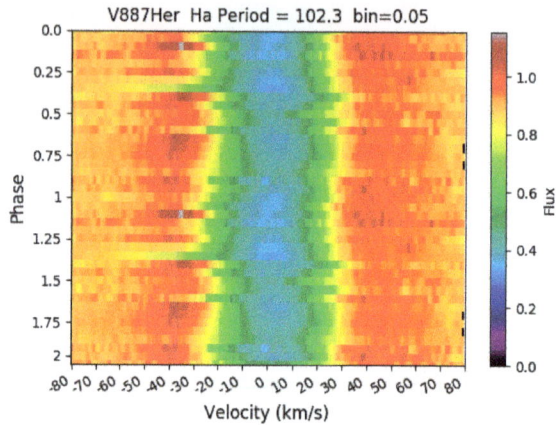

Figure 3. Dynamic Hα spectra of V887 Her, phased to 102.3 days and binned to 0.05, showing different phase-dependent variations in the line profiles. The velocity on the x-axis denotes the RV shift from the laboratory wavelength. The color indicates the continuum normalized flux.

Author Contributions: G.C.V.d.S.: obtained the data, did data analysis and data reduction, and wrote the paper B.J.H. obtained data, did data reduction and analysis, and helped writing up the paper. H.V.W., J.S., and D.B. contributed to the spectroscopic and/or photometric observations.

Funding: B.J.H. acknowledges ongoing support from the National Science Foundation (most recently AST 1009974, 1413660). J.S. acknowledge support of the Research Council of Lithuania under the grant MIP-085/2012. This research has been conducted in part based on funding from the Research Council of K.U. Leuven (GOA/13/012) and was partially funded by the Belgian Science Policy Office under contract BR/143/A2/STARLAB.

Acknowledgments: This research has made use of the SIMBAD database, operated at CDS, Strasbourg, France, and NASA's Astrophysical Data System. We also thank the many Valparaiso University undergraduate students and colleagues at the ROB, KUL, and ULB who obtained observations for the project.

Conflicts of Interest: The authors declare no conflict of interest.

References

1. Hrivnak, B.J.; Van de Steene, G.; Van Winckel, H.; Sperauskas, J.; Bohlender, D.; Lu, W. Where are the Binaries? Results of a Long-term Search for Radial Velocity Binaries in Proto-planetary Nebulae. *Astrophys. J.* **2017**, *846*, 96. [CrossRef]

2. Manick, R.; Van Winckel, H.; Kamath, D.; Hillen, M.; Escorza, A. Establishing binarity amongst Galactic RV Tauri stars with a disc. *Astron. Astrophys.* **2017**, *597*, A129. [CrossRef]

3. Van Winckel, H. Post-AGB binaries as tracers of stellar evolution. In *Planetary Nebulae: Multi-Wavelength Probes of Stellar and Galactic Evolution*; Liu, X., Stanghellini, L., Karakas, A., Eds.; IAU Symposium; International Astronomical Union: Paris, France, 2017; Volume 323, pp. 231–234. [CrossRef]

4. Min, M.; Jeffers, S.V.; Canovas, H.; Rodenhuis, M.; Keller, C.U.; Waters, L.B.F.M. The color dependent morphology of the post-AGB star HD 161796. *Astron. Astrophys.* **2013**, *554*, A15. [CrossRef]

5. Sahai, R.; Sánchez Contreras, C.; Morris, M.; Claussen, M. A Quadrupolar Preplanetary Nebula: IRAS 19475+3119. *Astrophys. J.* **2007**, *658*, 410–422. [CrossRef]

6. Ueta, T.; Meixner, M.; Bobrowsky, M. A Hubble Space Telescope Snapshot Survey of Proto-Planetary Nebula Candidates: Two Types of Axisymmetric Reflection Nebulosities. *Astrophys. J.* **2000**, *528*, 861–884. [CrossRef]

7. Lewis, B.M. On the Transience of Circumstellar Shells about |B| >=10deg OH/IR Stars. I. Basic Statistics. *Astrophys. J.* **2000**, *533*, 959–968. [CrossRef]

8. Raskin, G.; van Winckel, H.; Hensberge, H.; Jorissen, A.; Lehmann, H.; Waelkens, C.; Avila, G.; de Cuyper, J.P.; Degroote, P.; Dubosson, R.; et al. HERMES: A high-resolution fibre-fed spectrograph for the Mercator telescope. *Astron. Astrophys.* **2011**, *526*, A69. [CrossRef]

9. Van Winckel, H.; Jorissen, A.; Gorlova, N.; Dermine, T.; Exter, K.; Masseron, T.; Østensen, R.; Van Eck, S.; Van de Steene, G. Post-AGB binaries in an evolutionary perspective: A HERMES monitoring programme. *Memorie della Societa Astronomica Italiana* **2010**, *81*, 1022.

10. Hrivnak, B.J.; Lu, W.; Nault, K.A. Variability in Proto-planetary Nebulae. IV. Light Curve Analysis of Four Oxygen-rich, F Spectral Type Objects. *Astron. J.* **2015**, *149*, 184. [CrossRef]

11. Kochanek, C.S.; Shappee, B.J.; Stanek, K.Z.; Holoien, T.W.S.; Thompson, T.A.; Prieto, J.L.; Dong, S.; Shields, J.V.; Will, D.; Britt, C.; et al. The All-Sky Automated Survey for Supernovae (ASAS-SN) Light Curve Server v1.0. *Publ. Astron. Soc. Pac.* **2017**, *129*, 104502. [CrossRef]

12. Hrivnak, B.J.; Van de Steene, G.; Van Winckel, H.; Lu, W.; Sperauskas, J. Variability in Proto-Planetary Nebulae: V. Velocity and Light Curve Analyses of IRAS 17436+5003, 18095+2704, and 19475+3119. *arXiv* **2018**, arXiv:astro-ph.SR/1810.13037.

13. Hrivnak, B.J.; Lu, W.; Sperauskas, J.; Van Winckel, H.; Bohlender, D.; Začs, L. Studies of Variability in Proto-planetary Nebulae. II. Light and Velocity Curve Analyses of IRAS 22272+5435 and 22223+4327. *Astrophys. J.* **2013**, *766*, 116. [CrossRef]

14. Fokin, A.B.; Lèbre, A.; Le Coroller, H.; Gillet, D. Non-linear radiative models of post-AGB stars: Application to HD 56126. *Astron. Astrophys.* **2001**, *378*, 546–555. [CrossRef]

15. Klochkova, V.G. Circumstellar envelope manifestations in the optical spectra of evolved stars. *Astrophys. Bull.* **2014**, *69*, 279–295. [CrossRef]

16. Začs, L.; Musaev, F.; Kaminsky, B.; Pavlenko, Y.; Grankina, A.; Sperauskas, J.; Hrivnak, B.J. Spectroscopic Variability of IRAS 22272+5435. *Astrophys. J.* **2016**, *816*, 3. [CrossRef]

17. Bollen, D.; Van Winckel, H.; Kamath, D. Jet creation in post-AGB binaries: the circum-companion accretion disk around BD+46 442. *Astron. Astrophys.* **2017**, *607*, A60. [CrossRef]

galaxies

MDPI

Article

The Real-Time Evolution of V4334 Sgr

Peter A. M. van Hoof [1,*], Stefan Kimeswenger [2,3], Griet C. Van de Steene [1], Adam Avison [4], Albert A. Zijlstra [4], Lizette Guzman-Ramirez [5], Falk Herwig [6] and Marcin Hajduk [7]

1   Royal Observatory of Belgium, Ringlaan 3, B-1180 Brussels, Belgium; griet.vandesteene@oma.be
2   Instituto de Astronomía, Universidad Católica del Norte, Avenida Angamos 0610,
    Antofagasta 1240000, Chile; skimeswenger@ucn.cl
3   Institut für Astro- und Teilchenphysik, Universität Innsbruck, Technikerstr. 25/8, 6020 Innsbruck, Austria
4   School of Physics & Astronomy, University of Manchester, Oxford Road, Manchester M13 9PL, UK;
    adam.avison@manchester.ac.uk (A.A.); a.zijlstra@manchester.ac.uk (A.A.Z.)
5   Leiden Observatory, Leiden University, Niels Bohrweg 2, NL-2333 CA Leiden, The Netherlands;
    guzman@strw.leidenuniv.nl
6   Department of Physics and Astronomy, University of Victoria, P.O. Box 3055, Victoria, BC V8W 3P6, Canada;
    fherwig@uvic.ca
7   Space Radio-Diagnostics Research Centre, University of Warmia and Mazury, Prawochenskiego Str. 9,
    10-720 Olsztyn, Poland; marcin.hajduk@uwm.edu.pl
*   Correspondence: p.vanhoof@oma.be; Tel.: +32-2-3736787

Received: 19 June 2018; Accepted: 24 July 2018; Published: 26 July 2018

Abstract: V4334 Sgr (Sakurai's object) is an enigmatic evolved star that underwent a very late thermal pulse a few years before its discovery in 1996. It ejected a new hydrogen-deficient nebula in the process. The source has been observed continuously since, at many wavelengths ranging from the optical to the radio regime. In this paper we evaluate these data and discuss the evolution of this object. We reach the conclusion that we have seen no evidence for photoionization of the nebula yet and that the spectral features we see are caused either by shocks or by dust. These shocks are an integral part of the hydrodynamic shaping that is now producing a new bipolar nebula inside the old planetary nebula (PN), implying that we have a detailed observational record of the very early stages of the shaping of a bipolar nebula.

Keywords: stellar mass loss; stellar evolution; planetary nebulae; circumstellar dust

1. Introduction

V4334 Sgr (also referred to as Sakurai's object, after the Japanese amateur astronomer who discovered this object) is the central star of an old planetary nebula (PN) that underwent a very late thermal pulse (VLTP) a few years before its discovery in 1996 [1]. As a result of the thermal pulse, the star brightened considerably and became a very cool, born-again asymptotic giant branch star with a spectrum resembling a carbon star. During the VLTP, it ingested its remaining hydrogen-rich envelope into the helium-burning shell and ejected the processed material shortly afterwards to form a new, hydrogen-deficient nebula inside the old PN. After a few years, dust formation started in the new ejecta, and the central star became highly obscured. An [O III] image of the old PN overlaid with contours showing the first detection of free-free emission from the new ejecta in the center can be found in Hajduk et al. [2]. The contours were derived from 8.6 GHz radio continuum observations obtained with the Very Large Array (VLA). The VLA data traced the old PN quite reasonably, while the central emission has no counterpart in the [O III] image and was only marginally resolved.

Emission lines of ionized species were discovered as follows: first He I 1083 nm was discovered in 1998 [3], and later, in 2001, optical forbidden lines from neutral and singly ionized nitrogen, oxygen, and sulfur, as well as very weak Hα were found [4]. Chesneau et al. [5] observed Sakurai's object using the Very Large Telescope Interferometer (VLTI) at the European Southern Observatory (ESO) Paranal site. They detected a thick and dense dust disk with dimensions of 30 × 40 mas. This equates to 105 × 140 AU, assuming a distance of 3.5 kpc. Hinkle and Joyce [6] discovered bipolar lobes using deconvolved *Ks* images taken with the NIRI/Altair instrument on Gemini in 2010 and 2013 (see Figure 2 in their paper). The expansion of the bipolar structure was clearly visible. Also, the central star seems to be brightening in the near-infrared wavelength range. Note that the old PN has a round shape, while the new ejecta are bipolar.

2. Optical Observations

V4334 Sgr stunned the scientific community with its very fast evolution that was much faster than pre-discovery models predicted. To constrain the evolutionary models, we decided to derive the evolution of the central star temperature over time. The expectation was that the rising stellar temperature would result in forbidden lines from different ions emerging with time. We have been monitoring the evolution of the optical emission line spectrum since 2001 using low-resolution spectra taken with FORS1 and FORS2 on the ESO-VLT. A subset of these data was also discussed in earlier progress reports [7–12].

The optical lines initially showed an exponential decline in intensity and also a decreasing level of excitation (see Figure 1). This trend continued until 2007. Between 2001 and 2007, the optical spectrum was consistent with a shock that occurred before 2001 and started cooling and recombining afterwards. The low electron temperature derived from the [N II] lines in 2001 (3200–5500 K) and the [C I] lines in 2003 (2300–4300 K) was consistent with this [7]. The earliest evidence for this shock was the detection of the He I 1083 nm recombination line in 1998 [3]. This line was absent in 1997. Hence, the shock must have occurred around 1998 and must have stopped soon after, leaving cooling and recombining gas in its wake. In van Hoof et al. [7], the size of the [N II] emitting region was estimated to be 0.3–0.5 arcsec.

All line fluxes have been increasing since 2008. This is shown in Figure 1 for some selected strong lines. There are some minor exceptions for He I, [N I] and [O I] which are likely due to measurement errors and/or telluric contamination. This confirms the trend for the [C I] 982.4 and 985.0 nm doublet reported by Hinkle and Joyce [6]. Note that there was a strong discontinuous jump in the [O II] flux in 2008.

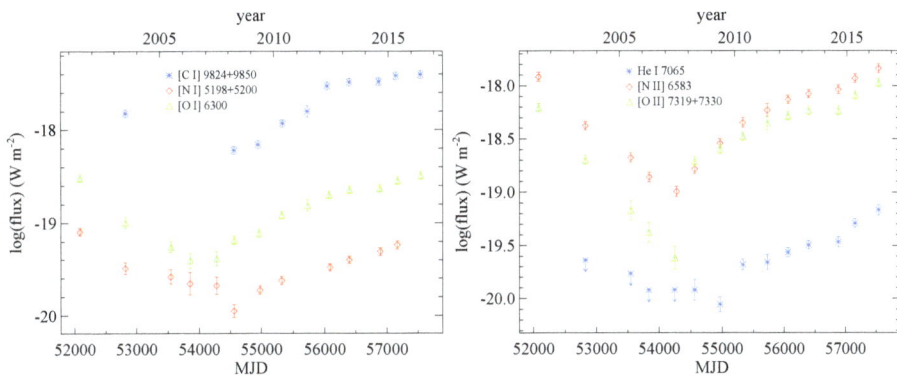

Figure 1. The flux evolution of selected lines in V4334 Sgr.

In 2015, we obtained a medium resolution echelle spectrum of V4334 Sgr using the Xshooter instrument at the ESO VLT. The slit was aligned along the bipolar axis (i.e., the position angle was PA = 13° east of north). In Figure 2, we show a position–velocity diagram of the [N II] 658.3 nm line. One can clearly see that the blue and red emissions come from different regions. The red-shifted and blue-shifted emission regions are +0.24 arcsec and −0.18 arcsec displaced with respect to the continuum source. This pattern is also seen in other forbidden and recombination lines (together referred to as nebular lines). From this, we conclude that the nebular emission lines originate in the bipolar lobes seen by Hinkle and Joyce [6].

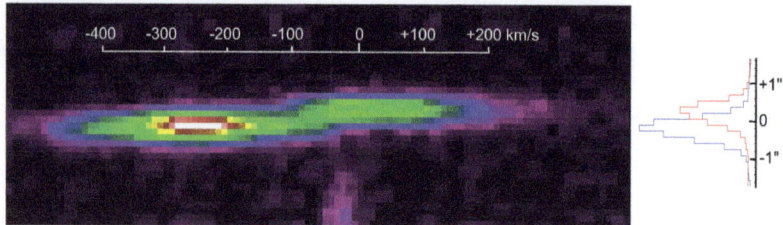

Figure 2. Position–velocity diagram of the [N II] 658.3 nm line taken at PA = 13°. Velocity runs along the x-axis, while position runs along the y-axis.

Since 2013, a complex of new lines has been emerging in the red. Many of these lines are still unidentified. The most prominent feature is shown in Figure 3. We have tentatively identified some of these as electronic transitions in CN (the 1,0 and 0,0 lines of A $^2\Pi_1 \rightarrow$ X $^2\Sigma^+$; the 0,0 lines would be the unidentified lines reported by Hinkle and Joyce [6]). We also identified the Na I doublet at 589.0 and 589.6 nm. The continuum is steadily rising, as was already reported by Hinkle and Joyce [6].

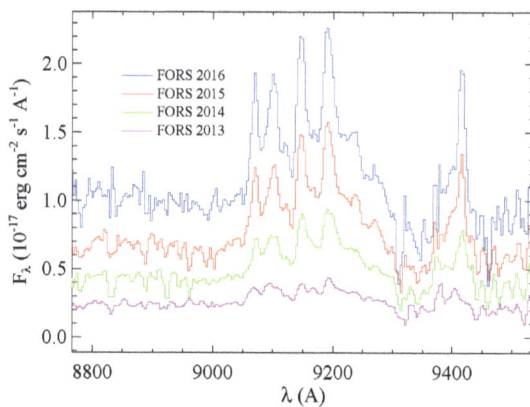

Figure 3. A complex of new emission lines that has been emerging since 2013. Note that the spectra have not been shifted.

3. Radio and ALMA Observations

Between 2004 and 2007, the 8 GHz radio flux was increasing. In van Hoof et al. [7], we interpreted this as evidence for the onset of photoionization of carbon. Since then, there has been a big gap in radio observations, but recently, we obtained new data that showed that the source has faded. The flux has dropped back to the levels of 2004 and 2005. This is inconsistent with photoionization of carbon, and the only plausible explanation is that the flux rise was due to a shock.

In 2015, we obtained Atacama Large Millimeter Array (ALMA) data of V4334 Sgr. The data set comprised a band 6 spectral scan, an incomplete band 7 spectral scan that was eventually rejected in quality control, and a continuum image. The band 6 spectral scan is shown in Figure 4 of van Hoof et al. [12]. We detected lines of CO, ^{13}CO, CN, likely ^{13}CN (blended), HC$_3$N, HC^{13}CCN, HCC^{13}CN, and possibly H^{13}CCCN. We also found CN absorption and a line at 239 GHz that was tentatively identified as CP.

We also created integrated images of the emission in each of the main emission lines. These are shown in Figure 4 where we also include the 2013 *Ks* image of Hinkle and Joyce [6] for comparison. One can see that the CN emission is spatially extended and coincides very nicely with the bipolar lobes detected by Hinkle and Joyce [6]. On the other hand, the CO, HC$_3$N, and the ALMA continuum image are nearly point-like and coincide with the central emission detected by Hinkle and Joyce [6]. This implies that these lines, as well as the dust emission, originate close to the central star, presumably from the disk detected by Chesneau et al. [5]. This implies that there is no dust emission detected in the bipolar lobes.

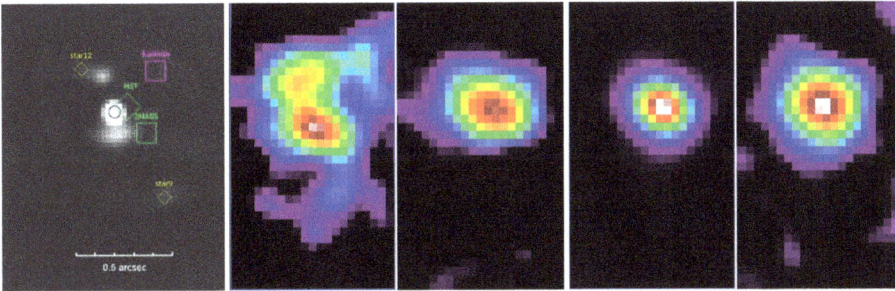

Figure 4. Panels from left to right: (1) the 2013 *Ks* image taken from Hinkle and Joyce [6] (© AAS, reproduced with permission); (2) the integrated CN emission observed with ALMA; (3) the integrated CO emission; (4) the continuum image; (5) the integrated HC$_3$N emission.

4. Preliminary Discussion

Using the data that have been presented in this paper, we reconstructed the following sequence of events. V4334 Sgr underwent a VLTP a few years before its discovery in 1996. It ejected a new hydrogen-deficient nebula in the process. The geometry of the source was clarified by Chesneau et al. [5] who discovered the presence of a dense and thick dust disk with dimensions of 30 × 40 mas using VLTI. All the dust was in the disk. The disk must have formed in the VLTP event (since there are no indications of the presence of hydrogen-rich material in the disk) and was already in place in 1997 (since it is responsible for obscuring the central star). It may be a Keplerian disk. Hinkle and Joyce [6] discovered the presence of bipolar lobes in the *Ks* band. These appear to be expanding. The total extent of these lobes along the major axis is approximately 0.4 arcsec.

Emission lines were first discovered in 1998 (He I 1083 nm) and 2001 (optical). The optical emission spectrum has been monitored since, initially showing an exponential decline in flux, while the level of excitation also dropped. We see this as evidence for a brief shock that occurred around 1998. A plausible explanation is that this is the fastest material ejected in the VLTP, hitting slower ejecta from the same event. Between 2005 and 2007, the 8 GHz radio emission showed a marked increase. The radio flux has returned to pre-2005 levels since. We see no counterpart for this behavior in the optical data. Our hypothesis is that this behavior was due to a shock in an obscured region. The optical line fluxes have started to increase again since 2008. The sudden jump in the [O II] flux in 2008 could point to a second shock as the cause of the change in behavior. Possibly this is the same shock that was already detected in radio emission and which now breaks out of the obscured region.

Our working hypothesis is that the central star wind is collimated into a bipolar jet and is now interacting with the lobes. So, the formation of the bipolar lobes may have started in 2008. The nebular lines were formed in the terminating bow shock. This was confirmed by Xshooter spectra. The optical spectrum shows new lines which have been emerging since 2013. Some have tentatively been identified as electronic transitions of CN and Na I. The optical CN lines, as well as the other lines that are emerging with them, formed close to the central star (based on Xshooter data that are not shown), possibly in the disk. If the optical CN lines are pumped by UV radiation from the central star, this is an indication that the reheating has started. Alternatively this could be a C-shock where the outflow is collimated by the disk into jets. In ALMA spectra, we detected the presence of CO, CN, HC_3N, and ^{13}C isotopologues. The CO and HC_3N (+isotopologues) emission is unresolved, so most likely comes from the disk. The ALMA CN and ^{13}CN emission was resolved and matched the bipolar lobes. A possible explanation is that CN is formed via shock-induced dissociation of HCN in the lobes. What is clear from these data is that we have a very detailed record of the very early stages of the hydrodynamic shaping of a bipolar nebula.

A more in-depth analysis of the available data will be presented in van Hoof et al. (A&A, in preparation).

Author Contributions: Conceptualization, P.A.M.v.H., S.K., G.C.V.d.S. and A.A.Z.; Data curation, P.A.M.v.H., S.K., G.C.V.d.S., A.A. and L.G.-R.; Funding acquisition, P.A.M.v.H. and M.H.; Investigation P.A.M.v.H., S.K., G.C.V.d.S., A.A.Z., F.H. and M.H.; Software, P.A.M.v.H. and F.H.; Writing—original draft, P.A.M.v.H.; Writing—review & editing, P.A.M.v.H., S.K., G.C.V.d.S., A.A.Z. and M.H.

Funding: P.A.M.v.H. acknowledges support from the Belgian Science Policy Office through the ESA PRODEX program and contract no. BR/154/PI/MOLPLAN. MH acknowledges the Polish MSHE for funding grants DIR/WK/2016/2017/05-1 and 220815/E-383/SPUB/2016/2 and the National Science Centre for grant No. 2016/23/B/ST9/01653.

Acknowledgments: We thank Kenneth Hinkle for kindly allowing us to reproduce the image from his paper.

Conflicts of Interest: The authors declare no conflict of interest.

References

1. Nakano, S.; Sakurai, Y.; Hazen, M.; McNaught, R.H.; Benetti, S.; Duerbeck, H.W.; Cappellaro, E.; Leibundgut, B. Novalike Variable in Sagittarius. Available online: http://www.cbat.eps.harvard.edu/iauc/06300/06322.html (accessed on 27 July 2018).
2. Hajduk, M.; Zijlstra, A.A.; Herwig, F.; van Hoof, P.A.M.; Kerber, F.; Kimeswenger, S.; Pollacco, D.L.; Evans, A.; Lopéz, J.A.; Bryce, M.; et al. The Real-Time Stellar Evolution of Sakurai's Object. *Science* **2005**, *308*, 231–233. [CrossRef] [PubMed]
3. Eyres, S.P.S.; Smalley, B.; Geballe, T.R.; Evans, A.; Asplund, M.; Tyne, V.H. Strong helium 10830-Å absorption in Sakurai's object (V4334 Sgr). *Mon. Not. R. Astron. Soc.* **1999**, *307*, L11–L15. [CrossRef]
4. Kerber, F.; Pirzkal, N.; De Marco, O.; Asplund, M.; Clayton, G.C.; Rosa, M.R. Freshly Ionized Matter around the Final Helium Shell Flash ObjectV4334 Sagittarii (Sakurai's Object). *Astrophys. J. Lett.* **2002**, *581*, L39–L42. [CrossRef]
5. Chesneau, O.; Clayton, G.C.; Lykou, F.; De Marco, O.; Hummel, C.A.; Kerber, F.; Lagadec, E.; Nordhaus, J.; Zijlstra, A.A.; Evans, A. A dense disk of dust around the born-again Sakurai's object. *Astron. Astrophys.* **2009**, *493*, L17–L20. [CrossRef]
6. Hinkle, K.H.; Joyce, R.R. The Spatially Resolved Bipolar Nebula of Sakurai's Object. *Astrophys. J.* **2014**, *785*, 146. [CrossRef]
7. Van Hoof, P.A.M.; Hajduk, M.; Zijlstra, A.A.; Herwig, F.; Evans, A.; Van de Steene, G.C.; Kimeswenger, S.; Kerber, F.; Eyres, S.P.S. The onset of photoionization in Sakurai's Object (V4334 Sagittarii). *Astron. Astrophys.* **2007**, *471*, L9–L12. [CrossRef]
8. Van Hoof, P.A.M.; Hajduk, M.; Zijlstra, A.A.; Herwig, F.; Van de Steene, G.C.; Kimeswenger, S.; Evans, A. Recent Observations of V4334 Sgr and V605 Aql. In *Hydrogen-Deficient Stars*; Werner, A., Rauch, T., Eds.; Astronomical Society of the Pacific Conference Series; Astronomical Society of the Pacific: San Francisco, CA, USA, 2008; Volume 391, p. 155.

Galaxies **2018**, *6*, 79

9.   Van Hoof, P.A.M.; Kimeswenger, S.; Van de Steene, G.C.; Zijlstra, A.A.; Hajduk, M.; Herwig, F. The Very Fast Evolution of the VLTP Object V4334 Sgr. In *19th European Workshop on White Dwarfs*; Dufour, P., Bergeron, P., Fontaine, G., Eds.; Astronomical Society of the Pacific: San Francisco, CA, USA, 2015; Volume 493, p. 95.

10.   Van Hoof, P.A.M.; Van de Steene, G.C.; Kimeswenger, S.; Zijlstra, A.A.; Hadjuk, M.; Herwig, F. The Very Fast Evolution of V4334 Sgr. *EAS Publ. Ser.* **2015**, *71–72*, 287–288. [CrossRef]

11.   Van de Steene, G.C.; van Hoof, P.A.M.; Kimeswenger, S.; Zijlstra, A.A.; Avison, A.; Guzman-Ramirez, L.; Hajduk, M.; Herwig, F. The very fast evolution of Sakurai's object. *Proc. Int. Astron. Union* **2016**, *12*, 380–381. [CrossRef]

12.   Van Hoof, P.A.M.; Herwig, F.; Kimeswenger, S.; Van de Steene, G.C.; Avison, A.; Zijlstra, A.A.; Hajduk, M.; Guzmán-Ramirez, L.; Woodward, P.R. The i process in the post-AGB star V4334 Sgr. *Mem. Soc. Astronom. Ital.* **2017**, *88*, 463.

galaxies

MDPI

Article

Revealing the True Nature of Hen 2-428

Nicole Reindl [1,*,†], Nicolle L. Finch [1,†], Veronika Schaffenroth [2], Martin A. Barstow [1], Sarah L. Casewell [1], Stephan Geier [2], Marcelo M. Miller Bertolami [3] and Stefan Taubenberger [4]

[1] Department of Physics and Astronomy, University of Leicester, University Road, Leicester LE1 7RH, UK; nlf7@le.ac.uk (N.L.F.); martin.barstow@leicester.ac.uk (M.A.B.); slc25@leicester.ac.uk (S.L.C.)

[2] Institute for Physics and Astronomy, University of Potsdam, Karl-Liebknecht-Str. 24/25, 14476 Potsdam, Germany; schaffenroth@astro.physik.uni-potsdam.de (V.S.); sgeier@astro.physik.uni-potsdam.de (S.G.)

[3] Instituto de Astrofísica de La Plata, UNLP-CONICET, La Plata, 1900 Buenos Aires, Argentina; mmiller@fcaglp.fcaglp.unlp.edu.ar

[4] European Southern Observatory, Karl-Schwarzschild-Str. 2, D-85748 Garching, Germany; staubenb@partner.eso.org

*   Correspondence: nr152@le.ac.uk; Tel.: +44-116-223-1385

† These authors contributed equally to this work.

Received: 19 June 2018; Accepted: 10 August 2018; Published: 14 August 2018

Abstract: The nucleus of Hen 2-428 is a short orbital period (4.2 h) spectroscopic binary, whose status as potential supernovae type Ia progenitor has raised some controversy in the literature. We present preliminary results of a thorough analysis of this interesting system, which combines quantitative non-local thermodynamic (non-LTE) equilibrium spectral modelling, radial velocity analysis, multi-band light curve fitting, and state-of-the art stellar evolutionary calculations. Importantly, we find that the dynamical system mass that is derived by using all available He II lines does not exceed the Chandrasekhar mass limit. Furthermore, the individual masses of the two central stars are too small to lead to an SN Ia in case of a dynamical explosion during the merger process.

Keywords: binaries: spectroscopic; stars: atmospheres; stars: abundances; supernovae

---

1. Introduction

The detection and study of progenitor systems of type Ia supernovae (SN Ia) are crucial to understand the exact explosion mechanism of these important cosmic distance indicators. Although there is a general consensus that only the thermonuclear explosion of a white dwarf can explain the observed features of those events, the nature of their progenitor systems still remains elusive. In the single-degenerate model, a white dwarf accretes material from a non-degenerate companion and explodes when it reaches the Chandrasekhar mass limit [1]. An alternative scenario is the double-degenerate model, in which the explosion is triggered during the merger process of two white dwarfs [2–4]. Identifying progenitors for the latter scenario is particularly interesting in view of the applicability of SN Ia as standardisable candles, as the merging system could exceed or fall below the Chandrasekhar limit significantly.

Santander-Garcia et al. [5] have claimed to have discovered the first definite double-degenerate, super-Chandrasekhar system that will merge within a Hubble time, namely the central stars of the planetary nebula (CSPN) Hen 2-428. They found a photometric period of 4.2 h and that He II λ 5411 Å is double lined and time variable. By fitting the radial velocities (RVs) and light curves, they concluded that the system consists of two pre-white dwarfs with equal masses of 0.88 $M_\odot$. In this case, the system would merge within 700 million years making, Hen 2-428 one of the best SN Ia progenitor candidates known.

This scenario has since been challenged by Garcia-Berro et al. [6], who criticized the strong mismatch between the luminosities and radii of both pre-white dwarf components as derived by [5] with the predictions from stellar evolution models [7]. In addition, Reference [6] suggested that the variable He II λ 5411 Å line might instead be a superposition of an absorption line plus an emission line, possibly arising from the nebula, the irradiated photosphere of a close companion, or a stellar wind. Since this would question the dynamical masses derived by [5], Reference [6] that repeated the light curve fitting and showed that the light curves of Hen 2-428 may also be fitted well by assuming an over-contact binary system that consists of two lower mass (i.e., masses of 0.47 $M_\odot$ and 0.48 $M_\odot$) stars. Thus, Reference [6] concludes that the claim that Hen 2-428 provides observational evidence for the double degenerate scenario for SN Ia is premature.

Given the potential importance of Hen 2-428 as a unique laboratory to study the double degenerate merger scenario, it is highly desirable to resolve this debate. Therefore, we use an improved approach for the analysis of this unique object by combining quantitative non-LTE spectral modelling, RV analysis, multi-band light curve fitting, and state-of-the art stellar evolutionary calculations.

2. Spectral Analysis

Our spectral analysis is based on the Very Large Telescope/FOcal Reducer and low dispersion Spectrograph 2 (VLT/FORS2) and Gran Telescopio Canarias/Optical System for Imaging and low-Intermediate-Resolution Integrated Spectroscopy (GTC/OSIRIS) spectra. FORS2 spectra were downloaded from the European Southern Observatory (ESO) archive (ProgIDs 085.D-0629(A), 089.D-0453(A)) and by reduced by using standard IRAF procedures. Calibrated GTC/OSIRIS spectra were obtained from the GTC PublicArchive (ProgID GTC41-13A). Since the relatively broad and deep absorption lines in the spectra of Hen 2-428 cannot be fitted assuming a single component, we conclude that Hen 2-428 is indeed a double lined spectroscopic binary. We employed the Tübingen Model Atmospheres Package (TMAP, [8–10]) to produce state-of-the-art non-LTE model atmospheres. Initially, our models contain only H and He. The model grid spans from T_{eff} = 30.0–70.0 kK (2.5 kK steps), $\log g$ = 4.25–6.00 (0.25 steps), and covers three He abundances ($\log (He/H) = -2, -1, 0$, number fractions). The first results of this analysis based on the FORS2 spectra only were presented in [11]. We extended this analysis by fitting the OSIRIS spectra and considering a variable flux contribution of the two stars in our fitting procedure. To constrain the parameters of the system, we used the XSPEC software [12,13], a χ^2 minimisation code, originally designed for X-ray spectra, but which has been adapted to work on optical data [14]. XSPEC determines the best fit model for the input parameters, which in our case are the effective temperatures (T_{eff}), surface gravities ($\log g [cm/s^2]$), He abundances, RVs, and the flux contribution of each star. First, the radial velocity of each component was found, and then these were fixed whilst deriving T_{eff}, $\log g$, $\log (He/H)$, and the flux contribution of each star simultaneously. For the first star, we find T_{eff} = 48 ± 7 kK, $\log g$ = 5.00 ± 0.1, $\log (He/H)$ = −1.1 and relative flux contribution of 46%. For the second star, we derive T_{eff} = 46 ± 7 kK, $\log g$ = 4.8 ± 0.1, $\log (He/H)$ = −1.0 and relative flux contribution of 54%. The reduced χ^2 value for these parameter is 1.04. In Figure 1, we show the best fit model (red) compared to an OSIRIS spectrum (grey). We obtain a good fit for He II $\lambda\lambda$ 4200, 4542, 5412 Å, and also the absorption wings of the He I and H I lines are reproduced nicely. The line cores of He II λ 4686 Å, however, appear too deep compared to the observation. This is a known problem when fitting the spectra of CSPNe and other hot stars with pure HHe models only. It has been shown that He II λ 4686 Å is particularly susceptible to metal line blanketing [15,16]. We expect the systematic effects introduced by neglecting metal-line blanketing effects to be of the order of 0.1 dex on $\log g$ and 1 kK on T_{eff} [17], but we will include metals in our models for future analysis of this system to overcome this problem.

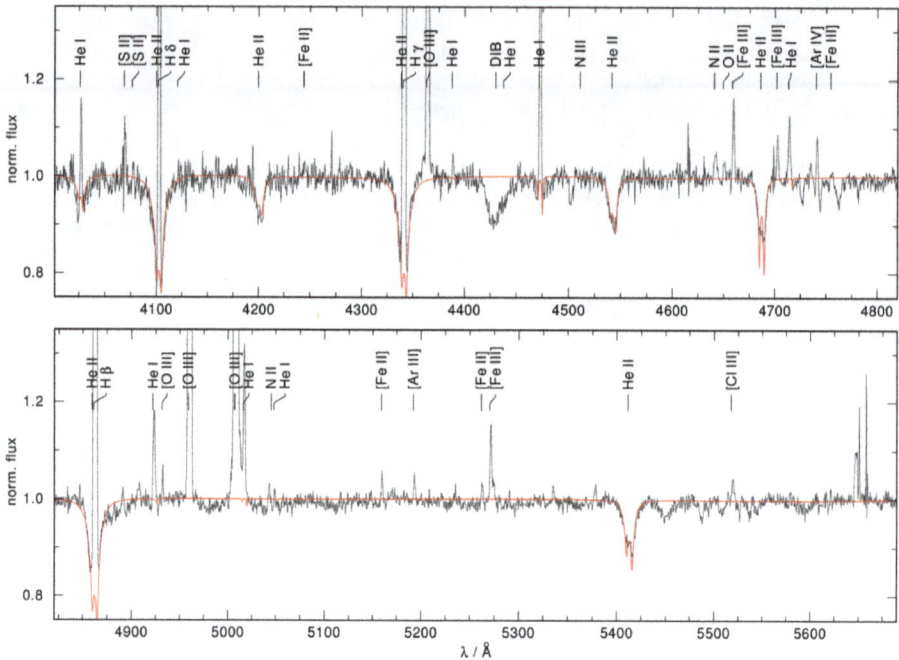

Figure 1. The best fit TMAP model (**red**) plotted against the OSIRIS spectrum (**grey**). Identified photospheric absorption lines, a diffuse interstellar band, and nebular emission lines are marked.

3. Light Curve Modelling

The analysis of the light curves (downloaded from [5]) was carried out simultaneously in B and i filter. For the analysis, we used MORO (Modified Roche Program, see [18]), which is based on the Wilson–Devinney code but is using a modified Roche model considering the influence of the radiation pressure on the shape of the stars. Due to the significant degeneracies in the parameters found in the analysis of the light curve, which result from not many independent parameters being used in the light curve analysis, we fixed the mass ratio of the system to the mass ratio, which was derived by the analysis of the radial velocity curve. Moreover, we used the temperatures derived by the spectral analysis as starting values. Initial attempts to reproduce the light curves with a contact system failed. The shape of the light curve could not be reproduced. Therefore, we tried to fit an over-contact system, which is assuming equal Roche potentials for both stars. We also considered a third light source since Hen 2-428 is known to exhibit a red-excess, which possibly results from the PN and a distant companion [19]. By varying the inclination, temperatures, Roche potentials, and luminosity ratio of both stars, the light curves could be reproduced nicely. Our preliminary best fit to the light curves is shown in Figure 2. We get a relative luminosity for the primary of 53.8% in B and 53.5% in i. We derived an additional constant flux component of 4.3% in B and 10.3% in i. We note that our fit reproduces the light curves better than the one of [6], and also slightly better than the model of [5].

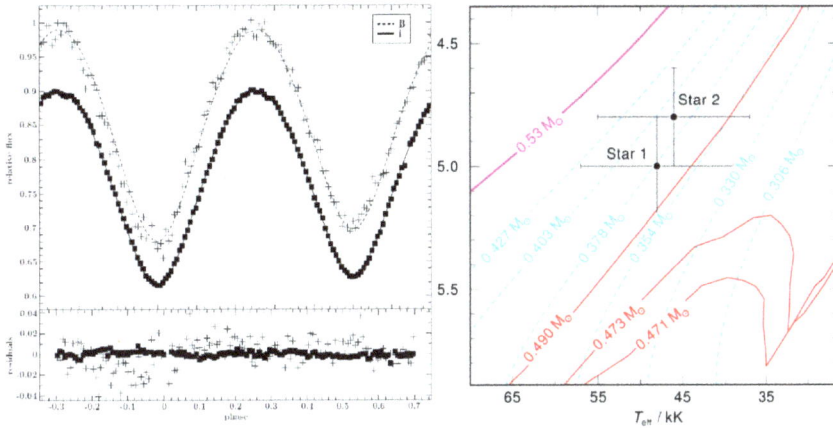

Figure 2. (**left**): Johnson B-band (dotted line) and Sloan i-band (solid line) light curves and LCURVE models; (**right**): the location of the two CSPNe of Hen 2-428 (black dots) in the T_{eff} -log g plane compared with a H-rich post-AGB evolutionary track (pink, [20]), post-RGB tracks (dashed blue, [21]), and post-EHB tracks (red, [22]). The tracks are labelled with stellar masses.

4. Dynamical Masses

The amplitudes of the RV curves (K_1, K_2) were obtained by sinusoidal fitting of the individual RV measurements obtained from the OSIRIS spectra of both components of the binary. This was done separately in fine steps over a range of test periods within the uncertainties of the orbital period (0.1758 ± 0.0005 d) derived from photometry by [5]. For each period, the χ^2 of the best fitting sine curve was determined and the solution with the lowest χ^2 was chosen. We can reproduce the results of [5] using their RVs as derived by merely Gaussian fitting of the He II λ 5412 Å absorption lines (fifth column in Table 1). Somewhat lower RV amplitudes are obtained when using our RVs as measured with XSPEC from our synthetic spectra for He II λ 5412 Å (forth column). Surprisingly, we find lower RV amplitudes when using only the RVs from He II λ 4686 Å (third column), and when all four He II lines were fitted simultaneously (first and second column). Since the inclination of the system ($i = 65.3 \pm 0.1$) can be constrained well from the light curve fitting, the dynamical masses of the two stars (bottom two rows in Table 1) can be calculated via the binary mass function. Importantly, we find that the combined dynamical mass obtained using all four He II lines is considerably lower than the one reported by [5] and does not exceed the Chandrasekhar mass limit.

Table 1. Dynamical and spectroscopic masses as derived in this work and by SG + 2015 [5].

	All Lines Syn. Spectra	He II λ 4686 Å Syn. Spectra	He II λ 5412 Å Syn. Spectra	He II λ 5412 Å Gaussians	SG + 2015 Gaussians
K_1 [km/s]	174 ± 16	171 ± 18	196 ± 19	203 ± 13	206 ± 8
K_2 [km/s]	174 ± 8	170 ± 10	191 ± 13	202 ± 12	206 ± 12
M_1 [$M_\odot$]	0.51 ± 0.15	0.48 ± 0.16	0.70 ± 0.23	0.81 ± 0.25	0.88 ± 0.13
M_2 [$M_\odot$]	0.51 ± 0.14	0.48 ± 0.15	0.71 ± 0.22	0.81 ± 0.24	0.88 ± 0.13

5. Discussion

Our spectral analysis is hindered by the limited number of unblended photospheric lines and the occurrences of metal line blanketing effects. Higher S/N spectra that cover other unblended photospheric lines as well as models that also include metal opacities would allow us to reduce the errors on the atmospheric parameters of the stars. Our effective temperatures are significantly larger than the ones reported by [5], who derived the stellar temperatures from light curve fitting. The narrow temperature range (30–40 kK) adopted by [5], however, is not valid as already pointed out by [6]. Ref. [5] established the upper limit of 40 kK based on the absence He II emission lines, but there are many CSPNe with even higher T_{eff} and also a lacking He II nebular lines. Our light curve analysis shows that the light curves can be fitted equally well with higher T_{eff} values compared to the ones derived by [5]. This suggests that, because of the light curve solution degeneracy, light curve modelling alone is not sufficient to derive the temperatures of the stars. Furthermore, our light curve modelling supports the idea that Hen 2-428 is an over-contact system, i.e., that both stars still share a common envelope. Therefore, should the PN be the result of the common envelope ejection, future studies of this system could reveal important insights on the common envelope ejection efficiency.

The dynamical masses are the main key to revealing the nature of Hen 2-428, i.e., to find out about the nature of the progenitor stars as well as its SNIa status. We found that the system masses vary depending on which He II lines were used, but agree within the error limits. Importantly, we find that the combined dynamical mass obtained using all four He II lines does not exceed the Chandrasekhar mass limit and that the individual masses of the two CSPNe are too small for a reasonable production of ^{56}Ni (which determines the explosion brightness) in case of a dynamical explosion during the merger process. Thus, the merging event of Hen 2-428 would not be identified as a SN Ia (see Figure 3 in [3]).

The reason why the analysis of the He II λ 5412 Å leads to a higher measurement of the RV amplitudes is unclear at the moment. Reference [19] reported an excess that shows up red-wards of 5000 Å, possibly originating from a late type companion. We speculate that this red-excess might impact the He II λ 5412 Å, but not the other three He II lines detected. Spectra extending toward longer wavelengths would help investigate the red excess further.

The lower system mass solution is also supported in view of the evolutionary time scales. The low surface gravities of both stars indicate that neither of the two stars has entered the white dwarf cooling sequence yet. The heating rate of post-asymptotic giant branch (AGB) stars is strongly mass dependent. For post-AGB stars with masses greater than about 0.7 $M_\odot$, the blue-ward evolution in the HRD is predicted to be so rapid that changes of T_{eff} would become noticeable within a human life span. Recent evolutionary models [20] predict that, for a 0.71 $M_\odot$ post-AGB star, it takes only 20 years to heat up from 30 kK (that is when the nebula becomes visible) to 50 kK (approximately the T_{eff} of both stars now). Since Hen 2-428 has been known for more than 50 years [23], this provides further evidence that the two CSPNe cannot both be such massive post-AGB stars. We also note that, if the two stars were in thermal non-equilibrium after the common-envelope was ejected, an even faster evolution would be predicted.

The spectroscopic masses of the two stars (Figure 2) agree with the dynamical masses from all four He II lines. We stress, however, that the spectroscopic masses should be treated with caution because it is not clear to what extent the evolutionary tracks are altered for over-contact systems. We also note that radii from the best light curve fit are significantly larger than the radii derived by the spectroscopic analysis due to the over-contact nature of the system. The errors on both the dynamical and spectroscopic masses of the two stars are relatively large. While the spectroscopic masses lead to the speculation that the two CSPNe might be AGB-manqué stars (stars that fail to evolve through the AGB), the dynamical masses do not allow us to distinguish whether one (or both) stars are post-AGB or post-red giant branch (RGB) stars. Potentially, one of them could be a post-extreme horizontal branch (EHB) star. More precise RVs and a more accurate orbital period would, therefore, be desirable to put stronger constraints on the masses.

6. Conclusions

Our preliminary results suggest that the dynamical system mass of Hen 2-428 that is derived by using all available He II lines does not exceed the Chandrasekhar mass limit. Furthermore, the individual masses of the two central stars are also too small to lead to an SN Ia in case of a dynamical

Galaxies **2018**, *6*, 88

explosion during the merger process. Further investigations on the red-excess and atmospheric models that consider opacities of heavy metals are mandatory for a reliable analysis this intriguing system.

Author Contributions: Writing—Original Draft Preparation, N.R.; Writing—Review and Editing, N.R., N.L.F, V.S., M.A.B., S.L.C., S.G., M.M.M.B., and S.T.; Visualization, N.R. and V.S.; Supervision, N.R., S.L.C., and M.A.B.; Data reduction: S.T.; Spectral Analysis: N.L.F. and N.R.; Light curve modeling: V.S.; RV analysis: N.L.F. and S.G.; mass determination: N.R.; Interpretation of the results: N.R., S.G., V.S., M.M.M.B, S.T.

Funding: N.R. is supported by a RC1851 research fellowship. N.L.F. is supported by an Science and Technology Facilities Council Studentship.

Acknowledgments: This work is based on observations collected at the European Organisation for Astronomical Research in the Southern Hemisphere under ESO programmes 085.D-0629(A) and 089.D-0453(A) based on data from the GTC PublicArchive at CAB (INTA-CSIC). The Tübingen Model-Atom Database (TMAD) service (http://astro-uni-tuebingen.de/~TMAD) used to compile atomic data for this work was constructed as part of the activities of the German Astrophysical Virtual Observatory. This research used the SPECTRE and ALICE High Performance Computing Facilities at the University of Leicester.

References

1. Whelan, J.; Iben, I., Jr. Binaries and Supernovae of Type I. *Astrophys. J.* **1973**, *186*, 1007–1014. [CrossRef]
2. Iben, I., Jr.; Tutukov, A.V. Supernovae of type I as end products of the evolution of binaries with components of moderate initial mass (M not greater than about 9 solar masses). *Astrophys. J. Suppl. Ser.* **1984**, *54*, 335–372. [CrossRef]
3. Shen, K.J. Every Interacting Double White Dwarf Binary May Merge. *Astrophys. J. Lett.* **2015**, *805*, L6. [CrossRef]
4. Webbink, R.F. Double white dwarfs as progenitors of R Coronae Borealis stars and Type I supernovae. *Astrophys. J.* **1984**, *277*, 355–360. [CrossRef]
5. Santander-Garcia, M.; Rodriguez-Gil, P.; Corradi, R.L.M.; Jones, D.; Miszalski, B.; Boffin, H.M.J.; Rubio-Díez, M.M.; Kotze, M.M. The double-degenerate, super-Chandrasekhar nucleus of the planetary nebula Henize 2-428. *Nature* **2015**, *519*, 63–65. [CrossRef] [PubMed]
6. Garcia-Berro, E.; Soker, N.; Althaus, L.G.; Ribas, I.; Morales, J.C. Is the central binary system of the planetary nebula Henize 2–C428 a type Ia supernova progenitor? *New Astron.* **2016**, *45*, 7–13. [CrossRef]
7. Blöcker, T.; Schönberner, D. A 7-solar-mass AGB model sequence not complying with the core mass-luminosity relation. *Astron. Astrophys.* **1991**, *244*, L43–L46.
8. Rauch, T.; Deetjen, J.L. Handling of Atomic Data. In Proceedings of the 19th European Workshop on White Dwarfs, Tübingen, Germany, 8–12 April 2002.
9. Werner, K.; Deetjen, J.L.; Dreizler, S.; Nagel, T.; Rauch, T.; Schuh, S.L. Model Photospheres with Accelerated Lambda Iteration. In Proceedings of the 19th European Workshop on White Dwarfs, Tübingen, Germany, 8–12 April 2002; pp. 31–50.
10. Werner, K.; Dreizler, S.; Rauch, T. *TMAP: Tübingen NLTE Model-Atmosphere Package*; University of Tübingen: Tübingen, Germany, 2012.
11. Finch, N.L.; Reindl, N.; Barstow, M.A.; Casewell, S.L.; Geier, S.; Bertolami, M.M.M.; Taubenberger, S. Spectral analysis of the binary nucleus of the planetary nebula Hen 2-428–first results. *Open Astron.* **2018**, *27*, 57–61. [CrossRef]
12. Arnaud, K.; Gordon, C.; Dorman, B. *An X-ray Spectral Fitting Package: User's Guide for Version 12.9.1*; National Aeronautics and Space Administration: Greenbelt, MD, USA, 2017.
13. Shafer, R.A.; Haberl, F.; Arnaud, K.A. *XSPEC: An X-Ray Spectral Fitting Package: Version 2 of the User's Guide*; National Aeronautics and Space Administration: Greenbelt, MD, USA, 1991.
14. Dobbie, P.D.; Pinfield, D.J.; Napiwotzki, R.; Hambly, N.C.; Burleigh, M.R.; Barstow, M.A.; Jameson, R.F.; Hubeny, I. Praesepe and the seven white dwarfs. *Mon. Not. R. Astron. Soc.* **2004**, *355*, L39–L43. [CrossRef]
15. Latour, M.; Fontaine, G.; Green, E.M.; Brassard, P. A non-LTE analysis of the hot subdwarf O star BD + 28–4211-II. The optical spectrum. *Astron. Astrophys.* **2015**, *579*, A39. [CrossRef]
16. Reindl, N.; Rauch, T.; Werner, K.; Kruk, J.W.; Todt, H. On helium-dominated stellar evolution: The mysterious role of the O(He)-type stars. *Astron. Astrophys.* **2014**, *566*, A116. [CrossRef]

17. Latour M., Fontaine G., Brassard P.; Chayer, P.; Green, E.M. A NLTE model atmosphere analysis of the pulsating sdO star SDSS J1600+0748. *Astrophys. Space Sci.* **2010**, *329*, 141–144. [CrossRef]

18. Drechsel, H.; Haas, S.; Lorenz, R.; Gayler, S. Radiation pressure effects in early-type close binaries and implications for the solution of eclipse light curves. *Astron. Astrophys.* **1995**, *294*, 723–743.

19. Rodriguez, M.; Corradi, R.L.M.; Mampaso, A. Evidence for binarity in the bipolar planetary nebulae A 79, He 2-428 and M 1-91. *Astron. Astrophys.* **2001**, *377*, 1042–1055. [CrossRef]

20. Miller Bertolami, M.M. New models for the evolution of post-asymptotic giant branch stars and central stars of planetary nebulae. *Astron. Astrophys.* **2016**, *588*, A25. [CrossRef]

21. Hall, P.D.; Tout, C.A.; Izzard, R.G.; Keller, D. Planetary nebulae after common-envelope phases initiated by low-mass red giants. *Mon. Not. R. Astron. Soc.* **2013**, *435*, 2048–2059. [CrossRef]

22. Dorman, B.; Rood, R.T.; O'Connell, R.W. Ultraviolet Radiation from Evolved Stellar Populations. *arXiv* **1993**, arXiv: astro-ph/9311022.

23. Henize, K.G. Observations of Southern Planetary Nebulae. *Astrophys. J. Suppl. Ser.* **1967**, *14*, 125. [CrossRef]

galaxies

MDPI

Article
Jsolated Stars of Low Metallicity

Efrat Sabach

Department of Physics, Technion—Israel Institute of Technology, Haifa 32000, Israel;
efrats@physics.technion.ac.il

Received: 15 July 2018; Accepted: 9 August 2018; Published: 15 August 2018

Abstract: We study the effects of a reduced mass-loss rate on the evolution of low metallicity Jsolated stars, following our earlier classification for angular momentum (J) isolated stars. By using the stellar evolution code MESA we study the evolution with different mass-loss rate efficiencies for stars with low metallicities of $Z = 0.001$ and $Z = 0.004$, and compare with the evolution with solar metallicity, $Z = 0.02$. We further study the possibility for late asymptomatic giant branch (AGB)—planet interaction and its possible effects on the properties of the planetary nebula (PN). We find for all metallicities that only with a reduced mass-loss rate an interaction with a low mass companion might take place during the AGB phase of the star. The interaction will most likely shape an elliptical PN. The maximum post-AGB luminosities obtained, both for solar metallicity and low metallicities, reach high values corresponding to the enigmatic finding of the PN luminosity function.

Keywords: late stage stellar evolution; planetary nebulae; binarity; stellar evolution

1. Introduction

Jsolated stars are stars that do not gain much angular momentum along their post main sequence evolution from a companion, either stellar or substellar, thus resulting with a lower mass-loss rate compared to non-Jsolated stars [1]. As previously stated in Sabach and Soker [1,2], the fitting formulae of the mass-loss rates for red giant branch (RGB) and asymptotic giant branch (AGB) single stars are set empirically by contaminated samples of stars that are classified as "single stars" but underwent an interaction with a companion early on, increasing the mass-loss rate to the observed rates. The mass-loss rate on the giant branches has extensive effects on stellar evolution and on the resulting planetary nebula (PN) in low and intermediate mass stars. The reduced mass-loss rate of Jsolated stars results in a larger AGB radii compared to the RGB and compared to the "traditional" evolution with the high mass-loss rate efficiency of non-Jsolated stars. The higher AGB radii reached for Jsolated stars can lead to possible late interaction with a low mass companion. If such a Jsolated star interacts late in its evolution with a companion, thus no longer qualifying as a Jsolated star afterwards, strong interaction might cause angular momentum gain, spin up, and increase in the mass-loss rate.

The role of low mass companions (brown dwarfs or planets) in shaping PNe has been long discussed over the past few decades and it has been suggested that most PNe result from binary interaction (e.g., [3–15]). As we have shown in Sabach & Soker [1] for solar type Jsolated stars (both in mass and in metallicity), such an interaction can occur during the AGB phase of evolution, where the companion is likely to be engulfed by the star. The engulfed companion will deposit angular momentum to the primary's envelope, increasing the mass-loss rate and by that later accelerating the post-AGB evolution. This late interaction can shape an elliptical bright PN. In addition, we further found under the Jsolated framework that as the sun is a Jsolated star it will most likely engulf the earth during the AGB rather than during the RGB.

We have also shown that such Jsolated stars have implications related to the puzzle of the bright end cut-off in the PN luminosity function (PNLF) of old stellar populations ([1,2]; for studies on the PNLF see, e.g., [16–22]). It was observed that both young and old populations have a steep bright end cut-off in the PNLF in [OIII] emission lines at $M^*_{5007} \simeq -4.5$ mag. This implies a more massive central star than expected in old populations, that reach post-AGB luminosities $L \geq 5000 \, L_\odot$ in order to ionize the observed bright nebulae to the desired level. Our previous results indicate that also for low mass stars the post-AGB luminosities of Jsolated stars are bright enough to account for the bright end cut-off in the PNLF of old stellar populations.

In Sabach & Soker [2] we focused on the implications of a reduced mass-loss rate on stellar evolution of solar-type stars and the shaping of elliptical PNe by a companion. In Sabach & Soker [1] we set the term *Jsolated stars* and studied the possible solution for the puzzling finding of bright PNe in old stellar populations, where the stellar mass is up to 1.2 $M_\odot$. Yet, we have only focused on Jsolated stars of solar metallicity, $Z = 0.02$. Here we continue the research and study the evolution of Jsolated star with low metallicities.

2. Results

We continue the study of Jsolated stars in old stellar populations by studying the evolution of low and intermediate mass stars with reduced mass-loss rates and with low metallicities of $Z = 0.001$ and $Z = 0.004$, compatible with the old population of the Small Magellanic Cloud (where the metallicity has been measured to be between $Z = 0.001$ and $Z = 0.004$; [23]). We conduct our simulations using the Modules for Experiments in Stellar Astrophysics (MESA, version 10398 [24]). We compare the evolution of stars with low metallicity to the evolution of stars with solar metallicity, $Z = 0.02$. Badenes et al. [25] find that most PN progenitors in the Large Magellanic Cloud correspond to stars of initial mass between 1 $M_\odot$ and 1.2 $M_\odot$. We here consider stars with an initial mass as low as $0.9 M_\odot$ for our study of Jsolated stars and the implications to the resulting PNe.

We focus on the effects of a reduced mass loss rate efficiency, as expressed by the empirical mass-loss formula for red giant stars of Reimers [26]

$$\dot{M} = \eta \times 4 \times 10^{-13} LM^{-1} R. \tag{1}$$

We repeat the procedure described in Sabach & Soker [1] and follow the evolution with several mass-loss rate efficiency parameters, from $\eta = 0.5$ (the "traditional" commonly used mass-loss rate efficiency; e.g., [27]) and as low as $\eta = 0.05$.

To examine whether an interaction can take place between a giant star and its low mass companion we focus on the condition for tidal capture (spiral-in of the low mass companion). Soker [28] expressed the maximum orbital separation at which tidal interaction is significant for brown dwarfs and planets. The maximum radius, scaled for a 10 M_J planet companion, is given by

$$a_m = 3.9R \left(\frac{\tau_{ev}}{6 \times 10^5 \text{yr}}\right)^{1/8} \left(\frac{L}{2000L_\odot}\right)^{1/24} \left(\frac{R}{200R_\odot}\right)^{-1/12} \left(\frac{M_{env}}{0.5M_1}\right)^{1/8} \left(\frac{M_{env}}{0.5M_\odot}\right)^{-1/24} \left(\frac{M_2}{0.01M_1}\right)^{1/8}, \tag{2}$$

where L, R and M are the luminosity, radius, and mass of the giant (RGB or AGB star), respectively, M_{env} is the giant's envelope mass, and τ_{ev} is the evolution time on the upper RGB or AGB. In other words, for a 10 M_J companion to be engulfed by the star the condition on the ratio between the giant's radius and the orbital separation is $R/a > 0.26$.

In Figures 1 and 2 we present the results of our simulations for a 0.9 $M_\odot$ star and a 1.2 $M_\odot$ star, since these masses bound the relevant mass range of PN progenitors of old stellar populations. In Figure 1 we present the final $\simeq 2.5 \times 10^5$ yr of the AGB and focus on 2 metallicities, $Z = 0.02$ and $Z = 0.001$, and three mass-loss rate efficiency parameters, $\eta = 0.5$ (the "traditional" commonly used mass-loss rate efficiency parameter), $\eta = 0.1$, and $\eta = 0.05$. When examining the effects of metallicity and reduced mass-loss on the evolution we focus on the values at the last AGB pulse. The plots are shifted so that they are centered around $t = 0$ where $M_{env} = 10^{-2} \, M_\odot$. We present (top to bottom) the mass, the mass loss rate, the radius and the luminosity. Since tidal interaction brings the planet into the envelope, and this interaction is highly sensitive to the ratio of $R(t)/a(t)$, We examine this quantity in details. We define the value of $L_{pAGB, max}$ by the maximum value of the luminosity during the last AGB pulse (around $t = 0$), disregarding the rapid rise in luminosity due to the helium flash. The reasoning behind choosing this value is that the short duration of the helium flash will have a small effect on observations whereas the "plateau" in the final pulse better reflects the observed luminosities. In Figure 2 we present the important values which have implication on the shaping and on the brightness of the resulting PN (if at all observed): the final mass (upper panel), the maximum value of R/a as expressed in Equation (2), both on the RGB and on the AGB, and the maximum value of the post-AGB luminosity.

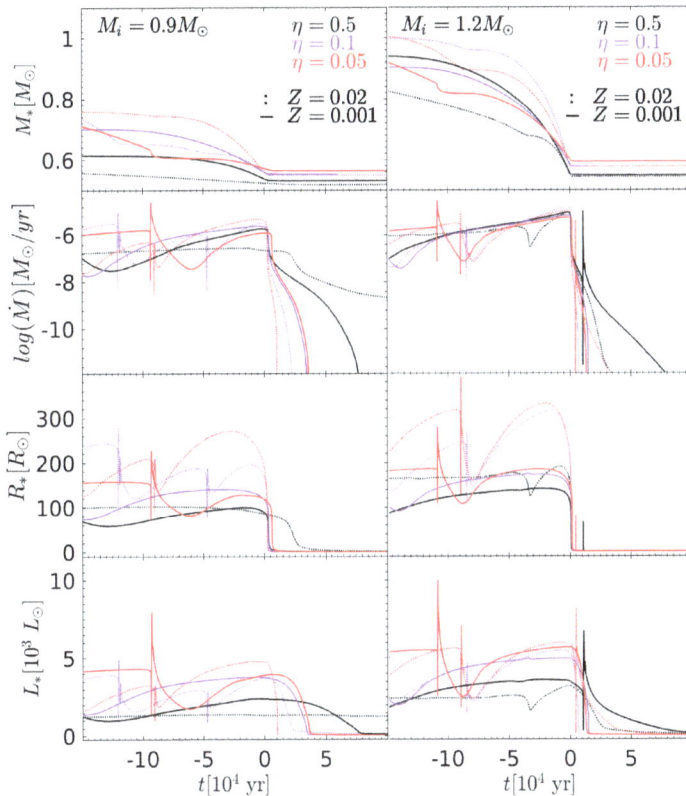

Figure 1. The evolution during the final $\simeq 2.5 \times 10^5$ yr of the asymptotic giant branch (AGB) of stars of initial mass 0.9 $M_\odot$ (**left plot**) and 1.2 $M_\odot$ (**right plot**). The graphs are shifted so that at $t = 0$ the envelope mass of the star is equal to 0.01 $M_\odot$. We present the evolution for 2 metallicities: $Z = 0.02$ (solar; dotted) and $Z = 0.001$ (old population; solid), and for three mass-loss rate efficiency parameters, $\eta = 0.5$ (the "traditional" commonly mass-loss rate efficiency; black), $\eta = 0.1$ (purple), and $\eta = 0.05$ (red). The panels depict, from top to bottom: the mass of the star, the mass loss rate (logarithmic of the absolute value), the stellar radius, and the stellar luminosity.

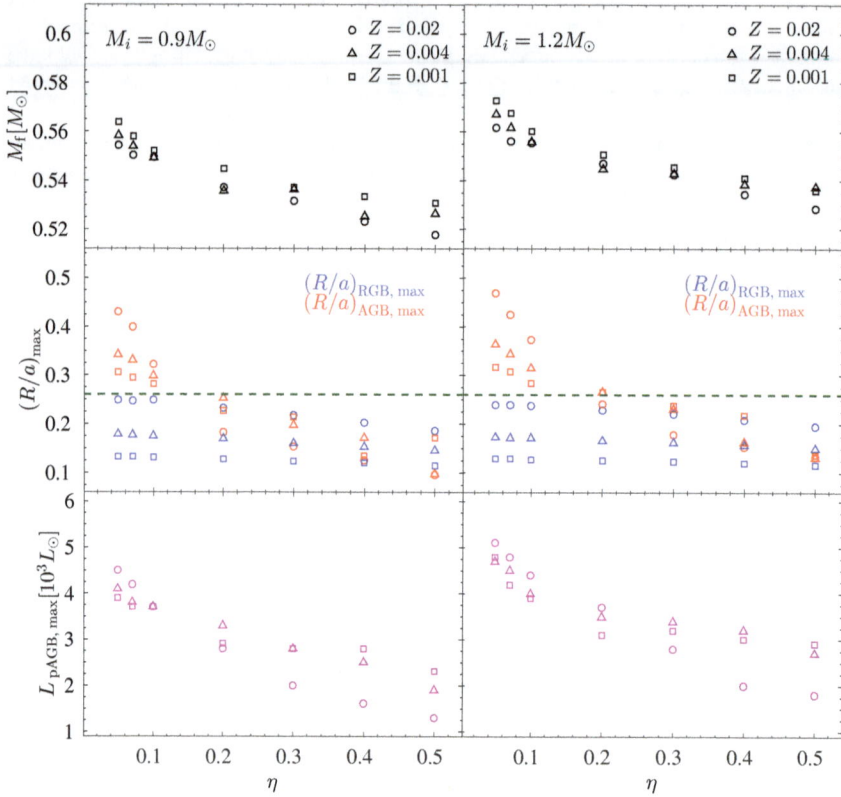

Figure 2. The summary of the evolution of a 0.9 $M_\odot$ star (**left plot**) and a 1.2 $M_\odot$ star (**right plot**). For each star we studied the evolution from zero age main sequence until the formation of a white dwarf, for several mass-loss rate efficiency parameters, from $\eta = 0.5$ (the "traditional" commonly mass-loss rate) to as low as $\eta = 0.05$. We present the evolution for 3 metallicities: $Z = 0.02$ (circles), $Z = 0.004$ (triangles) and $Z = 0.001$ (squares). The upper panels present the final white dwarf mass of each star. The middle panels present the maximum value of the stellar radius over the orbital separation, $(R/a)_{max}$, for both the red giant branch (RGB; blue) and the asymptotic giant branch (AGB; red). The companion mass is 10 M_J and the initial orbital separation taken is 3 AU. The green horizontal line indicates the capture condition above which planet engulfment can take place (Equation (2)). The lower panels present the maximum value of the luminosity on the post-AGB, $L_{pAGB, max}$.

3. Discussion

We studied the evolution of Jsolated stars with low metallicities ($Z = 0.001$ and $Z = 0.004$) and with a reduced mass-loss rate efficiency, from $\eta = 0.5$ (the commonly used value) to as low as $\eta = 0.05$. We compared the evolution of Jsolated stars with low metallicity with that of Jsolated stars with solar metallicity, $Z = 0.02$. Our conclusions are as follows.

1. A higher final mass (higher mass of the central star), M_f, implies a more luminous central star for the ionization of the PN. It can be seen for all metallicities that as the mass-loss rate efficiency parameter η decreases the value of M_f increases. There are very small differences between the values of M_f for different metallicities (yet with the same initial mass and the same value of η). Overall, by comparing $M_{f,\eta=0.5}$ and $M_{f,\eta=0.05}$ there is an increase of 5–8%. On the one hand this increase in M_f might be too small for dynamical effects, but on the other hand it has a large effect

on the luminosity, as we shall discuss below, since the luminosity is very sensitive to the central star mass.

2.   The ratio of the maximum stellar radius and the orbital separation, $(R/a)_{max}$, clearly increases both with the decrease in η and with increasing Z. Moreover, when examining Equation (2) for a planet companion of 10 M_J and at an initial orbital separation of 3 AU we reach the final conclusions as in Sabach & Soker [2]: For the RGB phase it is only marginal for the planet to be engulfed in all cases, hence the probability for an early interaction is at most very small. In addition, for the traditional evolution with a high mass-loss rate efficiency parameter of $\eta = 0.5$ the AGB value of $(R/a)_{\text{AGB, max}}$ is too small for planet engulfment for the low mass of 0.9 $M_\odot$, and is marginal for the larger mass of 1.2 $M_\odot$. However, when reducing the mass-loss rate efficiency to $\eta \leq 0.1$ it is clear from our results that there is a non-negligible range of separations for a planet to exist and be engulfed by the star during the AGB phase. This is because the AGB radius reaches higher values compared to the RGB and compared to the values obtained in "traditional" evolution. In other words, planet engulfment will take place on the AGB of Jsolated stars when the mass-loss rate efficiency parameter is $\eta \leq 0.1$, independent on the metallicity, for the representative cases studied here.

3.   Though we found that both for solar metallicity and for low metallicities planet engulfment is likely to take place on the AGB of Jsolated stars, it is interesting to note that the value of $(R/a)_{max}$ not only increases with an increasing value of Z, but the increase is also "stronger" as η reduces and the metallicity increases.

4.   Our results have implications on the bright end cut-off of the PNLF in old stellar populations, where luminosities of $\simeq$5000 $L_\odot$ and higher are needed for the central star to ionize the bright nebula. To examine this possibility of such a bright central star we focus on the post-AGB luminosities reached in our simulations. We find that the post-AGB luminosity also increases with the decrease of the mass-loss rate, reaching the high $\simeq$5000 $L_\odot$ luminosities needed to explain the brightest PNe in old stellar populations. Interestingly it seems that as the value of η decreases the values for different metallicities grow closer together. We point out that the value of $L_{\text{pAGB, max}}$ has a wide range since it is an approximate value taken at the final AGB phase (the maximum value of the luminosity around $t = 0$ in Figure 1), disregarding the sharp rise due to the helium shell flash. Indeed, we have shown that under the Jsolated framework we can explain the bright end cut-off of the PNLF also for low metallicity low mass stars.

Overall we reached for the case of low metallicity stars of $Z = 0.001$ and $Z = 0.004$ the same conclusions as in our previous works [1,2]. Jsolated stars, with a mass loss rate efficiency parameter $\eta \leq 0.1$ and in the mass range of $M_i = 0.9$–1.2 $M_\odot$, reach higher radii on the AGB compared to the RGB and also higher AGB radii compared to non-Jsolated stars, both for solar metallicity and low metallicities. The post-AGB luminosities of Jsolated stars also reach higher values for all metallicities studied. For the higher masses of 1.2 $M_\odot$ we find that the luminosities reached with a mass-loss rate efficiency parameter of $\eta = 0.1$ are $\simeq$5000 $L_\odot$ and might account for the PNLF in old stellar populations. For the lower mass stars of 0.9 $M_\odot$ we find that planet engulfment can take place on the AGB for a low mass loss rate efficiency parameter, but only for a very low efficiency parameter of $\eta = 0.05$ the luminosity exceeds 4000 $L_\odot$.

Together with a late interaction with a low mass companion such Jsolated stars could account for the shaping of elliptical PNe and possibly also the bright PNe in old stellar populations [1,2]. Once such an interaction takes place the star is no longer considered a Jsolated star. Recently, Giles et al. [29] found the longest period transiting planet candidate from radial velocity measurements, EPIC248847494b. The relevant system parameters are a planet of mass $1 - 10$ M_J at 4.5 ± 1.0 AU from a 0.9 ± 0.09 $M_\odot$ star. It will be interesting to examine the future of the system within our Jsolated framework.

We point out that new stellar evolution simulations find higher post-AGB luminosities compared to old calculations (e.g., [30–32]). Gesicki et al. [22] present new evolutionary tracks of low-mass stars and study the bright end cut-off of the PNLF. They use a different stellar evolution code and find for populations up to an age of 7 Gyr that the PNLF peak can be obtained by lower-mass stars than previously thought. The full answer to the puzzle of the PNLF might be a combination of our results of a low mass-loss rate of Jsolated stars and such new stellar evolution calculations.

Funding: This research was funded from the Israel Science Foundation and a grant from the Asher Space Research Institute at the Technion.

Acknowledgments: I thank Noam Soker and the referees for useful comments that helped improve the paper.

Conflicts of Interest: The author declares no conflicts of interest.

References

1. Sabach, E.; Soker, N. The Class of Jsolated Stars and Luminous Planetary Nebulae in old stellar populations. *Mon. Not. R. Astron. Soc.* **2018**, *479*, 2249–2255. [CrossRef]
2. Sabach, E.; Soker, N. Accounting for planet-shaped planetary nebulae. *Mon. Not. R. Astron. Soc.* **2018**, *473*, 286–294. [CrossRef]
3. Soker, N. What Planetary Nebulae Can Tell Us about Planetary Systems. *Astrophys. J. Lett.* **1996**, *460*, L53. [CrossRef]
4. Siess, L.; Livio, M. The accretion of brown dwarfs and planets by giant stars—I. Asymptotic giant branch stars. *Mon. Not. R. Astron. Soc.* **1999**, *304*, 925–937. [CrossRef]
5. Siess, L.; Livio, M. The accretion of brown dwarfs and planets by giant stars—II. Solar-mass stars on the red giant branch. *Mon. Not. R. Astron. Soc.* **1999**, *308*, 1133–1149. [CrossRef]
6. De Marco, O.; Moe, M. Common Envelope Evolution through Planetary Nebula Eyes. In Planetaty Nebulae as Astronomical Tools: International Conference on Planetaty Nebulae as Astronomical Tools, Gdańsk, Poland, 28 June–2 July 2005; Volume 804, pp. 169–172.
7. Soker, N.; Subag, E. A Possible Hidden Population of Spherical Planetary Nebulae. *Astron. J.* **2005**, *130*, 2717. [CrossRef]
8. Moe, M.; De Marco, O. Do Most Planetary Nebulae Derive from Binaries? I. Population Synthesis Model of the Galactic Planetary Nebula Population Produced by Single Stars and Binaries. *Astrophys. J.* **2006**, *650*, 916. [CrossRef]
9. Villaver, E.; Livio, M. Can Planets Survive Stellar Evolution? *Astrophys. J.* **2007**, *661*, 1192. [CrossRef]
10. Villaver, E.; Livio, M. The Orbital Evolution of Gas Giant Planets around Giant Stars. *Astrophys. J. Lett.* **2009**, *705*, L81. [CrossRef]
11. Nordhaus, J.; Spiegel, D.S.; Ibgui, L.; Goodman, J.; Burrows, A. Tides and tidal engulfment in post-main-sequence binaries: Period gaps for planets and brown dwarfs around white dwarfs. *Mon. Not. R. Astron. Soc.* **2010**, *408*, 631–641. [CrossRef]
12. De Marco, O.; Soker, N. The Role of Planets in Shaping Planetary Nebulae. *Publ. Astron. Soc. Pac.* **2011**, *123*, 402. [CrossRef]
13. Mustill, A.J.; Veras, D.; Villaver, E. Long-term evolution of three-planet systems to the post-main sequence and beyond. *Mon. Not. R. Astron. Soc.* **2014**, *437*, 1404–1419. [CrossRef]
14. Villaver, E.; Livio, M.; Mustill, A.J.; Siess, L. Hot Jupiters and Cool Stars. *Astrophys. J.* **2014**, *794*, 3. [CrossRef]
15. Meynet, G.; Eggenberger, P.; Privitera, G.; Georgy, C.; Ekström, S.; Alibert, Y.; Lovis, C. Star-planet interactions. IV. Possibility of detecting the orbit-shrinking of a planet around a red giant. *Astron. Astrophys.* **2017**, *602*, L7. [CrossRef]
16. Ciardullo, R.; Jacoby, G.H.; Ford, H.C.; Neill, J.D. Planetary nebulae as standard candles. II—The calibration in M31 and its companions. *Astrophys. J.* **1989**, *339*, 53–69. [CrossRef]
17. Jacoby, G. H. Planetary nebulae as standard candles. I—Evolutionary models. *Astrophys. J.* **1989**, *339*, 39–52. [CrossRef]
18. Ciardullo, R.; Sigurdsson, S.; Feldmeier, J.J.; Jacoby, G.H. Close Binaries as the Progenitors of the Brightest Planetary Nebulae. *Astrophys. J.* **2005**, *629*, 499. [CrossRef]
19. Van de Steene, G.C.; Jacoby, G.H.; Praet, C.; Ciardullo, R.; Dejonghe, H. Distance determination to NGC 55 from the planetary nebula luminosity function. *Astron. Astrophys.* **2006**, *455*, 891–896. [CrossRef]
20. Ciardullo, R. The Planetary Nebula Luminosity Function: Pieces of the Puzzle. *Publ. Astron. Soc. Aust.* **2010**, *27*, 149–155. [CrossRef]

21. Davis, B.D.; Ciardullo, R.; Feldmeier, J.J.; Jacoby, G.H. The Planetary Nebula Luminosity Function (PNLF): Contamination from Supernova Remnants. *Res. Notes Am. Astron. Soc.* **2018**, *2*, 32. [CrossRef]
22. Gesicki, K.; Zijlstra, A.A.; Miller Bertolami, M.M. The mysterious age invariance of the planetary nebula luminosity function bright cut-off. *Nat. Astron.* **2018**, *2*, 580. [CrossRef]
23. Diago, P.D.; Gutiérrez-Soto, J.; Fabregat, J.; Martayan, C. Pulsating B and Be stars in the Small Magellanic Cloud. *Astron. Astrophys.* **2008**, *480*, 179–186. [CrossRef]
24. Paxton, B.; Bildsten, L.; Dotter, A.; Herwig, F.; Lesaffre, P.; Timmes, F. Modules for Experiments in Stellar Astrophysics (MESA). *Astrophys. J. Suppl.* **2011**, *192*, 3. [CrossRef]
25. Badenes, C.; Maoz, D.; Ciardullo, R. Pulsating B and Be stars in the Small Magellanic Cloud. *Astrophys. J.* **2015**, *804*, L25. [CrossRef]
26. Reimers, D. The Progenitors and Lifetimes of Planetary Nebulae. *Memoires of the Societe Royale des Sciences de Liege* **1975**, *8*, 369.
27. Guo, J.; Lin, L.; Bai, C.; Liu, J. The effects of the Reimers η on the solar rotational period when our Sun evolves to the RGB tip. *Astrophys. Space Sci.* **2017**, *362*, 15. [CrossRef]
28. Soker, N. Energy and angular momentum deposition during common envelope evolution. *New Astron.* **2004**, *9*, 399–408. [CrossRef]
29. Giles, H.A.C.; Osborn, H.P.; Blanco-Cuaresma, S.; Lovis, C.; Bayliss, D.; Eggenberger, P.; Cameron, A.C.; Kristiansen, M.H.; Turner, O.; Bouchy, F.; et al. The longest period transiting planet candidate from K2. *arXiv* **2018**, arXiv:1806.08757.
30. Karakas, A.I. Helium enrichment and carbon-star production in metal-rich populations. *Mon. Not. R. Astron. Soc.* **2014**, *445*, 347–358. [CrossRef]
31. Miller Bertolami, M.M. New models for the evolution of post-asymptotic giant branch stars and central stars of planetary nebulae. *Astron. Astrophys.* **2016**, *588*, A25. [CrossRef]
32. Ventura, P.; Karakas, A.; Dell'Agli, F.; García-Hernández, D.A.; Guzman-Ramirez, L. Gas and dust from solar metallicity AGB stars. *Mon. Not. R. Astron. Soc.* **2018**, *475*, 2282–2305. [CrossRef]

galaxies

MDPI

Article

The Morphology of the Outflow in the Grazing Envelope Evolution

Sagiv Shiber

Physics Department, Technion—Israel Institute of Technology, Technion City, Haifa 3200003, Israel;
shiber@campus.technion.ac.il

Received: 24 July 2018 ; Accepted: 31 August 2018; Published: 6 September 2018

Abstract: We study the grazing envelope evolution (GEE), where a secondary star, which orbits the surface of a giant star, accretes mass from the giant envelope and launches jets. We conduct simulations of the GEE with different half-opening angles and velocities, and simulate the onset phase and the spiralling-in phase. We discuss the resulting envelope structure and the outflow geometry. We find in the simulations of the onset phase with narrow jets that a large fraction of the ejected mass outflows along the polar directions. The mass ejected at these directions has the fastest velocity and the highest angular momentum magnitude. In the simulations of the spiralling-in phase, a large fraction of the ejected mass concentrates around the orbital plane. According to our findings, the outflow with the highest velocity is closer to the poles as we launch narrower jets. The outflow has a toroidal shape accompanied by two faster rings, one ring at each side of the equatorial plane. The interaction of the jets with the giant envelope causes these outflow structures, as we do not include in our simulations the secondary star gravity and the envelope self-gravity.

Keywords: binaries: close; stars: AGB and post-AGB; stars: winds, outflows; ISM: jets and outflows

1. Introduction

Numerical simulations of the common envelope evolution (CEE) have been performed throughout the last thirty years [1–8]. The simulations attempt to answer fundamental questions regarding the CEE and aim to produce the observed close binary systems where at least one star is a white dwarf [9,10]. Due to the lack of symmetry axis, these simulations are three-dimensional and they usually include only the gravitational orbital energy of the in-spiralling binary.

Table 1 compares the results of several numerical simulations of the CEE. The second row lists the minimum final separation achieved between all the simulations presented in the same paper. The third row lists the maximal fraction of the envelope mass that was ejected from the system in each paper. Almost all simulations failed to achieve the envelope ejection and the observed small final separation. The exception is [3], where they included recombination energy as another source of energy. However, the role of recombination in common envelope removal is still in debate [11–14].

Another extra energy source that has been suggested to contribute to the envelope ejection comes from the accretion of envelope mass on to the more compact secondary star. In particular, when the secondary star accretes mass through an accretion disk and launches jets that interact with the envelope of the giant star. This possibility was first discussed in [15,16] for a neutron star, and then for other companions [17–19]. The jets themselves remove mass, angular momentum, and energy from the vicinity of the secondary star, and by that allow further accretion [20,21].

Table 1. Comparison between results from common envelope evolution simulations. a_f^{min} denotes the minimal final separation achieved in each paper and f_{env}^{max} denotes the maximal fraction of the envelope ejected in each paper. The fourth row states the numerical technique used in each paper, i.e., grid code or smoothed-particle hydrodynamics (SPH) code.

References	[6]	[8]	[7]	[5]	[4]	[3]	[1]
a_f^{min} ($R_\odot$)	4.3	4.4	8.6	5.9	4.2	0.43	16
f_{env}^{max}	0.1	0.31	0.26	0.15	0.06	1	0.16
Grid/SPH	SPH	Grid	Grid	Both	Moving Mesh	SPH	Both

Observational evidence supports mass accretion by the secondary star. For example, the observed chemically polluted secondary star in the Necklace nebula [22]. In several other planetary nebulae (PNe), an inflated secondary main sequence (MS) star is found in a close orbit with the central star [23]. The inflation is best explained by a recent accretion phase where the accretion rate needs to be high. Observations of Fleming 1 (PN G290.5+07.9) indicate that the formation of the jets precedes the ejection of the central nebula [24]. Accretion on to the secondary star, while the binary separation was still large, and, during a relatively long-lived period, caused the launching of the jets. Only later, a faster binary interaction created the central nebula.

In 2015, it was proposed that jets launched from the secondary star can prevent in some cases the system from entering the CEE [25]. The result is a new evolutionary phase that was termed the grazing envelope evolution (GEE). The system initial setup is of a nearly contacting binary where a secondary star already has an accretion disk as it approaches a giant star. The accretion disk launches two opposite jets that remove envelope mass and cause the giant radius to shrink simultaneously with the orbital separation. If jets are inefficient at removing mass, the system might enter a CEE.

First simulations of the GEE focused on simulating the initial evolution of the GEE [26]. A 0.5 $M_\odot$ MS star was placed on a circular orbit close to the surface of a 3.4 $M_\odot$ asymptotic giant branch (AGB) star. The simulations aimed to test the influence of the jets on the giant envelope, hence they did not include the secondary stellar gravity and the self-gravity of the envelope. The secondary star launched jets perpendicular to the orbital plane with a half-opening angle of 30 degrees while circling the giant. The jets were launched with a velocity of approximately the escape velocity from a low mass MS star, at a velocity of 700 km s^{-1}. The mass injected in the jets was taken to be between one to five percent of the Bondi–Hoyle–Lyttleton (BHL) accretion rate [27] on the giant surface. After eight orbital periods, the interaction of the jets with the envelope ejected between 0.05 $M_\odot$ to 0.25 $M_\odot$ of the envelope gas. Approximately 90 percent of the ejected gas was unbound.

In a later, more extended study the companion was set to spiral in to the envelope while it launches jets [28]. These simulations included different opening angles and velocities of the jets. The companion was set to move from the giant surface to two-thirds of the giant radius in a constant inward radial velocity while circling the giant in a tangential Keplerian velocity. The companion spirals-in in a time of three initial orbital periods, in a way that resembles orbits of some other CEE simulations. All the rest of the initial conditions (the giant profile, the secondary star mass, etc.) remained identical to the initial conditions of the onset of the GEE simulations. The results from this study indicate that, regardless of the different jets properties, the jets succeed at expelling mass from the spiraling-in secondary star vicinity and postpone the CEE.

Here, we focus on the morphological structure of nebulae formed by binary systems that have gone through the GEE. This is important because, in this way, we can differentiate between CEE and GEE while analyzing observations. The morphological structure we obtain in the numerical simulations can critically be compared to observations. Observers can link specific systems to this evolutionary phase which then could lead to predictions regarding the further evolution of such systems. In Section 2, we describe the results and, in Section 3, we discuss the results and summarize.

2. Results

The results from the simulations, both from the onset of the GEE and from the spiralling-in orbits, show several structures. We present new simulations of jets with a half-opening angle that equals 15 degrees, i.e., narrow jets. We describe the morphologies of the descendant nebulae from the onset of the GEE in Section 2.1 and from the spiralling-in phase in Section 2.2.

2.1. The Onset of the GEE

At the onset of the GEE, the secondary star orbits in a Keplerian circular motion on the surface of the giant star while launching jets. We simulated two new cases of jets with a half-opening angle of 15 degrees, which are different only in their jets velocity. We compare these two new simulations to a simulation with wider jets of 30 degrees half-opening angle. In all three cases, the jets inject mass at a rate of $\dot{M}_j = 0.001\ M_\odot$ year^{-1}. Figure 1 shows the mass ejected from the system and the outflow average speed as a function of angle from the equatorial plane. The legends show the properties of the simulations. The three first digits at each row represent the jet velocity in kilometres per second while the two last digits represent the jet's half-opening angle in degrees.

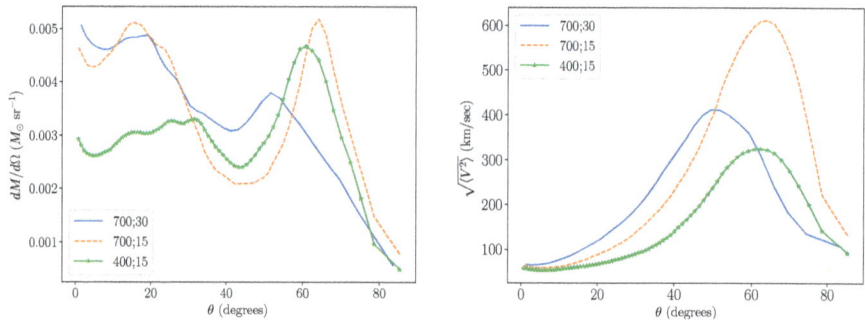

Figure 1. Left panel: The mass lost from the grid per unit solid angle as a function of the angle from the equatorial plane, for three different simulations of the onset of the grazing envelope evolution. The secondary star launches jets while it circles the asymptotic giant branch star on its equator in a Keplerian orbit with a period of 198.4 days. The simulations stop after 1620 days. The three first digits in the legend represent the jets velocity in kilometres per second while the two last digits represent the jets half-opening angle in degrees. The equator is at $\theta = 0$ and the poles are at $\theta = 90$. Simulation 700; 30 is from [26]. The amount of mass lost is that from the two hemispheres combined. In all three cases, the jets inject mass at a rate of $\dot{M}_j = 0.001\ M_\odot$ year^{-1}. **Right panel**: The average speed of the outflow as a function of the angle from the equatorial plane of these three simulations.

We find that narrower jets cause higher mass ejection at mid-latitude. In fact, we obtained a bipolar nebula. Moreover, according to [29], the fast post-AGB wind, which will later interact with the ejected envelope, will enhance the density contrast in the descendant planetary nebula (PN).

To further illustrate the effect of narrow jets, we focus on the simulation with jets velocity of 400 km s^{-1}. We sum over the mass and the kinetic energy that leave a sphere at the edge of the grid and derive the average speed of the outflow as function of the latitude and longitude. We also calculated the average specific angular momentum of the outflowing gas. Figure 2 shows maps of the average speed and of the average specific angular momentum in the z-direction (perpendicular to the orbital plane) of the outflow. The average specific angular momentum is divided by the initial specific orbital angular momentum of the binary system. The maximum in the outflow average velocity (the red area in the left panel of Figure 2) coincides with the angle of major mass ejection and with the maximum in negative average specific angular momentum (the blue area in the right panel of Figure 2).

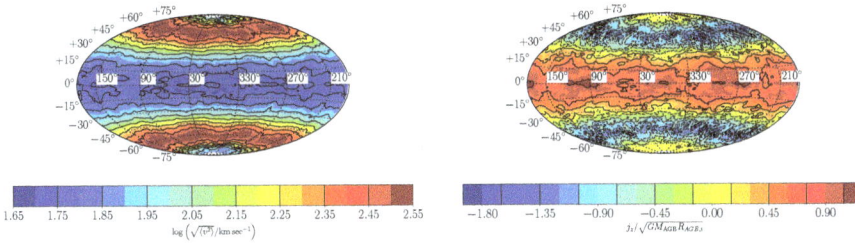

Figure 2. Left panel: The average speed of the outflow at a spherical shell of radius 4 AU for a simulation of the onset of the grazing envelope evolution with jet velocity of 400 km s^{-1} and jet's half-opening angle of 15°. **Right panel**: The average specific angular momentum in the z-direction (perpendicular to the orbital plane) in the same sphere of the same simulation. The average specific angular momentum is divided by the initial specific orbital angular momentum of the binary system. Zero latitude is the equatorial (orbital) plane and zero longitude (at the center) is the initial location of the companion. The companion is moving towards higher angles, namely from the right to the left.

Most of the outflowing gas concentrates in two expanding rings, one at each side of the orbital plane at latitudes of of ±70°, which contain the fastest outflowing gas. The magnitude of the angular momentum in these rings is the largest, *but in an opposite direction to the orbital angular momentum of the binary system*. The rings spin in an opposite direction due to the bending backwards of the jets by the dense envelope, an effect we have already seen and discussed in earlier studies [26,28]. The total angular momentum that leaves the computational domain is in an opposite direction to the orbital angular momentum of the binary system. As a result, the remaining bound envelope material mildly spins up.

We expect that the descendant nebula created by this outflow will exhibit a mirror symmetry. A mirror symmetry has been observed in many bipolar nebulae, such as nebula M2-9 [30]. The remnant of supernova (SN) 1987A shows a mirror symmetry with two quite symmetric outer rings that expand with a radial velocity of approximately 20 km s^{-1} [31], a slower velocity than the outflow velocity we obtain. This difference can be explained by the fact that we do not include radiative cooling in our simulations.

The results from the onset phase of the GEE can be compared to a class of post-AGB stars that have a companion with an intermediate orbital period, typically between one hundred to several thousands of days [32]. Their orbital periods lay in the gap of the traditional bimodal distribution of post-AGB binaries [33]. Several cases of jets were observed within these systems with velocities that indicate the presence of a MS companion [34]. Moreover, in one such system, BD+46°442, the half-opening angle was measured, assuming a certain inclination, to be roughly 60° [35]. Doppler tomography of this system reveals that the jet launching happens around the companion. Based on our results, we predict that post-AGB intermediate binaries (post-AGBIB) that launch narrow jets will produce bipolar nebulae. A known example for such a system is the red rectangle, HD 44179 [36].

2.2. GEE with Spiralling-in Orbits

The simulations with spiralling-in orbits show mass ejection mainly at low latitudes (see left panel in Figure 3), i.e., an equatorial dense outflow. The reason is the deflection of the jets towards the equatorial plane when the companion orbits inside the dense envelope. Conventional CEE simulations, which do not include jets, also lead to a dense equatorial outflow. The gravitational interaction of the envelope with the secondary star causes the equatorial outflow. We do not include the secondary star gravity in our simulations. We expect that the combined effect of the jets with the secondary gravity will produce a pre-PN and later a PN with a massive torus.

The right panel of Figure 3 presents the average speed of the outflow as a function of the latitude in the simulations of the GEE with spiralling-in orbits.

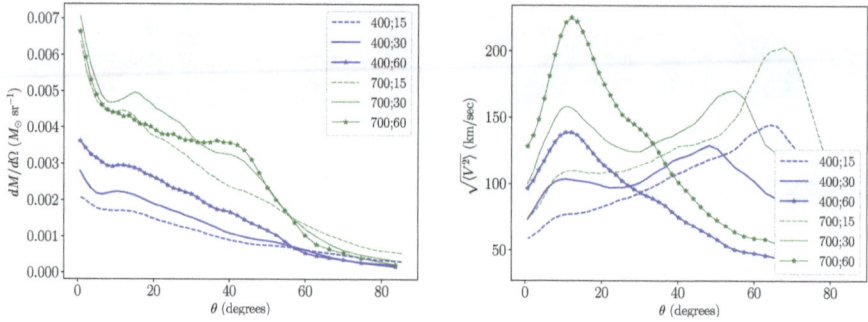

Figure 3. **Left panel**: The mass lost from the grid per unit solid angle as a function of the angle from the equatorial plane, for six different simulations with spiralling-in orbits. The secondary star launches jets while it spirals-in the asymptotic giant branch star envelope on its equator with a tangential Keplerian velocity and with a constant inward radial velocity. The initial orbital period equals to 198.4 days. The simulations stop after 595 days. The three first digits in the legend represent the jets velocity in kilometres per second while the two last digits represent the jets half-opening angle in degrees. The equator is at $\theta = 0$ and the poles are at $\theta = 90$. Simulations: 400;30, 400;60, 700;30, 700;60 are from [28], while simulations 400;15 and 700;15 are new. The amount of mass lost is that from the two hemispheres combined. In all six cases the jets inject mass at a rate of $\dot{M}_j = 0.001 \ M_\odot \ \text{year}^{-1}$. **Right panel**: The average speed of the outflow as a function of the angle from the equatorial plane of these six simulations.

The two simulations with the narrow jets are new, while the four simulations with wider jets are from [28]. We see that narrow jets shift the highest velocity outflow towards the poles. Simulations of jets with a half-opening angle of 60 degrees have a definite peak in the outflow speed at low latitude, while simulations with narrow jets (dashed lines in the right panel of Figure 3) have a definite peak in the outflow speed close to the poles. The escape velocity from the giant star at the boundary of the grid (a distance of 4 AU) equals to approximately 40 km s^{-1}, thus, as seen in Figure 3, a large fraction of the ejected gas is unbound.

We plot in Figure 4 the average speed of the outflow as a function of the latitude and longitude for the case of jets with a half-opening angle of 30 degrees and jets velocity of 400 km s^{-1}.

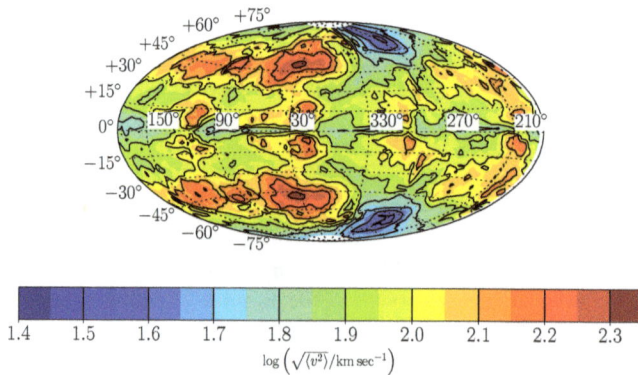

Figure 4. The average speed of the outflow at a spherical shell of radius 4 AU for a simulation with spiralling-in orbit with jets velocity of 400 km s^{-1} and jets half-opening angle of 30°. Zero latitude is the equatorial (orbital) plane and zero longitude (at the center) is the initial location of the companion. The companion is moving towards higher angles, namely from the right to the left.

While there are sporadic peaks near the equatorial plane, there are extended peaks at around ±45° latitude, which lead to the formation of arcs of high velocity outflow, one at each hemisphere. The sporadic peaks in the outflow average speed near the equatorial plane together with the clumpy mass ejection found near the equatorial plane [28] can create the knotty waists observed in the Necklace nebula and in some other similar PNe, such as Hen 2-161 [22,23]. We note that the equatorial plane in SN 1987A also shows knotty features.

To conclude, the morphology of the descendant nebula in the spiralling-in simulations is of a massive torus surrounding the central binary. Fast rings or arcs of material expand below and above the torus. The size of the rings, their number, and even their distance at a given time are the imprints of the jets interaction with the giant envelope.

3. Discussion

We examined the outflow morphology of the GEE as a result of jets that a secondary star launches when it accretes mass from the envelope of a giant star. We conducted hydrodynamical simulations of two phases, the onset of the GEE, when the secondary star orbits the giant star in a circular Keplerian motion at the giant surface, namely, it grazes the envelope from outside, and the spiralling-in phase, when the secondary star grazes the envelope from inside as it spirals-in.

We presented simulations with jets of different half-opening angles and velocities. The simulations with narrow jets, of a half opening angle of 15 degrees, are presented for the first time for the GEE. Simulations of the onset of the GEE with narrow jets show massive ejection near the poles (left panel of Figure 1). Jets with a larger half-opening angle of 30 degrees during the onset phase, and all of the jets that we simulated in the spiralling-in phase, interact with the envelope and create a dense equatorial outflow. This will later form a nebula with a dense expanding torus, which might have knotty structures. Rings (right panel of Figure 3) or arcs (Figure 4) of fast outflow appear at mid-latitudes below and above the central torus (left panel of Figure 3).

The structure of a knotty equatorial torus and a pair of faster rings, one at each side of the equatorial plane and at mid latitudes, is similar to the structure of the three rings of SN 1987A [31]. We suggest that the three rings of SN 1987A were formed by a GEE, with a companion that later entered the envelope and collided with the core of the progenitor of SN 1987A. The three rings expand in a relatively slow speed, much slower than the velocities we obtain. It is possible that the inclusion of radiative cooling into our simulations will yield lower outflow velocities.

We postulate that the observed post-AGBIBs are systems that experienced in the past or are experiencing at present the GEE. Our simulated orbital periods are consistent with the short end of the post-AGBIBs orbital periods. Indeed, a recent study used the MESA BINARY code to show that systems experiencing a significant mass ejection by jets can avoid the CEE [37].

At this stage, we conducted three-dimensional hydrodynamical simulations only of the onset phase of the GEE and of the spiralling-in phase of the GEE. After the onset of the GEE, the system can either continue in the GEE or in the CEE. The exact conditions that determine the evolutionary route are still unclear. To test the transition from the onset phase of the GEE to a long-lasting GEE or to a CEE, simulations of the GEE that include the secondary gravity and the envelope self-gravity should be performed. This will be the subject of a forthcoming study.

Funding: This research received funding from the Israel Science Foundation.

Acknowledgments: I thank the referees for useful comments that helped improve the paper. I thank Noam Soker for useful discussions. This work was supported by the Cy-Tera Project, which is co-funded by the European Regional Development Fund and the Republic of Cyprus through the Research Promotion Foundation.

Conflicts of Interest: The author declares no conflict of interest.

References

1.   Iaconi, R.; De Marco, O.; Passy, J.-C.; Staff, J. The effect of binding energy and resolution in simulations of the common envelope binary interaction. *Mon. Not. R. Astron. Soc.* **2018**, *477*, 2349–2365. [CrossRef]

2. Livio, M.; Soker, N. The common envelope phase in the evolution of binary stars. *Astrophys. J.* **1988**, *329*, 764–779. [CrossRef]

3. Nandez, J.L.A.; Ivanova, N. Common envelope events with low-mass giants: Understanding the energy budget. *Mon. Not. R. Astron. Soc.* **2016**, *460*, 3992–4002. [CrossRef]

4. Ohlmann, S.T.; Röpke, F.K.; Pakmor, R.; Springel, V. Hydrodynamic Moving-mesh Simulations of the Common Envelope Phase in Binary Stellar Systems. *Astrophys. J. Lett.* **2016**, *816*, L9. [CrossRef]

5. Passy, J.-C.; De Marco, O.; Fryer, C.L.; Herwig, F.; Diehl, S.; Oishi, J.S.; Mac Low, M.-M.; Bryan, G.L.; Rockefeller, G. Simulating the Common Envelope Phase of a Red Giant Using Smoothed-particle Hydrodynamics and Uniform-grid Codes. *Astrophys. J.* **2012**, *744*, 52. [CrossRef]

6. Rasio, F.A.; Livio, M. On the Formation and Evolution of Common Envelope Systems. *Astrophys. J.* **1996**, *471*, 366. [CrossRef]

7. Ricker, P.M.; Taam, R.E. An AMR Study of the Common-envelope Phase of Binary Evolution. *Astrophys. J.* **2012**, *746*, 74. [CrossRef]

8. Sandquist, E.L.; Taam, R.E.; Chen, X.; Bodenheimer, P.; Burkert, A. Double Core Evolution. X. Through the Envelope Ejection Phase. *Astrophys. J.* **1998**, *500*, 909. [CrossRef]

9. De Marco, O.; Izzard, R.G. Dawes Review 6: The Impact of Companions on Stellar Evolution. *Publ. Astron. Soc. Aust.* **2017**, *34*, e001. [CrossRef]

10. Ivanova, N.; Justham, S.; Chen, X.; De Marco, O.; Fryer, C.L.; Gaburov, E.; Ge, H.; Glebbeek, E.; Han, Z.; Li, X.-D.; et al. Common envelope evolution: Where we stand and how we can move forward. *Astron. Astrophys. Rev.* **2013**, *21*, 59. [CrossRef]

11. Clayton, M.; Podsiadlowski, P.; Ivanova, N.; Justham, S. Episodic mass ejections from common-envelope objects. *Mon. Not. R. Astron. Soc.* **2017**, *470*, 1788–1808. [CrossRef]

12. Grichener, A.; Sabach, E.; Soker, N. The limited role of recombination energy in common envelope removal. *Mon. Not. R. Astron. Soc.* **2018**, *478*, 1818–1824. [CrossRef]

13. Ivanova, N. On the Use of Hydrogen Recombination Energy during Common Envelope Events. *Astrophys. J. Lett.* **2018**, *858*, L24. [CrossRef]

14. Sabach, E.; Hillel, S.; Schreier, R.; Soker, N. Energy transport by convection in the common envelope evolution. *Mon. Not. R. Astron. Soc.* **2017**, *472*, 4361–4367. [CrossRef]

15. Armitage, P.J.; Livio, M. Black Hole Formation via Hypercritical Accretion during Common-Envelope Evolution. *Astrophys. J.* **2000**, *532*, 540. [CrossRef]

16. Chevalier, R.A. Common Envelope Evolution Leading to Supernovae with Dense Interaction. *Astrophys. J.* **2012**, *752*, L2. [CrossRef]

17. López-Cámara, D.; De Colle, F.; Moreno Méndez, E. self-regulating jets during the Common Envelope phase. *arXiv* **2018**, arXiv:1806.11115.

18. Moreno Méndez, E.; López-Cámara, D.; De Colle, F. Dynamics of jets during the common-envelope phase. *Mon. Not. R. Astron. Soc.* **2017**, *470*, 2929–2937. [CrossRef]

19. Soker, N. Energy and angular momentum deposition during common envelope evolution. *New Astron.* **2004**, *9*, 399. [CrossRef]

20. Chamandy, L.; Frank, A.; Blackman, E.G.; Carroll-Nellenback, J.; Liu, B.; Tu, Y.; Nordhaus, J.; Chen, Z.; Peng, B. Accretion in common envelope evolution. *arXiv* **2018**, arXiv:1805.03607.

21. Shiber, S.; Schreier, R.; Soker, N. Binary interactions with high accretion rates onto main sequence stars. *Res. Astron. Astrophys.* **2016**, *16*, 117. [CrossRef]

22. Miszalski, B.; Boffin, H.M.J.; Corradi, R.L.M. A carbon dwarf wearing a Necklace: First proof of accretion in a post-common-envelope binary central star of a planetary nebula with jets. *Mon. Not. R. Astron. Soc.* **2013**, *428*, L39. [CrossRef]

23. Jones, D.; Boffin, H.M.J.; Rodríguez-Gil, P.; Wesson, R.; Corradi, R.L.M.; Miszalski, B.; Mohamed, S. The post-common envelope central stars of the planetary nebulae Henize 2-155 and Henize 2-161. *Astron. Astrophys.* **2015**, *580*, A19. [CrossRef]

24. Boffin, H.M.J.; Miszalski, B.; Rauch, T.; Jones, D.; Corradi, R.L.M.; Napiwotzki, R.; Day-Jones, A.C.; Koeppen, J. An Interacting Binary System Powers Precessing Outflows of an Evolved Star. *Science* **2012**, *338*, 773–775. [CrossRef] [PubMed]

25. Soker, N. Close Stellar Binary Systems by Grazing Envelope Evolution. *Astrophys. J.* **2015**, *800*, 114. [CrossRef]

26. Shiber, S.; Kashi, A.; Soker, N. Simulating the onset of grazing envelope evolution of binary stars. *Mon. Not. R. Astron. Soc.* **2017**, *465*, L54. [CrossRef]

27. Bondi, H. On spherically symmetrical accretion. *Mon. Not. R. Astron. Soc.* **1952**, *112*, 195–204. [CrossRef]

28. Shiber, S.; Soker, N. Simulating a binary system that experiences the grazing envelope evolution. *Mon. Not. R. Astron. Soc.* **2018**, *477*, 2584–2598. [CrossRef]

29. Frank, A.; Chen, Z.; Reichardt, T.; De Marco, O.; Blackman, E.; Nordhaus, J. Planetary Nebulae Shaped By Common Envelope Evolution. *arXiv* **2018**, arXiv:1807.05925.

30. Livio, M.; Soker, N. The "Twin Jet" Planetary Nebula M2-9. *Astrophys. J.* **2001**, *552*, 685. [CrossRef]

31. Burrows, C.J.; Krist, J.; Hester, J.J.; Sahai, R.; Trauger, J.T.; Stapelfeldt, K.R.; Gallagher, J.S., III; Ballester, G.E.; Casertano, S.; Clarke, J.T.; et al. Hubble Space Telescope Observations of the SN 1987A Triple Ring Nebula. *Astrophys. J.* **1995**, *452*, 680. [CrossRef]

32. Manick, R.; Van Winckel, H.; Kamath, D.; Hillen, M.; Escorza, A. Establishing binarity amongst Galactic RV Tauri stars with a disc. *Astron. Astrophys.* **2017**, *597*, A129. [CrossRef]

33. Nie, J.D.; Wood, P.R.; Nicholls, C.P. Predicting the fate of binary red giants using the observed sequence E star population: Binary planetary nebula nuclei and post-RGB stars. *Mon. Not. R. Astron. Soc.* **2012**, *423*, 2764–2780. [CrossRef]

34. Van Winckel, H. Post-AGB binaries as tracers of stellar evolution. In *Planetary Nebulae: Multi-Wavelength Probes of Stellar and Galactic Evolution*; Cambridge University Press: Cambridge, UK, 2016; Volume 12, pp. 231-234.

35. Bollen, D.; Van Winckel, H.; Kamath, D. Jet creation in post-AGB binaries: the circum-companion accretion disk around BD+46°442. *Astron. Astrophys.* **2017**, *607*, A60. [CrossRef]

36. Van Winckel, H. Why is the Red Rectangle Unique? *Diffuse Interstellar Bands* **2014**, *297*, 180. [CrossRef]

37. Abu-Backer, A.; Gilkis, A.; Soker, N. Orbital Radius during the Grazing Envelope Evolution. *Astrophys. J.* **2018**, *861*, 136. [CrossRef]

galaxies

MDPI

Article

Planetary Nebulae Shaped by Common Envelope Evolution

Adam Frank [1,*], Zhuo Chen [1], Thomas Reichardt [2], Orsola De Marco [2], Eric Blackman [1] and Jason Nordhaus [3]

[1] Department of Physics and Astronomy, University of Rochester, Rochester, NY 14627, USA; zhuo.chen@rochester.edu (Z.C.); blackman@pas.rochester.edu (E.B.)

[2] Department of Physics & Astronomy, Macquarie University, Sydney, NSW 2109, Australia; thomas.reichardt@mq.edu.au (T.R.); orsola.demarco@mq.edu.au (O.D.M.)

[3] National Technical Institute for the Deaf, Rochester Institute of Technology, Rochester, NY 14623, USA; nordhaus@astro.rit.edu

* Correspondence: afrank@pas.rochester.edu

Received: 16 July 2018; Accepted: 2 October 2018; Published: 26 October 2018

Abstract: The morphologies of planetary nebula have long been believed to be due to wind shaping processes in which a "fast wind" from the central star impacts a previously ejected envelope. It is assumed that asymmetries existing in the "slow wind" envelope would lead to inertial confinement, shaping the resulting interacting wind flow. We present new results demonstrating the effectiveness of Common Envelope Evolution (CEE) at producing aspherical envelopes which, when impinged upon by a spherical fast stellar wind, produce highly bipolar, jet-like outflows. We have run two simple cases using the output of a single PHANTOM SPH CEE simulation. Our work uses the Adaptive Mesh Refinement code AstroBEAR to track the interaction of the fast wind and CEE ejecta allows us to follow the morphological evolution of the outflow lobes at high resolution in 3-D. Our two models bracket low and high momentum output fast winds. We find the interaction leads to highly collimated bipolar outflows. In addition, the bipolar morphology depends on the fast wind momentum injection rate. With this dependence comes the initiation of significant symmetry breaking between the top and bottom bipolar lobes. Our simulations, though simplified, confirm the long-standing belief that CEE can plan a major role in PPN and PN shaping. These simulations are intended as an initial exploration of the post-CE/PPN flow patterns that can be expected from central source outflows and CE ejecta.

Keywords: post-AGB stars; pre-PN hydrodynamic models; planetary nebulae: Common Envelope; planetary nebulae: individual (OH231+8+04.2)

1. Introduction

Planetary nebulae ("PN") are formed via gas ejected from highly evolved stars starting within a few thousand years before their final state as white dwarfs. The interaction between a fast wind ejected during the post-Asymtotic Giant Branch (post-AGB) evolutionary state of the star and the previously ejected AGB slow material drives both Pre-PN ("PPN") and mature PN [1]. Given the bipolar nature of many PPN and PN, a "Generalized Interacting Stellar Wind" model has long been thought to be a dominant mechanism for understanding their shaping. The Interacting Stellar Winds model began with the seminal work of [2] who proposed that fast, line driven winds from the hot Central Star of a PN (CSPN) expanding at velocities of order $v_f \sim 1000$ km/s would drive shocks into the heavy and slowly expanding AGB wind ejected earlier ($v_f \sim 10$ km/s). The bright rims of PN were then the ionized shells of swept up AGB material. The generalization of this model, to explain aspherical PN, began with suggestions by [1,3] in which they proposed that slow AGB winds could include a pole-to-equator density contrast. Once a spherical fast wind begins impinging on this aspherical slow wind, "inertial confinement" leads the resulting shocks to take on the elliptical or bipolar configurations. In this Generalized Interacting Stellar Wind model (GISW), the run from mildly elliptical PN to strongly bipolar "butterfly" shaped nebula depends on the pole-to-equator density contrast in the AGB slow wind (e = ρ_p/ρ_e) and its aspherical morphology. Analytic solutions [3] and later simulations [4–6] verified the ability of the GISW model to capture observed PN morphologies. With time, other shaping processes have been considered including the role of magnetic fields [7,8], clumps [9] and the possibility that the fast wind was already collimated in the form of jets [10]. In addition, new observations indicated that much of the shaping process for PN may actually occur before the central star heats up enough to produce ionizing radiation (the post-AGB/PPN phases [11]). While this recognition, along with the other mechanisms such as MHD, have contributed greatly to the understanding of PN shaping, the GISW model often remains in the background. For example, MHD models may still require a strong pole to equator density contrast to produce certain kinds of wasp-waisted nebula. Thus, in spite of the evolution of PN shaping studies since Kwok, Balick and Icke's initial ground-breaking studies, the existence of a pre-existing toroidal density distribution still figures centrally in our understanding of the PN shaping process.

Of course, the assumption of a torioidal circumstellar slow wind has always begged the question of its formation. While single star models for the production of a pole-to-equator density contrast have been proposed [12], most of these rely on degrees of rotation that have proven difficult to obtain for slow rotating AGB stars. Thus binary stars have been seen as essential to modeling PN shaping for some time. Early work of Livio and Soker [13,14] provided key insights into how binary interactions could produce aspherical circumstellar environments. More recent 3-D simulation work has shown how even detached binary interactions can produce fertile ground for PPN shaping [15].

Throughout the extensive discussion of the role that binary stars may play in creating PN, Common Envelope Evolution (CEE) has long been seen as a primary means for generating AGB environments with a high pole to equator density contrast [16]. Common Envelope Evolution occurs when a more compact companion plunges into the envelope of an RGB or AGB star [17]. The release of gravitational energy during the rapid orbital decay is expected to unbind some, or all, of the AGB envelope, creating an expanding toroidal flow that can then serve as the aspherical slow wind for GISW models. The ability to calculate CEE out to the point where the expanding envelope could serve as input for GISW models has, in the past, been hampered by the complexity of CEE 3-D flows. Over the last decade, however, a number of new simulation platforms have been developed for CEE including smooth particle methods and fixed grid [18], AMR mesh methods [19,20] as well as moving mesh methods [21].

In this contribution, we take on the problem of PPN evolution from CEE systems by combining a SPH CEE simulation with an AMR grid-based wind interaction simulation. Our goal in these initial studies is to map out the basic flow patterns emerging from GISW interactions with CEE initial conditions, using fully 3-D, high-resolution simulations.

2. Model and Methods

Our simulations begin with the output of CEE models calculated using the SPH code PHANTOM. The model tracks a binary system with a M_1 = 0.88$M_\odot$ primary and M_2 = 0.6$M_\odot$ companion. The envelope of the primary holds $M_{1,e}$ = 0.48$M_\odot$ of mass. Thus, there is approximately 1.1$M_\odot$ in the two stellar cores that will eventually form a tight binary or merged object. From the simulations, the final separation of the binary has stabilized at approximately R = 20$R_\odot$ (for more details on the simulation including the EOS used, see [22,23].

The results in terms of 3-D flow variable distributions then become the initial conditions for our AstroBEAR AMR simulation (for details of the AstroBEAR AMR multi-physics code see [24]). The numerical results from the SPH simulation (density, 3-D velocity and temperature) are then sampled at regularized points in a grid. This grid of points is then mapped onto an Eulerian mesh appropriate for an AMR simulation. The sampling occurs at the highest level of a 7 level AMR tree and is then restricted using the AMR tolerances for gradients within the flow to the less resolved grids in the AMR hierarchy. The base grid of the mesh has a physical dimension of $(1000R_\odot)^3$ and the 7th level grid has a physical dimension of $(7.8R_\odot)^3$. The simulation domain is cropped to $(16,000R_\odot)^2 \times (128,000R_\odot)$. The origin of the simulation is set at the center of the domain.

After reading in the original SPH data and mapping it to the Eulerian mesh, a point particle with mass 1 M_0 is placed at the grid origin. The point particle's gravity is imposed on the gas, however, the self-gravity of the gas is not considered. A spherical inner "wind" boundary is created around the point particle with $r = 46.9R\_sun$ (spanning 6 of finest mesh cells). The fast wind conditions are injected into the grid through this boundary. Note that the binary orbit at the end of the SPH CE simulation fits within this wind boundary region. Thus, we do not resolve the binary's evolution any further or consider its effect on the fast wind that is considered a spherical outflow in our models.

After preparing the AMR grid with the SPH CEE conditions, we allow for a quiescent period of evolution where the CEE ejecta is simply allowed to expand for some time before the fast wind is initiated. We do this to allow the expanding CE flow to relax on the grid. We note that while we do not include self-gravity in the calculation, our tests during the quiescent period show no significant morphological evolution of the CE ejecta during this time. The most important feature of the quiescent period is that some innermost CE material falls back onto the inner boundary condition (which during this period is set to allow for infall). The fast wind is then turned on instantaneously. For the fast wind interaction with the CE ejecta we used a polytropic index of $\gamma = 5/3$. In fact, cooling is likely too important in these flows, however, for these initial experiments we chose to explore the fundamental hydrodynamic response of the wind/CE interactions. We ran two different models for the fast wind: a low and a high momentum flux case. We have listed both types of condition in Table 1.

Table 1. Parameters for Runs.

	Case A: High Momentum Flux	Case B: Low Momentum Flux
ρ_{FW} (g cm^{-3})	1.0×10^{-11}	5.0×10^{-13}
V_{FW} (km s^{-1})	300	300
Quiescent Phase Δt (days)	500	6000
Mass Loss Rate ($M_\odot$ y^{-1})	6.4×10^{-4}	3.2×10^{-5}
T_{FW} ($^\circ$ak)	30,000	30,000

Our simulations occurred just a few years after the onset of the CEE. At this point, the nature of the central object driving the central outflow was quite uncertain. It could have been a wind from a disk surrounding either the the the core of the primary [25] or the secondary [26] or It could have also been a magnetically powered flow from the primary via a Poynting–Flux driven "magnetic tower" [27]. Thus, linking this immediate post-CE flow to the radiative flux driving single post-AGB stars is only one of a number of possibilities. Most importantly, however, are the results of [11] which showed that many PPN showed "momentum excesses" in their outflows that were orders of magnitude above what stellar photons could provide. Thus, it is particularly unlikely that the central outflow is a "standard" post-AGB radiative driven wind. Note that the timescales for the launching of the outflow in the Bujarrabal et al. study were generally shown to be $t_l \sim 100$ year. This, along with the measured mass in their flows, implies that brief periods (or order a 100 years or less) with mass loss rates of 10^{-5} or 10^{-4} solar masses/year may not be uncommon in PPN flows. Note that this "momentum excess" problem has only been strengthened by results since the 2001 paper. It has been the motivation for a number of MHD driving models for the winds of post-CE objects.

3. Results

In the GISW model, two shocks formed as the fast wind interacted with the previously deposited slow-moving material. An outer shock, facing radially outward from the central wind source, was created as slow wind material was accelerated and swept up via the driving action of the interior fast wind. In addition, an inner shock formed facing back to the wind source via the deceleration of the fast wind. The inner shock converts wind kinetic energy into thermal energy of post-shock fast wind material. If both the fast and slow wind are spherical, the two shocks will be spherical. If, however, the slow wind is aspherical, the inner shock takes on a convex semi-elliptical shape. This distortion of the inner shock is important because radial fast wind streamlines no longer encounter the shock normally but instead strike it at an angle. Streamlines passing through a shock at an acute angle are refracted away from the shock normal. In this way, an aspherical slow wind in GISW models lead to a "shock focused inertial confinement" of the fast wind producing well-collimated jets [5]. The stronger the pole to equator contrast in the slow wind, the more effective is the shock focusing of fast wind material into bipolar jets.

- Model A: In Figure 1 we show slices of the X-Z plane from our AMR AstroBEAR simulation taken at two different times. The first slice is taken 780 days after the initialization of the fast wind and the second is taken at 1600 days.

We first note that the CEE ejecta began with a high pole to equator density contrast (e > 100) and a "cored apple" morphology. During the quiescent evolution of the CEE ejecta, we did not observe this morphology to change significantly. Once the fast wind began, we saw that the high densities of the CEE led to strong inertial confinement in which the momentum injection via the fast wind was unable to move ejecta material in the equatorial direction. In the polar regions, however, fast wind material was able to drive through the ejecta, leading to two highly collimated shocked wind flows which, together, created an overall bipolar outflow.

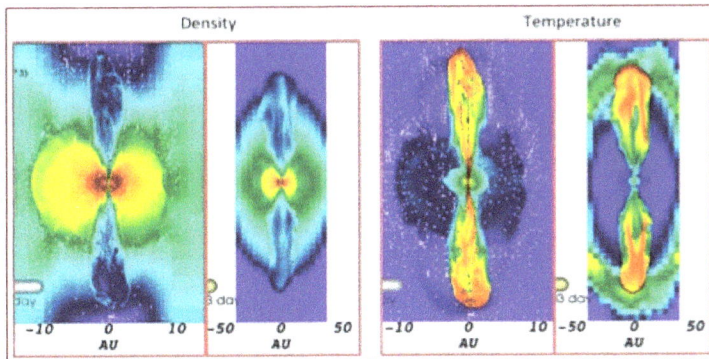

Figure 1. 2-D slices in the X-Z plane of log density (**right**) and log temperature (**left**) from 3-D AMR simulations of a high momentum fast wind being driven into CEE ejecta. We showed two different times for the simulations. Each left sub-panel is at t = 780 days after initiation of fast wind. Each right sub-panel is at t = 1600 days after initiation of fast wind. Note the change of scale between the two times as the nebular lobes expand.

Consideration of the flow at t = 780 days shows that the lobes are essentially hollow, bounded externally by shocked, swept up CEE ejecta and filled internally with low density shocked fast wind material. The shocked fast wind leads, as one would expect, to high temperatures (T ~10^6 K) in the lobe interiors.

As the lobes expand, their basic morphology remains relatively steady with a length to radius ratio of order L_{lobe}/R_{lobe} ~7. What is noteworthy, however, is the difference between the upper and lower lobes in the two time slices. We emphasize that the flows associated with CEE are inherently and fully 3-D. In particular, the binary star's movement through the CE gas sets up strong shearing shocks that lead to KH instabilities which drive local regions of turbulence. The complexity of the flows after repeatedly being shocked by the subsequent driving of unstable modes during each orbit creates local flow patterns that does not need to be axi-symmetric or top-bottom symmetric. We note that such perturbations are seen not just in SPH simulations but also in AMR grid-based simulations ([19], Figure 4, [28], Figure 1). We note that while it's possible that the initial perturbations driving instabilities may have numerical seeds, the development of turbulent, non-symmetric local flow conditions is to be expected.

Because the evolution of the CE ejecta need not respect top/bottom symmetry relative to the orbital plane, it presents a different environment for fast wind interaction above and below the equatorial regions. Thus, we see the details of density and temperature inhomogeneities in the top and bottom lobes differing significantly. This natural cause of symmetry breaking in the two lobes holds great promise in explaining the rich morphologies seen in high-resolution images of PNe.

- Model B: In Figure 2, we show slices from a simulation with a fast wind momentum injection rate that is 20 times lower than Model A. In this model, it takes considerable time for the bipolar lobes to work their way out of the aspherical CEE ejecta as it is demonstrated by the time at which the two sets of images are taken ($t = 6440$ and 6770 days, respectively). The two times shown are relatively close, as compared to what was shown in Model A. This is because before $t = 6440$, the lobes have yet to break out from the central regions of the CEE ejecta. Once the lobes escape the central regions, they begin a rapid expansion in the z-direction (as defined by the binary orbital angular momentum vector).

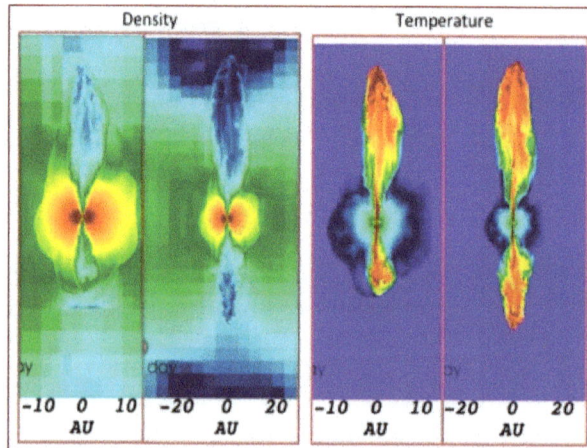

Figure 2. 2-D slices of the X-Z plane of log density (**right**) and log temperature (**left**) from 3-D AMR simulations of a low momentum fast wind being driven into CEE ejecta. We present two different times for the simulations. Each left sub-panel is at $t = 6440$ days after initiation of fast wind. Each right sub-panel is at $t = 6770$ days after initiation of fast wind. Note the change of scales between the two times as the nebular lobes expand.

What is most noteworthy for this model is the strong asymmetry between the top and bottom lobes. While Model A showed differences in details of density/temperature inhomogeneities, Model B provides evidence for a very different evolutionary history between the two lobes. While the top lobe has advanced to almost $Z = 60$ AU from the central source, the bottom lobe has only made it to $Z = 40$ AU. Inspection of the evolution of the Model B shows that the difference between the top and bottom lobe lies in sensitivity to the wind interaction in the CEE polar "channels" (the cored-apple structures in density). In the low momentum case, shocked fast wind accumulates in central regions of CEE ejecta. Inhomogeneities in the "walls" of the CEE ejecta, which are not symmetric about the equatorial plane, lead to the shocked fast wind being becoming trapped/diverted in local density pockets in the lower half of the domain compared to the upper half. Thus, while the shocked fast wind is capable of pushing through the upper CEE channel to form an extended bipolar lobe, the fast wind in the lower channel is blocked for some time. By the time it escapes from the lower channel to form a jet, it significantly lags behind the upper lobe. As discussed above, these global scale asymmetries in the outflow result from local scale asymmetries in the "cored apple" region of the CE ejecta which acts as a nozzle or funnels for collimating the spherical wind into a collimated, bipolar outflow.

Note that the combination of the temperature and density maps show that once they form, the lobes are, just as in Model A, composed of high temperature, low density shocked fast wind material.

4. Conclusions

Using the full CEE simulations as input, we have performed AMR simulations that track the interaction of a fast wind from a central source (the compact binary) and a strongly aspherical CEE ejecta. We find that CEE ejecta provides the strong aspherical density environment needed in classic GISW models to produce strongly bipolar outflows such as those seen in some PPN and PN. Thus, our models confirm and extend the original emphasis on CEE in PN shaping.

We note that one significant new result is the ability of full 3-D CEE models to drive symmetry breaking in GISW bipolar outflows. Our models not only show the differences between top and bottom lobes in terms of inhomogeneous, but they also showed a difference in the size and extent of the upper and lower lobe. If this result continues to be obtained in a more detailed and extensive modeling, then it may offer a natural explanation for objects like OH231+8+04.2 and other highly asymmetric bipolar nebulae.

Finally we note that 2-D fixed grid studies of this problem have recently been completed by [29] using a different set of CE inputs. These results along with the work discussed here show that the field is finally in a position to calculate the PN morp $R_\odot$ hologies directly from models of the evolution of ejecta from the host binary star system.

Author Contributions: Z.C. and T.R. ran the simulations shown in this paper. A.F., E.B., J.N. and O.D.M. helped in the set-up and analysis of the models.

Funding: This paper is supported the Extreme Science and Engineering Discovery Environment (XSEDE), supported by National Science Foundation grant number ACI-1548562. (through XSEDE allocation TG-AST120060) and with funding by Department of Energy grant GR523126, the National Science Foundation grant GR506177, the Space Telescope Science Institute grant GR528562 and NASA grants HST-15044 and HST-14563. **Acknowledgments:** We thank Bruce Balick for many helpful discussions.

Conflicts of Interest: The authors declare no conflict of interest.

References

1. Balick, B. The evolution of planetary nebulae. I-Structures, ionizations, and morphological sequences. *Astron. J.* **1987**, *94*, 671–678. [CrossRef]
2. Kwok, S.; Purton, C.R.; Fitzgerald, P.M. On the origin of planetary nebulae. *Astrophys. J.* **1978**, *219*, 125–127. [CrossRef]
3. Icke, V. Blowing bubbles. *Astron. Astrophys.* **1988**, *202*, 177–188.
4. Frank, A.; Balick, B.; Icke, V.; Mellema, G. Astrophysical gasdynamics confronts reality-The shaping of planetary nebulae. *Astrophys. J.* **1993**, *404*, L25–L27. [CrossRef]
5. Icke, V.; Mellema, G.; Balick, B.; Eulderink, F.; Frank, A. Collimation of astrophysical jets by inertial confinement. *Nature* **1992**, *355*, 524–526. [CrossRef]
6. Icke, V.; Balick, B.; Frank, A. The hydrodynamics of aspherical planetary nebulae. II—Numerical modelling of the early evolution. *Astron. Astrophys.* **1992**, *253*, 224–264.

7. Garcia-Segura, G.; Langer, N.; Rozyczka, M.; Franco, J. Shaping Bipolar and Elliptical Planetary Nebulae: Effects of Stellar Rotation, Photoionization Heating, and Magnetic Fields. *Astrophys. J.* **1999**, *517*, 767–781. [CrossRef]

8. Garcia-Segura, G.; Lopez, J.A.; Franco, J. Magnetically Driven Winds from Post-Asymptotic Giant Branch Stars: Solutions for High-Speed Winds and Extreme Collimation. *Astrophys. J.* **2005**, *618*, 919–931. [CrossRef]

9. Steffen, W.; Lopez, J.A. On the Velocity Structure in Clumpy Planetary Nebulae. *Astrophys. J.* **2004**, *612*, 319–331.

10. Lee, C.-F.; Sahai, R. Magnetohydrodynamic Models of the Bipolar Knotty Jet in Henize 2-90. *Astrophys. J.* **2004**, *606*, 483–496.

11. Bujarrabal, V.; Castro-Carrizo, A.; Alcolea, J.; Sanchez Contreras, C. Mass, linear momentum and kinetic energy of bipolar flows in protoplanetary nebulae. *Astron. Astrophys.* **2001**, *377*, 868–872. [CrossRef]

12. Bjorkman, J.E.; Cassinelli, J.P. Equatorial disk formation around rotating stars due to Ram pressure confinement by the stellar wind. *Astrophys. J.* **1993**, *409*, 429–449. [CrossRef]

13. Soker, N.; Livio, M. Interacting winds and the shaping of planetary nebulae. *Astrophys. J.* **1989**, *339*, 268–278.

14. Livio, M. Planetary Nebulae with Binary Nuclei. In *IAU Symposium No 155, Planetary Nebulae*; Weinberger, R., Acker, A., Eds.; Kluwer Academic Publishers: Dordrecht, The Netherlands, 1993.

15. Chen, Z.; Nordhaus, J.; Frank, A.; Blackman, E.G.; Balick, B. Three-dimensional hydrodynamic simulations of L2 Puppis. *Mon. Not. R. Astron. Soc.* **2016**, *460*, 4182–4187. [CrossRef]

16. Livio, M.; Soker, N. The common envelope phase in the evolution of binary stars. *Astrophys. J.* **1988**, *329*, 764–799.

17. Paczynski, B. Common envelope binaries. *Proc. Int. Astron. Union* **1976**, *73*, 75–80. [CrossRef]

18. Passy, J.-C.; De Marco, O.; Fryer, C.; Herwig, F.; Diehl, S.; Oishi, J.; Mac Low, M.; Bryan, G.; Rockefeller, G. Simulating the Common Envelope Phase of a Red Giant Using Smoothed-particle Hydrodynamics and Uniform-grid Codes. *Astrophys. J.* **2012**, *744*, 52–79. [CrossRef]

19. Chamandy, L.; Frank, A.; Blackman, E.G.; Carroll-Nellenback, J.; Liu, B.; Tu, Y.; Nordhaus, J.; Chen, Z.; Peng, B. Accretion in common envelope evolution. *Mon. Not. R. Astron. Soc.* **2018**, *480*, 1898–1911. [CrossRef]

20. Ricker, P.M.; Taam, R.E. An AMR Study of the Common-envelope Phase of Binary Evolution. *Astrophys. J.* **2012**, *746*, 74.

21. Ohlmann, S.T.; Ropke, F.K.; Pakmor, R.; Springel, V. Hydrodynamic Moving-mesh Simulations of the Common Envelope Phase in Binary Stellar Systems. *Astrophys. J.* **2016**, *816*, L9–15.

22. Iaconi, R.; Reichardt, T.; Staff, J.; De Marco, O.; Pass, J.; Price, D.; Wurster, J.; Herwig, F. The effect of a wider initial separation on common envelope binary interaction simulations. *Mon. Not. R. Astron. Soc.* **2017**, *464*, 4028–4044. [CrossRef]

23. Reichardt, T.; De Marco, O.; Iaconi, R.; Tout, C.; Price, D. Extending Common Envelope Simulations from Roche Lobe Overflow to the Nebular Phase. *Mon. Not. R. Astron. Soc.* **2018**, in press.

24. Carroll-Nellenback, J.J.; Shroyer, B.; Frank, A.; Ding, C.J. Efficient Parallelization for AMR MHD Multiphysics Calculations. *Comput. Phys.* **2013**, *236*, 461–476. [CrossRef]

25. Blackman, E.G.; Frank, A.; Markiel, J.A.; Thomas, J.H.; Van Horn, H.M. Dynamos in asymptotic-giant-branch stars as the origin of magnetic fields shaping planetary nebulae. *Nature* **2001**, *409*, 485–487. [CrossRef] [PubMed]

26. Blackman, E.G.; Frank, A.; Welch, C. Magnetohydrodynamic Stellar and Disk Winds: Application to Planetary Nebulae. *Astrophys. J.* **2001**, *546*, 288–298. [CrossRef]

27. Matt, S.; Frank, A.; Blackman, E.G. Astrophysical Explosions Driven by a Rotating, Magnetized, Gravitating Sphere. *Astrophys. J. Lett.* **2006**, *647*, L45–L48. [CrossRef]

28. MacLeod, M.; Ramirez-Ruiz, E. Asymmetric accretion flows within a common envelope. *Astrophys. J.* **2015**, *803*, 41–62. [CrossRef]

29. Garcia-Segura, G.; Ricker, P.M.; Taam, R.E. Common Envelope Shaping of Planetary Nebulae. *Astrophys. J.* **2018**, *860*, 19–33. [CrossRef]

galaxies

MDPI

Article

Planetary Nebulae Embryo after a Common Envelope Event

Natalia Ivanova [1,*] and Jose L. A. Nandez [1,2]

[1]   Department of Physics, University of Alberta, Edmonton, AB T6G 2E7, Canada; jnandez@sharcnet.ca
[2]   SHARCNET, Faculty of Science, University of Western Ontario, London, ON N6A 5B7, Canada
*   Correspondence: nata.ivanova@ualberta.ca

Received: 7 June 2018; Accepted: 9 July 2018; Published: 19 July 2018

Abstract: In the centers of some planetary nebulae are found close binary stars. The formation of those planetary nebulae was likely through a common envelope event, which transformed an initially-wide progenitor binary into the currently observed close binary, while stripping the outer layers away. A common envelope event proceeds through several qualitatively different stages, each of which ejects matter at its own characteristic speed, and with a different degree of symmetry. Here, we present how typical post-common envelope ejecta looks kinematically a few years after the start of a common envelope event. We also show some asymmetric features we have detected in our simulations (jet-like structures, lobes, and hemispheres).

Keywords: stellar evolution; binarity; planetary nebulae

---

1. Introduction

About 20 percent of planetary nebulae (PN) were likely formed as a result of a common envelope (CE) event that took place in the progenitor binary [1]. A CE event is an interaction during which one of the stars of the binary (the donor) shares its outer layers with its companion star (see for a review [2]). One of the possible outcomes of a CE event is the formation of a binary that is more compact than the initial binary, where one of the stars is the hot core of the initial donor star. In the case of binary formation, the shared CE is ejected; if the stripped core remains hot, the expelled CE may appear in the future as a PN.

About 50 post-CE PN with central stars are known [1,3,4]; CE events could be responsible for most non-spherical PNe [5]. Even though many post-CE PN are bipolar [4], the morphology of the post-CE PN is not uniform. The observed morphology may depend on the observation angle—e.g., an equatorial view favors a narrow-waist shape while a pole-on view favors a ring-type shape. Observed post-CE nebulae sometimes have jets and are elliptical in a shape (see the collection of various observed shapes of PNe in [1]). Not all of the observed non-spherical PNe have to be explained by a CE interaction, some could be a result of a triple interaction or be shaped by planets [6,7]. On the other hand, it has also been shown that a bipolar PN, which could be the most common shape for non-spherical PNe (see, e.g., [4]), could not have been formed during a single star's Asymptotic Giant Branch (AGB) evolution [8,9], where a binary is needed to provide the angular momentum to shape the final PN. This all hints that understanding of strong binary interactions during the envelope removal is important for an overall understanding of PN morphology.

Proper simulation of PN formation due to a CE event must include a self-consistent three-dimensional simulation of the CE event with complete ejection, and the follow-up evolution during which the ejecta becomes a PN. The latter is expected to include a long-term interaction with a pre-CE interstellar medium (presumably shaped by the previous episodes of mass loss by stellar winds), radiative cooling of the expanding envelope, stages of a fast wind, and re-ionization by a hot central star. Recent simulations by Garcia-Segura et al. [10] have shown that, indeed, a bipolar shape can be achieved as a result of considering the interaction of the envelope and a binary. The outer lobes are usually bipolar and the inner lobes are elliptical, bipolar or barrel-type. That study gave detailed attention to the late stages of post-CE evolution. For the initial conditions, they used CE simulations presented in [11]. These CE simulations were performed using the adaptive mesh refinement three-dimensional code FLASH 2.4 [12]. We note that the CE simulations used there as the initial conditions did not end with a complete CE ejection, and there is still some substantial envelope mass left behind, with an unknown fate.

In this contribution, we do the opposite: we provide the kinematic picture of very young, but complete, post-CE ejecta obtained using the smoothed particle hydrodynamics three-dimensional code StarSmasher [13,14]. A typical setup and description of successful CE events can be found in [15,16], but the evolution of the ejecta is limited to about three years. We stress that the expansion age of the ejecta that we can model is very small. It is much smaller not just compared to the age of a typical PNe, $\geq$10,000 years, but also compared to the expansion age of the so-called protoplanetary, or preplanetary, nebulae [17], which is of the order of 3000 years [18]. We term therefore the post-CE ejecta a few years after a CE event as a PN embryo.

2. The Stages of a CE Event, and the Associated Mass Outflows

The outcome of a CE event is usually found using the energy formalism, first introduced by Webbink [19], Livio and Soker [20]. The original energy formalism balances the released orbital energy as an energy source, $E_{orb,ini} - E_{orb,fin}$, with the required energy to dispel the bound envelope, $E_{bind,env}$, as the energy sink. It can be noted that, in this balance, the post-CE ejecta should have null velocity at infinity. Recently, the energy formalism was updated to take into account what fraction of the released orbital energy is taken away by the ejected envelope at infinity, α_{unv}^{∞}, as well as the potentially available recombination energy, E_{rec} [16]:

$$(E_{orb,ini} - E_{orb,fin})(1 - \alpha_{unb}^{\infty}) + E_{bind,env} + E_{rec} = 0 \qquad (1)$$

Recombination energy is a "potential" energy in the sense that it is not readily available to drive an envelope expansion at the start of the CE event, but, after the recombination is triggered, it can be useful at the late stages of a CE event. Recombination energy can remove all the remaining bound parts of the envelope if the envelope material starts to recombine after it expands beyond the "recombination" radius (for the definition, see [21]). The discussion on whether the recombination energy can be used for driving the expansion of a CE, or it would be lost as radiation remains active. Recently, it was shown that neither of the two energy transfer mechanisms in stars—normal sub-sonic stellar convection or radiative diffusion—are capable of transferring the released recombination energy to the surface in most of the cases of regular (as opposed to self-regulated) CE events [22]. A regular CE event takes place on a dynamical timescale; during a regular CE event radiative losses from the surface are negligible compared to the energy released from shrinking orbit (or other energy sources inclusive of recombination). During a self-regulated stage of a CE event, first identified by Meyer and Meyer-Hofmeister [23], the energy loss via surface luminosity matches the energy source. During a self-regulated stage, the ionization profile by mass remains either constant, or propagates too slowly for recombination to affect the CE event energetics and outcomes.

In the published simulations of completed CE events, the ejecta was found to take away about 20–50% of the released orbital energy [16]. The corresponding mass-averaged velocity of the ejecta at infinity is in the range of 40–80 km/s. This is in principle higher than the observed expansion velocities, which are typically <40 km/s [24–26], but this does not yet account for future deceleration due to interactions with previously lost matter in the stellar wind, or with the interstellar medium.

The velocity of the ejected material is not constant across the ejecta–see the typical direction-averaged profile in Figure 1. Ivanova and Nandez [21] identified four kinds of matter ejection processes during a CE event. First is an "initial ejection", before the plunge (for exact definitions of various stages, see [21]). In the initial ejecta, most of the angular momentum is taken away, and the ejected material has a high velocity. The second ejecta are formed during the plunge, when the energy the CE material receives is mainly the mechanical energy transferred from the shrinking binary orbit. Once the envelope is puffed up and is no longer tidally coupled with the shrunken binary orbit, the plunge is slowed down. The envelope ejection is now driven primarily by the recombination energy and is termed as a recombination outflow. This fraction of the ejecta has almost uniform velocity; the rate of the mass loss in the analyzed simulations is 0.2–2 $M_\odot$ year^{-1}, but can be expected to vary in general. Finally, as the slowly expanding but still bound envelope evolves on its new parabolic trajectory, the bottom part of the envelope may fall back and trigger an additional dynamical ejection, termed as a "shell-triggered" ejection (not shown in Figure 1).

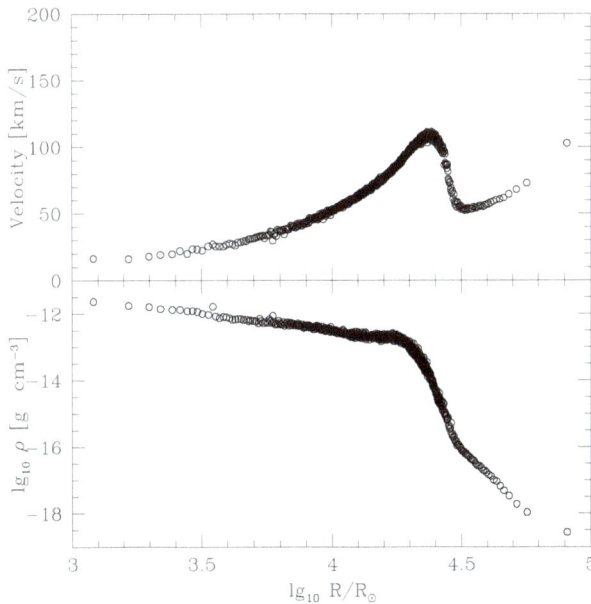

Figure 1. A typical profile of CE ejecta, weighted-by-mass averaged over all directions, shown using the CE simulation of an initial binary consisting of a 1.2 $M_\odot$ red giant with a 0.32 $M_\odot$ core and a 0.32 $M_\odot$ white dwarf companion. The top panel shows velocity and the bottom panel shows density. The profile is shown 800 days after the plunge. There is no envelope material left within the inner 1000 $R_\odot$.

If the rate of the envelope removal by recombination outflows is slow or ineffective, the system may evolve into a self-regulated spiral-in. During the self-regulated spiral-in, the envelope continues to lose material, by the combined work of recombination and pulsations [27,28]; a CE can also experience shell-triggered ejections similar to the case of a dynamical CE event. The mass loss is mainly driven during pulsations. This can only be modeled now using one-dimensional codes, and hence the kinematic profile of the ejecta during a self-regulated spiral-in is not entirely clear.

3. Morphology of the CE Ejecta

The variety—in terms of donor masses, radii at the start of a CE event, the companion masses, how well the stars' rotation are synchronized with the orbit, etc.—of the three-dimensional simulations performed to date with successful CE ejection is very limited, due to the heavy computational demand of those simulations. Besides, the existing three-dimensional simulations were performed for the case of red giant donors, while the most likely donors in the case of the observed PNe are AGB stars. In this respect, the results presented below are limited, and should not be expected to explain all plausible PNe morphology, but they may give us some clues. Here, we use the simulations that were described in [16]; the data presented here have not yet been published elsewhere. Simulations with a very low mass companion were performed for this contribution, using the same numerical methods as described in [16], using 200,000 particles for the donor star.

3.1. Typical Shape—Spherical Symmetry

The most typical shape of a post-CE ejecta, as found in three-dimensional simulations with red giant donors, is spherically symmetric. As a slice in density, it can be said to have a ring shape (see Figure 2), although one has to keep in mind that it is a two-dimensional representation of otherwise rather spherically symmetric object. The ring contains most of the mass, although the entire ejecta propagates well beyond the shown domain. This shape visually resembles the images of such post-CE nebulae as NGC 6337 and Sp 1, which were found to be post-CE PNe [29,30], or Hf 38, also argued to have two central stars (see the image in [5]). This visual symmetry of the ejecta is mainly due to the symmetry of recombination outflows. It is, however, difficult to fully separate how much symmetry was produced by the recombination outflows, and how much was already imprinted in the plunge-driven ejecta (for more discussion on the symmetry of angular momentum inside the CE during the plunge, see [21]). We note that in simulations that did not end with the complete removal of the envelope, e.g. in the simulations of the merger event V1309 Sco [31], the outflow was found to be less spherically symmetric, and resembles an actual ring, which would be seen as a circular ring only from a pole.

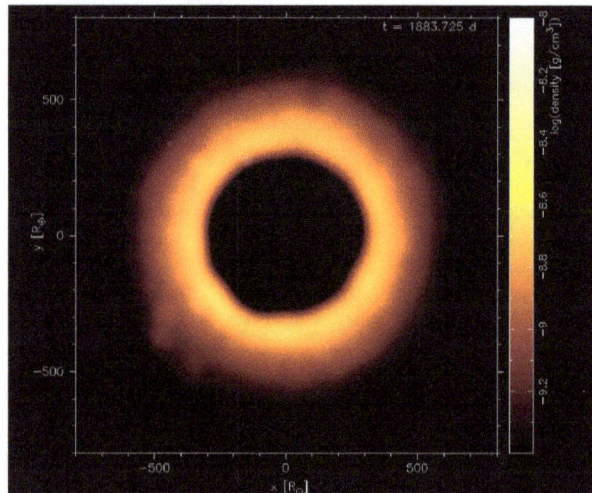

Figure 2. A typical post-CE nebula from a three-dimensional simulation. Here, we show the case of a 1.6 $M_\odot$ red giant with a 0.32 $M_\odot$ core and a 0.36 $M_\odot$ white dwarf companion. Only material with densities above $10^{-9.3}$ is shown (the shown intensity scale starts at $10^{-9.4}$ g cm^{-3} and ends at 10^{-8} g cm^{-3}). In the entire domain of simulation, there is no material with a higher density than used for the intensity scale, a lower density material surrounds the shown object (see also Figure 3). This is about 1000 days after the plunge. The shown domain is 1600 $R_\odot$ by 1600 $R_\odot$. This figure was created using ParaView [32].

3.2. Deviations from Spherical Symmetry

We find that deviations from a spherically symmetric shape are produced in the earlier stages of a CE event, either during the initial stage when most of the angular momentum is lost, or during the plunge. Noticeable deviations can include a jet-like formation in a polar direction (see Figures 3 and 4), accompanied by an underdense lobe cleared by the jet, and one or more hemispheric bubbles (see Figures 4 and 5).

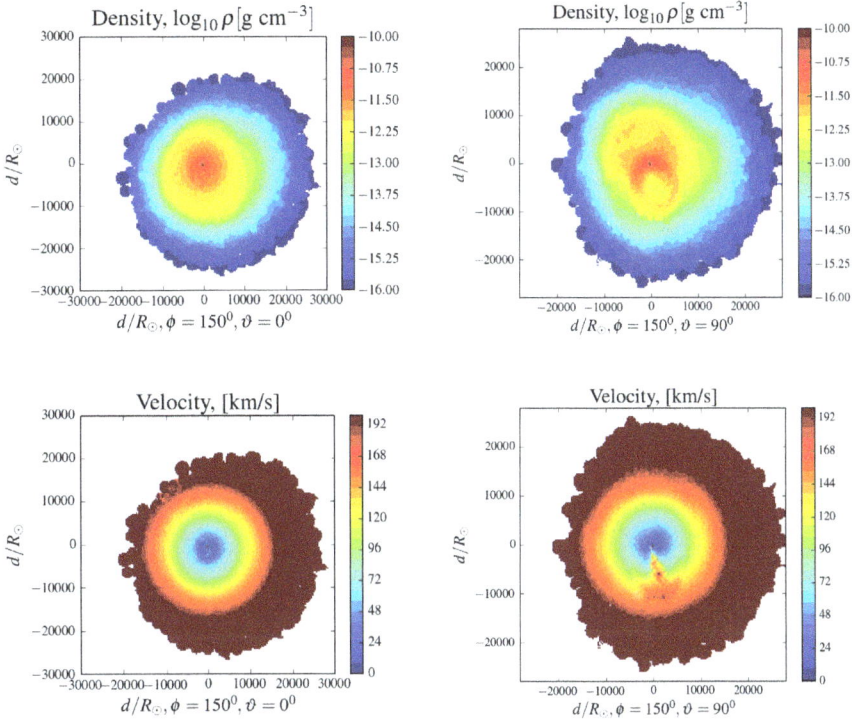

Figure 3. Density and velocity, in the equatorial xy ($\theta = 0°$) and vertical $x'z$ planes ($\theta = 90°$). For these "slices", the plane goes through the origin, and then is first rotated by an angle ϕ around the axis z to produce new axis x', and then by an angle θ around the axis x'. The thickness of the plane is 0.1 of the size. ϕ rotation is chosen to bring the maximum symmetry in $x'y$ plane for $\theta = 0$. The matter with density below 10^{-16} g cm^{-3} is not shown. This CE simulation uses an initial binary consisting of a 1.2 $M_\odot$ red giant with a 0.32 $M_\odot$ core and a 0.32 $M_\odot$ white dwarf companion. These slices are shown about 1000 days after the plunge.

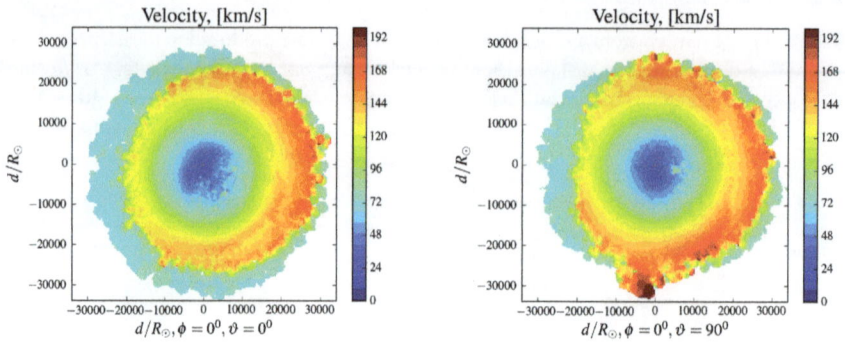

Figure 4. Velocity, in the equatorial xy ($\theta = 0°$) and vertical $x'z$ planes ($\theta = 90°$), constructed similarly to Figure 3. This CE simulation uses an initial binary consisting of a 1.2 $M_\odot$ red giant with a 0.32 $M_\odot$ core and a 0.4 $M_\odot$ white dwarf companion. This is about 800 days after the plunge.

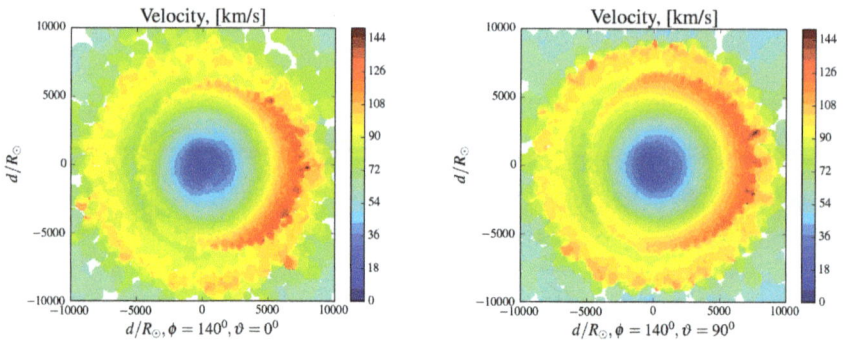

Figure 5. Velocity, in the equatorial xy ($\theta = 0°$) and vertical $x'z$ planes ($\theta = 90°$), constructed similarly to Figure 3. This CE simulation uses an initial binary consisting of a 1.25 $M_\odot$ red giant with a 0.41 $M_\odot$ core and a 0.067 $M_\odot$ white dwarf companion. This is about 1000 days after the plunge.

4. Discussion

Analyzing the simulations of successful CE events with red giant donors, we find that post-CE ejecta are not uniform: neither in the average expansion velocity across the ejecta, nor in their density. Earlier, faster and lower-density ejecta is less spherically symmetric and can feature jet-like formations, spherical shells, hemispherical bubbles, and lobes. Later, slower but more dense ejecta is more spherically symmetric. Its symmetrical features are likely due to the inherent symmetry of the recombination outflows that drive it, or they can be produced at a late stage of the plunge. An asymmetry in the late ejecta could be produced by shell-triggered ejections.

The considered ejecta, regarding their expansion ages, are just embryos of possible future PNe. During the considered early stages of the post-CE expansion, we do not detect the bipolar shape typical of many observed PNe. The bipolar shape could be produced if additional physics is taken into account during subsequent stages of ejecta evolution (see, e.g., [10]). In addition, the cases modeled so far are not directly related to CE events in binaries with AGB donors, where the envelopes are less bound than the donors we have considered, and hence these results should be taken with a grain of salt.

An essential point is that kinematic properties of the post-CE ejecta are imprinted by several stages of the CE event that formed it, as well as by the total energy that the ejecta takes away. We hope, therefore, that kinematic studies of the observed PNe and preplanetary nebulae could help us to substantially constrain the physics of CE events.

Author Contributions: N.I. and J.N. conceived and designed the simulations; J.N. performed the simulations; and N.I. analyzed the data and wrote the paper.

Funding: N.I. acknowledges support from CRC program, and funding from an NSERC Discovery Grant.

Acknowledgments: This research was enabled by the use of computing resources provided by WestGrid, Sharcnet and Compute/Calcul Canada.

Conflicts of Interest: The authors declare no conflict of interest.

Abbreviations

The following abbreviations are used in this manuscript:

AGB   Asymptotic Giant Branch
PN   planetary nebula
CE   common envelope

References

1. De Marco, O.; Reichardt, T.; Iaconi, R.; Hillwig, T.; Jacoby, G.H.; Keller, D.; Izzard, R.G.; Nordhaus, J.; Blackman, E.G. Post-common envelope PN, fundamental or irrelevant? In *Planetary Nebulae: Multi-Wavelength Probes of Stellar and Galactic Evolution*; Liu, X., Stanghellini, L., Karakas, A., Eds.; International Astronomical Union: Paris, France, 2017; Volume 323, pp. 213–217. [CrossRef]
2. Ivanova, N.; Justham, S.; Chen, X.; De Marco, O.; Fryer, C.L.; Gaburov, E.; Ge, H.; Glebbeek, E.; Han, Z.; Li, X.D.; et al. Common envelope evolution: Where we stand and how we can move forward. *Astron. Astrophys. Rev.* **2013**, *21*, 59. [CrossRef]
3. Miszalski, B.; Acker, A.; Moffat, A.F.J.; Parker, Q.A.; Udalski, A. Binary planetary nebulae nuclei towards the Galactic bulge. I. Sample discovery, period distribution, and binary fraction. *Astron. Astrophys.* **2009**, *496*, 813–825. [CrossRef]
4. Miszalski, B.; Acker, A.; Parker, Q.A.; Moffat, A.F.J. Binary planetary nebulae nuclei towards the Galactic bulge. II. A penchant for bipolarity and low-ionisation structures. *Astron. Astrophys.* **2009**, *505*, 249–263. [CrossRef]
5. Barker, H.; Zijlstra, A.; De Marco, O.; Frew, D.J.; Drew, J.E.; Corradi, R.L.M.; Eislöffel, J.; Parker, Q.A. The binary fraction of planetary nebula central stars—III. the promise of VPHAS$^+$. *Mon. Not. R. Astron. Soc.* **2018**, *475*, 4504–4523. [CrossRef]
6. Bear, E.; Soker, N. Planetary Nebulae that Cannot Be Explained by Binary Systems. *Astrophys. J. Lett.* **2017**, *837*, L10. [CrossRef]
7. Sabach, E.; Soker, N. Accounting for planet-shaped planetary nebulae. *Mon. Not. R. Astron. Soc.* **2018**, *473*, 286–294. [CrossRef]
8. García-Segura, G.; Villaver, E.; Langer, N.; Yoon, S.C.; Manchado, A. Single Rotating Stars and the Formation of Bipolar Planetary Nebula. *Astrophys. J.* **2014**, *783*, 74. [CrossRef]
9. García-Segura, G.; Villaver, E.; Manchado, A.; Langer, N.; Yoon, S.C. Rotating Stars and the Formation of Bipolar Planetary Nebulae. II. Tidal Spin-up. *Astrophys. J.* **2016**, *823*, 142. [CrossRef]
10. Garcia-Segura, G.; Ricker, P.M.; Taam, R.E. Common Envelope Shaping of Planetary Nebulae. *ArXiv* **2018**, arXiv:astro-ph.SR/1804.09309.
11. Ricker, P.M.; Taam, R.E. An AMR Study of the Common-envelope Phase of Binary Evolution. *Astrophys. J.* **2012**, *746*, 74. [CrossRef]
12. Fryxell, B.; Olson, K.; Ricker, P.; Timmes, F.X.; Zingale, M.; Lamb, D.Q.; MacNeice, P.; Rosner, R.; Truran, J.W.; Tufo, H. FLASH: An Adaptive Mesh Hydrodynamics Code for Modeling Astrophysical Thermonuclear Flashes. *Astrophys. J. Suppl. Ser.* **2000**, *131*, 273–334. [CrossRef]

13. Gaburov, E.; Lombardi, J.C., Jr.; Portegies Zwart, S. On the onset of runaway stellar collisions in dense star clusters—II. Hydrodynamics of three-body interactions. *Mon. Not. R. Astron. Soc. Lett.* **2010**, *402*, 105–126. [CrossRef]

14. Gaburov, E.; Lombardi, J.C., Jr.; Portegies Zwart, S.; Rasio, F.A. *StarSmasher: Smoothed Particle Hydrodynamics Code for Smashing Stars and Planets*; Astrophysics Source Code Library: Houghton, MI, USA, 2018.

15. Nandez, J.L.A.; Ivanova, N.; Lombardi, J.C. Recombination energy in double white dwarf formation. *Mon. Not. R. Astron. Soc. Lett.* **2015**, *450*, L39–L43. [CrossRef]

16. Nandez, J.L.A.; Ivanova, N. Common envelope events with low-mass giants: understanding the energy budget. *Mon. Not. R. Astron. Soc.* **2016**, *460*, 3992–4002. [CrossRef]

17. Volk, K.M.; Kwok, S. Evolution of protoplanetary nebulae. *Astrophys. J.* **1989**, *342*, 345–363. [CrossRef]

18. Sahai, R.; Sánchez Contreras, C.; Morris, M. A Starfish Preplanetary Nebula: IRAS 19024+0044. *Astrophys. J.* **2005**, *620*, 948–960. [CrossRef]

19. Webbink, R.F. Double white dwarfs as progenitors of R Coronae Borealis stars and Type I supernovae. *Astrophys. J.* **1984**, *277*, 355–360. [CrossRef]

20. Livio, M.; Soker, N. The common envelope phase in the evolution of binary stars. *Astrophys. J.* **1988**, *329*, 764–779. [CrossRef]

21. Ivanova, N.; Nandez, J.L.A. Common envelope events with low-mass giants: understanding the transition to the slow spiral-in. *Mon. Not. R. Astron. Soc.* **2016**, *462*, 362–381. [CrossRef]

22. Ivanova, N. On the Use of Hydrogen Recombination Energy during Common Envelope Events. *Astrophys. J. Lett.* **2018**, *858*, L24. [CrossRef]

23. Meyer, F.; Meyer-Hofmeister, E. Formation of cataclysmic binaries through common envelope evolution. *Astron. Astrophys.* **1979**, *78*, 167–176.

24. Weinberger, R. A catalogue of expansion velocities of Galactic planetary nebulae. *Astron. Astrophys. Suppl. Ser.* **1989**, *78*, 301–324.

25. Gesicki, K.; Zijlstra, A.A. Expansion velocities and dynamical ages of planetary nebulae. *Astron. Astrophys.* **2000**, *358*, 1058–1068.

26. Arnaboldi, M.; Doherty, M.; Gerhard, O.; Ciardullo, R.; Aguerri, J.A.L.; Feldmeier, J.J.; Freeman, K.C.; Jacoby, G.H. Expansion Velocities and Core Masses of Bright Planetary Nebulae in the Virgo Cluster. *Astrophys. J. Lett.* **2008**, *674*, L17. [CrossRef]

27. Ivanova, N.; Justham, S.; Podsiadlowski, P. On the role of recombination in common-envelope ejections. *Mon. Not. R. Astron. Soc.* **2015**, *447*, 2181–2197. [CrossRef]

28. Clayton, M.; Podsiadlowski, P.; Ivanova, N.; Justham, S. Episodic mass ejections from common-envelope objects. *Mon. Not. R. Astron. Soc.* **2017**, *470*, 1788–1808. [CrossRef]

29. García-Díaz, M.T.; Clark, D.M.; López, J.A.; Steffen, W.; Richer, M.G. The Outflows and Three-Dimensional Structure of NGC 6337: A Planetary Nebula with a Close Binary Nucleus. *Astrophys. J.* **2009**, *699*, 1633–1638. [CrossRef]

30. Hillwig, T.C.; Jones, D.; De Marco, O.; Bond, H.E.; Margheim, S.; Frew, D. Observational Confirmation of a Link Between Common Envelope Binary Interaction and Planetary Nebula Shaping. *Astrophys. J.* **2016**, *832*, 125. [CrossRef]

31. Nandez, J.L.A.; Ivanova, N.; Lombardi, J.C., Jr. V1309 Sco—Understanding a Merger. *Astrophys. J.* **2014**, *786*, 39. [CrossRef]

32. Ayachit, U. *The ParaView Guide: A Parallel Visualization Application*; Kitware, Inc.: New York, NY, USA, 2015.

galaxies

MDPI

Article

Simulations and Modeling of Intermediate Luminosity Optical Transients and Supernova Impostors

Amit Kashi

Department of Physics, Ariel University, P.O. Box 3, Ariel 40700, Israel; kashi@ariel.ac.il; Tel.: +972-3-914-3046

Received: 23 June 2018; Accepted: 30 July 2018; Published: 1 August 2018

Abstract: More luminous than classical novae, but less luminous than supernovae, lies the exotic stellar eruptions known as Intermediate luminosity optical transients (ILOTs). They are divided into a number of sub-groups depending on the erupting progenitor and the properties of the eruption. A large part of the ILOTs is positioned on the slanted Optical Transient Stripe (OTS) in the Energy-Time Diagram (ETD) that shows their total energy vs. duration of their eruption. We describe the different kinds of ILOTs that populate the OTS and other parts of the ETD. The high energy part of the OTS hosts the supernova impostors—giant eruptions (GE) of very massive stars. We show the results of the 3D hydrodynamical simulations of GEs that expose the mechanism behind these GEs and present new models for recent ILOTs. We discuss the connection between different kinds of ILOTs and suggest that they have a common energy source—gravitational energy released by mass transfer. We emphasize similarities between Planetary Nebulae (PNe) and ILOTs, and suggest that some PNe were formed in an ILOT event. Therefore, simulations used for GEs can be adapted for PNe, and used to learn about the influence of the ILOT events on the central star of the planetary nebula.

Keywords: stellar evolution; late stage stellar evolution; binarity: transients: planetary nebulae

---

1. Introduction

Intermediate luminosity optical transients (ILOTs) are exotic transients which fall in between the luminosities of novae and supernovae (SN). The group consists of many different astronomical eruptions that, at first, appear to look different from one another but are found to have shared properties. The ILOTs are discussed in this paper and many more are classified according to their total energy and eruption timescale (see Section 2) using a tool named the energy-time diagram (ETD). Most ILOTs reside on the optical transient stripe (OTS) on the ETD that gives us information about the power involved in the eruption and its magnitude. Figure 1 shows the ETD with many ILOTs positioned on the OTS. An extended version can be found on http://phsites.technion.ac.il/soker/ilot-club/. The different kinds of ILOTs are described in Section 3. In Section 4, we discuss simulations for the massive kind of ILOTs and its relevance to the study of Planetary Nebulae (PNe). Our summary and discussion appear in Section 5.

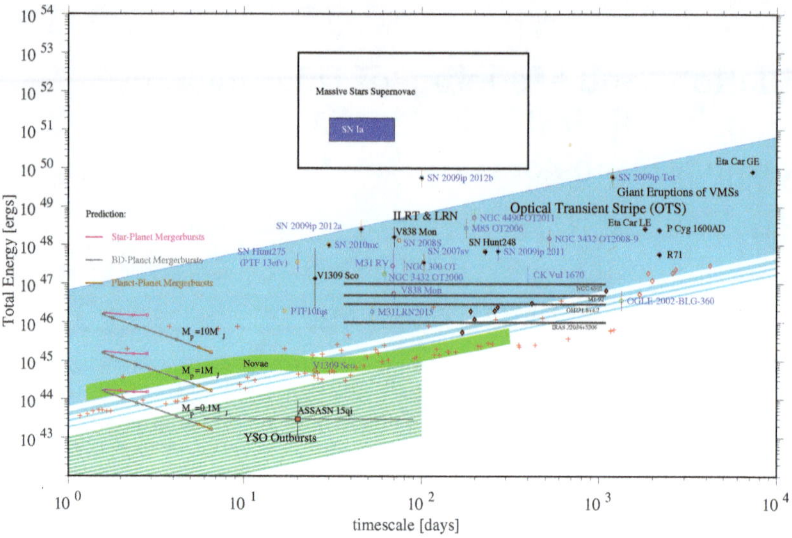

Figure 1. Intermediate luminosity optical transients (ILOTs) on the energy-time diagram (ETD). The abscissa represents the eruption timescale, usually the time for $\Delta V = -3$ mag. The ordinate represents the eruption energy. The blue empty circles represent the total (radiated + kinetic) energy. The black asterisk represents the total available energy (modeled value for some ILOTs). The optical transient stripe is populated by ILOT events powered by gravitational energy or complete merger events or vigorous mass transfer events. Novae models are marked with a green line [1] with red crosses [2] or with diamonds [3]. The four horizontal lines show Planetary Nebulae (PNe) and Proto Planetary Nebulae (PPNe). that might have been formed by ILOT events [4]. Merger models of a planet with a planet or a brown dwarf (BD) star [5] are presented on the left, along with models we added for smaller merging planet with a mass of 0.1 M_J. The lower-left part (hatched in green) is our extension for younger objects [6], including ASASSN-15qi (red square), where the planets are of lower density and can more easily undergo tidal destruction.

2. Common Properties of ILOTs

In reference [7], we noticed that when scaling the time axis for ILOTs of the three kinds, there is an amazing similarity in the light curve surfaces. From the peak or peaks of the light cuve, the decline in the optical bands with time is similar over almost four magnitudes. This could not be a coincidence, and a common physical mechanism must have been involved in all ILOTs, one that resulted in a similar decline.

The high-accretion-powered ILOT (HAPI) model is a model that aims to focus on the shared properties of many of the ILOTs. In its present state, the model accounts for the source and the amount of energy involved in the events and their timescales. The step of obtaining the exact decline rates of the events has not yet been performed and is currently in progress.

The HAPI model was built on the premise that the luminosity of an ILOT comes from the gravitational energy of accreted mass that is partially channeled to radiation. The radiation is eventually emitted in the visible spectrum, possibly after scattering and/or absorption and re-emission.

To quantitatively obtain the power of the transients, we started by defining M_a and R_a as the mass and radius (respectively) of star 'a', which accretes the mass. Star 'b' is the one that supplies the mass to the accretion; it is possibly a destructed MS star, as in Luminous Red Novae (LRN), or alternatively, an evolved star in an unstable evolutionary phase during which it loses a huge amount of mass, as in the giant eruptions (GEs) of η Car. The average total gravitational power is obtained by multiplying the average accretion rate and the potential well of the accreting star:

$$L_G = \frac{GM_a \dot{M}_a}{R_a}. \tag{1}$$

The accreted mass may form an accretion disk or a thick accretion belt around star 'a'. In the case of a merger event, this belt consists of the destructed star. The accretion time should be longer than the viscosity time scale for the accreted mass to lose its angular momentum, so it can be actually accreted. The viscosity timescale is scaled according to

$$t_{visc} \simeq \frac{R_a^2}{\nu} \simeq 73 \left(\frac{\alpha}{0.1}\right)^{-1} \left(\frac{H/R_a}{0.1}\right)^{-1} \left(\frac{C_s/v_\phi}{0.1}\right)^{-1} \left(\frac{R_a}{5\,R_\odot}\right)^{3/2} \left(\frac{M_a}{8\,M_\odot}\right)^{-1/2} \text{days}, \tag{2}$$

where H is the thickness of the disk, C_s is the sound speed, α is the disk viscosity parameter, $\nu = \alpha\, C_s H$ is the viscosity of the disk, and v_ϕ is the Keplerian velocity. M_a and R_a are scaled in Equation (2) according to the parameters of V838 Mon [8]. For these parameters, the ratio of viscosity to the Keplerian timescale is $\chi \equiv t_{visc}/t_K \simeq 160$.

The accreted mass is determined by the details of the binary interaction process and varies for different objects. It is scaled by by $M_{acc} = \eta_a M_a$. As we have learnt from modeled systems (e.g., V838 Mon, V 1309 Sco, η Car), $\eta_a \lesssim 0.1$ with a large variation. The value of $\eta_a \lesssim 0.1$ can be understood as follows. If the MS star collides with a star and tidally disrupts it, as in the model for V838 Mon [9,10], the destructed star is likely to be less massive than the accretor $M_{acc} < M_b \lesssim 0.3 M_a$. In another possible case, an evolved star loses a huge amount of mass, but the accretor gains only a small fraction of the ejected mass, as in the great eruption of η Car.

The viscosity time scale gives an upper limit for the rate of accretion:

$$\dot{M}_a < \frac{\eta_a M_a}{t_{visc}} \simeq 4 \left(\frac{\eta_a}{0.1}\right) \left(\frac{\alpha}{0.1}\right) \left(\frac{H/R_a}{0.1}\right) \left(\frac{C_s/v_\phi}{0.1}\right) \left(\frac{R_a}{5\,R_\odot}\right)^{-3/2} \left(\frac{M_a}{8\,M_\odot}\right)^{3/2} M_\odot\,\text{yr}^{-1}. \tag{3}$$

The maximum gravitational power is, therefore,

$$L_G < L_{max} = \frac{GM_a \dot{M}_a}{R_a} \simeq 7.7 \times 10^{41} \left(\frac{\eta_a}{0.1}\right) \left(\frac{\chi}{160}\right)^{-1} \left(\frac{R_a}{5\,R_\odot}\right)^{-5/2} \left(\frac{M_a}{8\,M_\odot}\right)^{5/2} \text{erg s}^{-1}, \tag{4}$$

where the parameters of the viscosity time scale are replaced with the ratio of the viscosity to the Keplerian time (χ).

The top line of the OTS is calculated according to Equation (4), and describes a supper-Eddington luminosity. The top line might be crossed by some ILOTs if the accretion efficiency (η) is higher or if the parameters of the accreting star are different. For most of the ILOTs, the accretion efficiency is lower; hence, they are located below that line, giving rise to the relatively large width of the OTS. The uncertainty in η_a is not small, but rarely crosses 1. Therefore, one does not expect to find objects above the upper limit indicated by the top line of the OTS very frequently. We also note that the above estimate was performed with the expressions relevant to a thin disk, and a thick disk may need different treatment. More accurate treatment requires hydrodynamic simulations together with radiation transfer to obtain the radiation emitted in each waveband.

3. Types of ILOTs

In recent years, the literature has been far from consistent in referring to transient events. Since many ILOTs are being discovered, the time has arrived for everyone to converge on one naming scheme that will eliminate any ambiguity. In reference [11], we picked up the gauntlet and suggested a complete set of names for the new types of transients. Since then, there have been developments in the field, and new types have been suggested. Therefore, we hereby update the classification scheme of ILOTs.

A. Type I ILOTs. Type I ILOTs is the term for the combination of the three groups listed below: ILRT, LRN, and SN impostors (GEs of LBV). These events share many common physical processes. In particular, they are powered by gravitational energy released in a high-accretion rate event, according to the high-accretion-powered ILOT (HAPI) model discussed below. The condition for an ILOT to be classified as type I is that the observing direction is such that the ILOT is not obscured from the observer by an optically thick medium.

1. **ILRT: Intermediate Luminosity Red Transients.** Events involving evolved stars, such as Asymptotic Giant Branch (AGB) stars and similar objects, such as stars on the Red Giant Branch (RGB). The scenario which leads to these events is most probably a companion which accretes mass and the gravitational energy of the accreted mass supplies the energy of the eruption. Examples include NGC 300 OT, SN 2008S, and M31LRN 2015 (note the self-contradiction in the names of the last two transients).
2. **GEs: Giant Eruptions.** Eruptive events of Luminous Blue Variables (LBVs) or other kinds of very massive stars (VMS), also known as SN impostors. We note that the weaker eruptions of LBVs, known as S Dor eruptions, are not included. From all ILOTs, these GEs are the ones with the highest energy. The energy released in one or in a sequence of these GEs can reach a few $\times 10^{49}$ erg. Examples include the seventeenth century GE of P Cyg, the nineteenth century GEs of η Car, and the pre-explosion eruptions of SN 2009ip. ILRTs are the low mass relatives of LBV GEs.
3. **LRN (or RT): Luminous Red Novae or Red Transients or Merger-Bursts.** Refers to a quite diverse group. These transients are powered by a complete merger of two stars. The eruption can be preceded by a characteristic merger light-curve, as was observed for V1309 Sco. During the eruption, the observer sees a process of destruction of the less compressed star onto the denser star which is accompanied by the release of gravitational energy that powers the transient. Examples for LRN include V838 Mon and V1309 Sco and possibly, the more massive eruption NGC 4490-OT.

B. Type II ILOTs. Since most ILOTs are non-spherically-symmetric eruptions, it is possible that the same event would be observed differently only because it has a different orientation in the sky. In a recent work, reference [12], a new type of binary-powered ILOT was suggested, referred to as Type II ILOT. For Type II ILOTs, the line of sight to the observer intersects a thick dust torus or shell which hides or severely attenuates a direct view of the binary system or the photosphere of the merger product. Type II ILOT may take place in the presence of a strong binary interaction such as a periastron passage in an eccentric orbit causing strong tidal effects that trigger the eruption of a star which is unstable it its outer layers. The interaction leads to an axisymmetrical mass ejection which significantly departs from spherical symmetry. This is much like the morphologies of many planetary nebulae (e.g., references [13–17]). In most cases, the obscuring matter would reside in equatorial directions. The binary system will be obscured to the observer in the optical and IR bands as long as the dust has not dissipated. The type II ILOT is accompanied by some polar mass ejection that may also form dust. The dust in the polar directions reprocesses the radiation that arrives from the central source, and by doing so, enables the observation of the type II ILOT which becomes much fainter. A possible example is the outburst observed from the red supergiant N6946-BH1 in 2009 [18].

C. Proposed scenarios for ILOTs. Other types of ILOTs have been suggested to exist and populate empty regions of the ETD.

1. **Weaker Mergerburst between a planet and a brown dwarf (BD).** It was suggested that in such a scenario the planet is shredded into a disk around the BD, and the energy from accretion lead to an outburst [5]. The destruction of the planet before it plunges into the BD may occur since its average density is smaller than the average density of the BD it encounters. The planet must enter the tidal radius of the BD for the scenario to be applicable. That may occur if the planet

is in a highly eccentric orbit and gets perturbed. Once the planet is destructed as a result of the sequence of events, the remnant of the merger will resamble other LRNs, but on a shorter time scale and smaller energy. Nevertheless, this process is super Eddington. Mergerbursts between a planet and a BD occupy the lower part of the OTS on the ETD.

2. **Weak Outburst of a Young Stellar Object (YSO).** In reference [6], it was suggested that the unusual outburst of the YSO ASASSN-15qi [19] is an ILOT event, similar in many respects to LRN events, such as V838 Mon, but much fainter and of lower total energy. The erupting system was young, but unlike the LRN, the secondary object that was tidally destroyed onto the primary main sequence (MS) star that was suggested to be a Saturn-like planet instead of a low mass MS companion. Such ILOTs are unusual in the sense that they have low power and reside below the OTS. These outbursts are related to FUor outbursts (e.g., reference [20]), which are pre-MS that experience an extreme change in magnitude with slow (years) decline and spectral type. They can be regarded as the more energetic counterparts of the EXor class of outburst, of which EX Lupis is a prototype [21]. These are pre-MS variables that show flares of a few months to a few years and of several magnitudes of amplitude as a result of episodic mass accretion. Another transient, ASASSN-13db [22], may also be a related object.

3. **ILOT which created a Planetary Nebulae (PN).** In reference [4] we identified some intriguing similarities between PNe and ILOTs: (a) a linear velocity–distance relation; (b) a bipolar structure; (c) a total kinetic energy of $\approx 10^{46}$–10^{49} erg. We therefore suggest that some PNe may have formed in an ILOT event lasting a few months (a short "lobe-forming" event). The power source is similar, namely, mass accretion onto a MS companion from the AGB (or ExAGB). The velocity of the fastest gas parcels in such an outburst will be in the order of the escape speed of the MS star, namely a few $\times 100$ km s^{-1}, though most of the gas is expected to interact with the AGB wind and slow down to a few $\times 10$ km s^{-1}. Examples include the PN NGC 6302, the pre-PNe OH231.8+4.2, M1-92, and the IRAS 22036+5306. This process was demonstrated in simulations [23], in which a very short impulsive mass ejection event from a binary system, namely an ILOT, developed a PN with clumpy lobes, quite similar to NGC 6302.

4. GEs in Very Massive Stars

As mentioned above, the most energetic ILOTs are the GEs of LBVs. There is some controversy about the identification of the stars which undergo GE with LBVs, so for safety, we will refer to them as Very Massive Stars or VMS.

GEs in VMSs may occur as a result of an interaction with a binary [24]. A similar process is involved in other types of ILOTs, as discussed above. GE in VMSs may create an accretion disk around the companion, which, in turn, blows jets that shape the ambient gas into bipolar lobes. The mechanism that causes the eruption has an interesting similarity to asymmetric PNe. Many PNe are formed by binary interactions and jets, a process that was suggested two decades ago [25–27]. This process has been observed in PPNe, with examples ranging from the discovery of the butterfly-shaped PPN IRAS 17106-3046 that shows a disk and a bipolar outflow [28], and Hen 2-90 with bipolar morphology and indication of a disk-jet operation [29], to recent detailed observations of the PPN IRAS 16342-3814 that allowed constraints to be out on the mass accretion rate and the accretion channel [30]. Many more examples can be found in reference [31]. Numerical simulations have shown that jets from a companion can interact with the wind of the AGB to form a bipolar PN [32], or a multi-polar PN [33], and more modern simulations have even demonstrated the formation of a 'messy' PN lacking any type of symmetry, also known as a highly irregular PN, and other simulations have emphasized the large variety of morphological features that can be formed by jets [34,35].

In reference [36], we used the hydro code FLASH [37] to model the response of a massive star to a high mass loss episode. The hydro simulation started with the results of a run of a modified version of the 1D stellar evolution code MESA [38–41] in which we obtained a model of an evolved VMS. The MESA obtained stellar model had properties similar to those of η Car. We hydrodynamically simulated a GE with the FLASH code using two approaches: (1) manually removal of a layer from the star, taking the energy required for the process inner layers; and (2) transferring energy from the VMS core to outer layers, which, in turn, caused a spontaneous mass loss. We found that the star developed a strong wind, powered by pulsation in the inner parts of the star. The strong eruptive mass loss phase lasted for a few years, followed by centuries of continually weakening mass loss. Figure 2 shows the resulting mass loss rate over time after the initiation of the GE. The three different lines indicate different simulations with different amount sof mass removed from the VMS. After about two centuries, the mass loss rate of the star declined quite dramatically. The explanation for this behavior is the change in the stellar structure as a response to the huge mass loss in the simulated GE and the two hundred years of high mass loss that followed. At a certain time, the structure became such that the mechanism that accelerated the wind—non-adiabatic κ-mechanism pulsations induced near the iron-bump—stopped being efficient. At that point, the rate of mass loss decreased. The decline in the observed rate of mass loss rate has been observed in η Car over the last two decades. Variations in the spectral properties of the star, especially near the periastron passage and across the spectroscopic event, have taught us that this change-of-state has been happening [42–44]. However the reason for it was, at first, unknown until the simulations in reference [36] revealed the physical mechanisms and demonstrated its work.

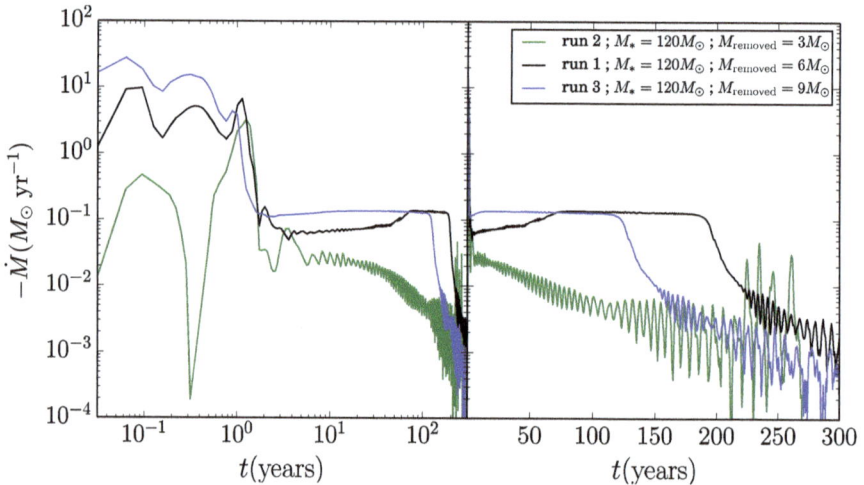

Figure 2. Mass-loss rate as a function of time for runs using the 120 $M_\odot$ VMS model and approaching 1 for initiating a GE. The three different lines indicate different simulations with different amounts of mass removed from the VMS. Two of the runs (the blue and black lines) show a change of state in the mass loss rate of about ≈130 and ≈190 years (respectively) after the GE started. The simulations reproduced the decline in mass loss observed in η Car in previous years, (~175 years after its great eruption), during which the rate of mass loss has been declining by a factor of 2–4 (and possibly, continues to decline).

In follow up simulations we were able to use super-high resolution to simulate a GE in 3D for the first time. The purpose of doing so was to investigate multidimensional effects that cannot be completely or at all addressed otherwise, namely, convection and mixing, rotation, turbulence, hydrodynamical instabilities, meridional currents, tides, and especially, multi-dimensional pulsation. These effects are known to considerably influence the stellar properties, and therefore, we expect them to have significant impacts on how a VMS recovers from a GE.

Figure 3 shows the stellar properties as a function of time after the GE. In the 3D simulations, we found that the pulsations behaved much more chaotically and were much less coherent than in the

1D simulations. The three spatial degrees of freedom in the 3D simulation engendered destructive interference of the pulsations (the chance of creating a constructive interference is low) that damped the waves before they reached the surface and ejected mass. As a result, the mass-loss rate obtained after the GE was smaller (Figure 2), and during the period spanning the first few years after the eruption, the VMS reached an almost stable hydrostatic equilibrium.

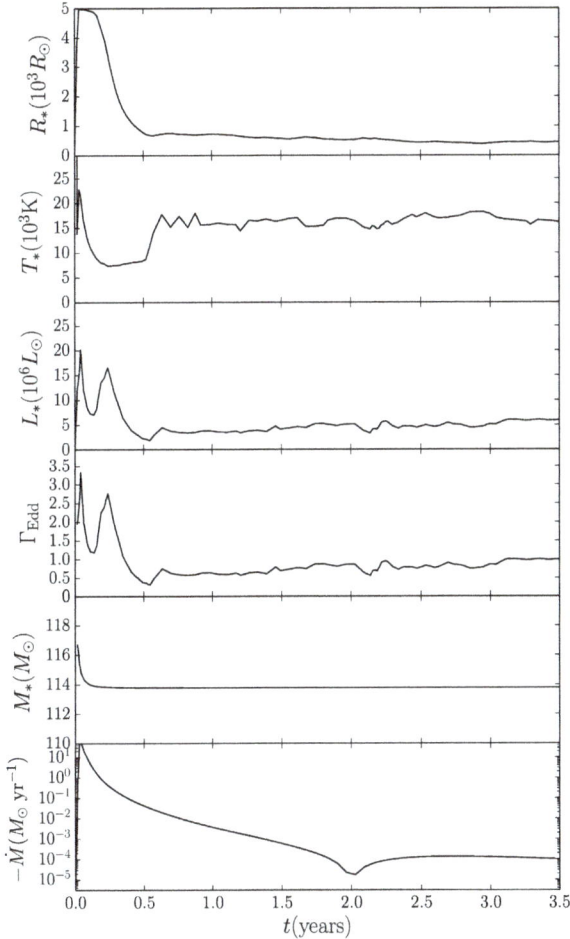

Figure 3. Results of one of our 3D simulations showing the recovery of a non-rotating VMS of 120 $M_\odot$ after removing an outer layer of 6 $M_\odot$ according to approach 2 (as a result of extracting energy from the core). From top to bottom, the panels show the stellar radius, effective temperature, luminosity, mass, Eddington ratio, and mass-loss rate. The temperature, radius, and luminosity were calculated at an optical depth of $\tau = 3$. This figure can be compared to the right panel of Figure 3 in [36] which shows the equivalent 1D simulation. In this 3D simulation, the first few years after the eruption bring the star to an almost stable hydrostatic equilibrium. The mass-loss rate is roughly two orders of magnitude smaller than that in the 1D simulation. Although the results of this preliminary work are still under study, they are probably due to the lower coherency of the pulsations in the 3D, compared to the 1D, simulation. The three spatial degrees of freedom engender destructive interference of the pulsations that damp the waves before they reach the surface and eject mass.

In reference [24], we suggested that major LBV eruptions are triggered by binary interactions. The possibly most problematic example to be considered as a refute of our claim is P Cygni, as it was previously believed to be a single star which underwent such an GE in the seventeenth century. However, reference [45] showed that even the GEs of P Cygni presented evidence of a binary interaction, printed in the varying time gaps between its consequent eruptions in the seventeenth century. Recently, reference [46] used observations of P Cygni spanning over seven decades along with signal processing methods to identify a periodicity in the stellar luminosity. The period they found is a possible indication for the presence of the companion suggested by reference [45] on the basis of theoretical arguments. This gives support to the conjecture that probably all major eruptions in VMSs are triggered by interactions with a secondary star.

5. Summary and Discussion

We discussed different types of ILOTs and categorized them in a manner that gives order to the confused nomenclature in the field. We reviewed commonalities between these types and the HAPI model that suggest they are all gravitationally powered. We showed simulations of the most massive and energetic outbursts that consist of the ILOTs group, the GEs in VMSs. The simulations explained the mechanism for this as being the strong mass loss after the GEs, which accounts for the change of state in the mass loss more than 100 years after the eruption.

Since the physical process for GE in VMSs and many asymmetric PNe—accretion onto a secondary star and jet launching—is similar, hydrodynamical simulations performed for one of them, can relatively easy be adapted and used for the other. The simulations presented here for the recovery of a VMS after a GE can be modified to simulate how the AGB star responds to a large mass loss event as a result of interaction with a companion star. Simulations have so far only modeled the AGB wind, or, at best, the outer layers of the AGB only, and not the entire star. Since, for VMSs, the timescales of the GEs are years to decades, and the recovery time is in the order of centuries, it would be plausible to predict that AGB stars also experience a recovery face that affects their centers, later to become the central star of Planetary Nebulae.

Funding: This research received no external funding.

Acknowledgments: I thank Noam Soker, Amir Michaelis and anonymous referees for helpful comments. I acknowledge Support from the the Authority for Research & Development and the Rector of Ariel University. This work used the Extreme Science and Engineering Discovery Environment (XSEDE) TACC/Stampede2 through allocation TG-AST150018. This work used resources of the Cy-Tera Project, co-funded by the European Regional Development Fund and the Rep. of Cyprus through the Research Promotion Foundation.

Conflicts of Interest: The authors declare no conflict of interest.

Abbreviations

The following abbreviations are used in this manuscript:

AGB	Asymptotic Giant Branch
ASASSN	All-Sky Automated Survey for Supernovae
BD	Brown Dwarf
CSPN	Central Star of Planetary Nebula
ETD	Energy-Time Diagram
GE	Giant Eruption
HAPI	High Accretion Powered ILOT
ILOT	Intermediate Luminosity Optical Transient
ILRT	Intermediate Luminosity Red Transient
LBV	Luminous Blue Variable
LRN	Luminous Red Nova
MESA	Modules for Experiments in Stellar Astrophysics

MS   Main Sequence
OTS   Optical Transient Stripe
PN   Planetary Nebulae
SN   Supernova
VMS   Very Massive Star
YSO   Young Stellar Object

References

1. Della Valle, M.; Livio, M. The Calibration of Novae as Distance Indicators. *Astrophys. J.* **1995**, *452*, 704–709. [CrossRef]
2. Yaron, O.; Prialnik, D.; Shara, M.M.; Kovetz, A. An Extended Grid of Nova Models. II. The Parameter Space of Nova Outbursts. *Astrophys. J.* **2005**, *623*, 398. [CrossRef]
3. Shara, M.M.; Yaron, O.; Prialnik, D.; Kovetz, A.; Zurek, D. An Extended Grid of Nova Models. III. Very Luminous, Red Novae. *Astrophys. J.* **2010**, *725*, 831. [CrossRef]
4. Soker, N.; Kashi, A. Formation of Bipolar Planetary Nebulae by Intermediate-luminosity Optical Transients. *Astrophys. J.* **2012**, *746*, 100. [CrossRef]
5. Bear, E.; Kashi, A.; Soker, N. Mergerburst transients of brown dwarfs with exoplanets. *Mon. Not. R. Astron. Soc.* **2011**, *416*, 1965–1970. [CrossRef]
6. Kashi, A.; Soker, N. An intermediate luminosity optical transient (ILOTs) model for the young stellar object ASASSN-15qi. *Mon. Not. R. Astron. Soc.* **2017**, *468*, 4938–4943. [CrossRef]
7. Kashi, A.; Frankowski, A.; Soker, N. NGC 300 OT2008-1 as a Scaled-down Version of the Eta Carinae Great Eruption. *Astrophys. J. Lett.* **2010**, *709*, L11. [CrossRef]
8. Tylenda, R.; Soker, N.; Szczerba, R. On the progenitor of V838 Monocerotis. *Astron. Astrophys.* **2005**, *441*, 1099–1109. [CrossRef]
9. Soker, N.; Tylenda, R. Violent stellar merger model for transient events. *Mon. Not. R. Astron. Soc.* **2006**, *373*, 733–738. [CrossRef]
10. Tylenda, R.; Soker, N. Eruptions of the V838 Mon type: Stellar merger versus nuclear outburst models. *Astron. Astrophys.* **2006**, *451*, 223–236. [CrossRef]
11. Kashi, A.; Soker, N. Operation of the jet feedback mechanism (JFM) in intermediate luminosity optical transients (ILOTs). *Res. Astron. Astrophys.* **2016**, *16*, 99. [CrossRef]
12. Kashi, A.; Soker, N. Type II intermediate-luminosity optical transients (ILOTs). *Mon. Not. R. Astron. Soc.* **2017**, *467*, 3299. [CrossRef]
13. Balick, B. The evolution of planetary nebulae. I—Structures, ionizations, and morphological sequences. *Astron. J.* **1987**, *94*, 671–678. [CrossRef]
14. Corradi, R.L.; Schwarz, H.E. Morphological populations of planetary nebulae: Which progenitors? I. Comparative properties of bipolar nebulae. *Astron. Astrophys.* **1995**, *293*, 871–888.
15. Manchado, A.; Guerrero, M.A.; Stanghellini, L.; Serra-Ricart, M. *The IAC Morphological Catalog of Northern Galactic Planetary Nebulae*; Instituto de Astrofisica de Canarias: La Laguna, Spain, 1996.
16. Parker, Q.A.; Bojičić, I.S.; Frew, D.J. HASH: The Hong Kong/AAO/Strasbourg Hα planetary nebula database. *J. Phys. Conf. Ser.* **2016**, *728*, 032008. [CrossRef]
17. Sahai, R.; Morris, M.R.; Villar, G.G. Young Planetary Nebulae: Hubble Space Telescope Imaging and a New Morphological Classification System. *Astron. J.* **2011**, *141*, 134. [CrossRef]
18. Adams, S.M.; Kochanek, C.S.; Gerke, J.R.; Stanek, K.Z.; Dai, X. The search for failed supernovae with the Large Binocular Telescope: Confirmation of a disappearing star. *Mon. Not. R. Astron. Soc.* **2017**, *468*, 4968–4981. [CrossRef]
19. Herczeg, G.J.; Dong, S.; Shappee, B.J.; Chen, P.; Hillenbrand, L.A.; Jose, J.; Kochanek, C.S.; Prieto, J.L.; Stanek, K.Z.; Kaplan, K.; et al. The Eruption of the Candidate Young Star ASASSN-15QI. *Astron. J.* **2016**, *831*, 133. [CrossRef]
20. Audard, M.; Ábrahám, P.; Dunham, M.M.; Green, J.D.; Grosso, N.; Hamaguchi, K.; Kastner, J.H.; Kóspál, A.; Lodato, G.; Romanova, M.M.; et al. Episodic Accretion in Young Stars. In *Protostars Planets VI*; University of Arizona Press: Tucson, AZ, USA, 2014; pp. 387–410.
21. Herbig, G.H. EX Lupi: History and Spectroscopy. *Astron. J.* **2007**, *133*, 2679. [CrossRef]

22. Sicilia-Aguilar, A.; Oprandi, A.; Froebrich, D.; Fang, M.; Prieto, J.L.; Stanek, K.; Scholz, A.; Kochanek, C.S.; Henning, T.; Gredel, R.; et al. The 2014–2017 outburst of the young star ASASSN-13db. A time-resolved picture of a very-low-mass star between EXors and FUors. *Astron. Astrophys.* **2017**, *607*, A127. [CrossRef]

23. Akashi, M.; Soker, N. Impulsive ejection of gas in bipolar planetary nebulae. *Mon. Not. R. Astron. Soc.* **2013**, *436*, 1961–1967. [CrossRef]

24. Kashi, A.; Soker, N. Periastron Passage Triggering of the 19th Century Eruptions of Eta Carinae. *Astrophys. J.* **2010**, *723*, 602. [CrossRef]

25. Livio, M.; Pringle, J.E. Wobbling Accretion Disks, Jets, and Point-symmetric Nebulae. *Astrophys. J.* **1997**, *486*, 835. [CrossRef]

26. Soker, N. Collimated Fast Winds in Wide Binary Progenitors of Planetary Nebulae. *Astrophys. J.* **2001**, *558*, 157. [CrossRef]

27. Soker, N.; Livio, M. Disks and jets in planetary nebulae. *Astrophys. J.* **1994**, *421*, 219–224. [CrossRef]

28. Kwok, S.; Hrivnak, B.J.; Su, K.Y. Discovery of a Disk-collimated Bipolar Outflow in the Proto-Planetary Nebula IRAS 17106-3046. *Astrophys. J. Lett.* **2000**, *544*, L149. [CrossRef]

29. Sahai, R.; Nyman, L.-Å. Discovery of a Symmetrical Highly Collimated Bipolar Jet in Hen 2-90. *Astrophys. J. Lett.* **2000**, *538*, L145. [CrossRef]

30. Sahai, R.; Vlemmings, W.H.; Gledhill, T.; Contreras, C.S.; Lagadec, E.; Nyman, L.Å.; Quintana-Lacaci, G. ALMA Observations of the Water Fountain Pre-planetary Nebula IRAS 16342-3814: High-velocity Bipolar Jets and an Expanding Torus. *Astrophys. J. Lett.* **2017**, *835*, L13. [CrossRef] [PubMed]

31. Kwok, S. On the Origin of Morphological Structures of Planetary Nebulae. *Galaxies* **2018**, *6*, 66. [CrossRef]

32. García-Arredondo, F.; Frank, A. Collimated Outflow Formation via Binary Stars: Three-Dimensional Simulations of Asymptotic Giant Branch Wind and Disk Wind Interactions. *Astrophys. J.* **2004**, *600*, 992. [CrossRef]

33. Blackman, E.G.; Frank, A.; Welch, C. Magnetohydrodynamic Stellar and Disk Winds: Application to Planetary Nebulae. *Astrophys. J.* **2001**, *546*, 288. [CrossRef]

34. Akashi, M.; Soker, N. Bipolar rings from jet-inflated bubbles around evolved binary stars. *Mon. Not. R. Astron. Soc.* **2016**, *462*, 206–216. [CrossRef]

35. Akashi, M.; Soker, N. Shaping planetary nebulae with jets in inclined triple stellar systems. *Mon. Not. R. Astron. Soc.* **2017**, *469*, 3296–3306. [CrossRef]

36. Kashi, A.; Davidson, K.; Humphreys, R.M. Recovery from Giant Eruptions in Very Massive Stars. *Astrophys. J.* **2016**, *817*, 66. [CrossRef]

37. Fryxell, B.; Olson, K.; Ricker, P.; Timmes, F.X.; Zingale, M.; Lamb, D.Q.; MacNeice, P.; Rosner, R.; Truran, J.W.; Tufo, H. FLASH: An Adaptive Mesh Hydrodynamics Code for Modeling Astrophysical Thermonuclear Flashes. *Astrophys. J. Suppl. Ser.* **2000**, *131*, 273. [CrossRef]

38. Paxton, B.; Bildsten, L.; Dotter, A.; Herwig, F.; Lesaffre, P.; Timmes, F. Modules for Experiments in Stellar Astrophysics (MESA). *Astrophys. J. Suppl. Ser.* **2011**, *192*, 3. [CrossRef]

39. Paxton, B.; Cantiello, M.; Arras, P.; Bildsten, L.; Brown, E.F.; Dotter, A.; Mankovich, C.; Montgomery, M.H.; Stello, D.; Timmes, F.X.; et al. Modules for Experiments in Stellar Astrophysics (MESA): Planets, Oscillations, Rotation, and Massive Stars. *Astrophys. J. Suppl. Ser.* **2013**, *208*, 4. [CrossRef]

40. Paxton, B.; Marchant, P.; Schwab, J.; Bauer, E.B.; Bildsten, L.; Cantiello, M.; Dessart, L.; Farmer, R.; Hu, H.; Langer, N.; et al. Modules for Experiments in Stellar Astrophysics (MESA): Binaries, Pulsations, and Explosions. *Astrophys. J. Suppl. Ser.* **2015**, *220*, 15. [CrossRef]

41. Paxton, B.; Schwab, J.; Bauer, E.B.; Bildsten, L.; Blinnikov, S.; Duffell, P.; Farmer, R.; Goldberg, J.A.; Marchant, P.; Sorokina, E.; et al. Modules for Experiments in Stellar Astrophysics (MESA): Convective Boundaries, Element Diffusion, and Massive Star Explosions. *Astrophys. J. Suppl. Ser.* **2018**, *234*, 34. [CrossRef]

42. Davidson, K.; Ishibashi, K.; Martin, J.C.; Humphreys, R.M. Eta Carinae's Declining Outflow Seen in the UV, 2002–2015. *Astrophys. J.* **2018**, *858*, 109. [CrossRef]

43. Davidson, K.; Martin, J.; Humphreys, R.M.; Ishibashi, K.; Gull, T.R.; Stahl, O.; Weis, K.; Hillier, D.J.; Damineli, A.; Corcoran, M.; et al. A Change in the Physical State of η Carinae? *Astron. J.* **2005**, *129*, 900. [CrossRef]

44. Mehner, A.; Davidson, K.; Humphreys, R.M.; Walter, F.M.; Baade, D.; De Wit, W.J.; Martin, J.; Ishibashi, K.; Rivinius, T.; Martayan, C.; et al. Eta Carinae's 2014.6 spectroscopic event: Clues to the long-term recovery from its Great Eruption. *Astron. Astrophys.* **2015**, *578*, A122. [CrossRef]
45. Kashi, A. An indication for the binarity of P Cygni from its 17th century eruption. *Mon. Not. R. Astron. Soc.* **2010**, *405*, 1924–1929. [CrossRef]
46. Michaelis, A.M.; Kashi, A.; Kochiashvili, N. Periodicity in the light curve of P Cygni-Indication for a binary companion? *New Astron.* **2018**, *65*, 29. [CrossRef]

Article

Close Binaries and the Abundance Discrepancy Problem in Planetary Nebulae

R. Wesson [1,2,*], **D. Jones** [3,4], **J. García-Rojas** [3,4], **H. M. J. Boffin** [5] **and R. L. M. Corradi** [3,6]

[1] Department of Physics and Astronomy, University College London, Gower St, London WC1E 6BT, UK
[2] European Southern Observatory, Alonso de Córdova 3107, Casilla, Santiago 19001, Chile
[3] Instituto de Astrofísica de Canarias, E-38205 La Laguna, Tenerife, Spain; djones@iac.es (D.J.);
     jogarcia@iac.es (J.G.-R.); romano.corradi@gtc.iac.es (R.L.M.C.)
[4] Departamento de Astrofísica, Universidad de La Laguna, E-38206 La Laguna, Tenerife, Spain
[5] European Southern Observatory, Karl-Schwarzschild-Str. 2, 85738 Garching bei München, Germany;
     hboffin@eso.org
[6] Gran Telescopio Canarias S.A., c/ Cuesta de San José s/n, Breña Baja, E-38712 Santa Cruz de Tenerife, Spain
* Correspondence: rw@nebulousresearch.org

Received: 27 July 2018; Accepted: 15 October 2018; Published: 19 October 2018

Abstract: Motivated by the recent establishment of a connection between central star binarity and extreme abundance discrepancies in planetary nebulae, we have carried out a spectroscopic survey targeting planetary nebula with binary central stars and previously unmeasured recombination line abundances. We have discovered seven new extreme abundance discrepancies, confirming that binarity is key to understanding the abundance discrepancy problem. Analysis of all 15 objects with a binary central star and a measured abundance discrepancy suggests a cut-off period of about 1.15 days, below which extreme abundance discrepancies are found.

Keywords: planetary nebulae; binarity; abundances; stellar evolution

1. Introduction

Heavy element abundances in planetary nebulae (PNe) may be calculated from bright, easily-observed collisionally-excited lines (CELs; typical fluxes of 10–1000, where F(Hβ) = 100) or from the much fainter recombination lines (RLs; typical fluxes of 0.01–1 on the same scale). For 70 years, it has been known that RL abundances exceed those from CELs, with the so-called abundance discrepancy factor (*adf*) ranging from 2–3 in the majority of cases, up to nearly three orders of magnitude in the most extreme cases (see, e.g., [1–3]). Many mechanisms have been proposed to account for the discrepancy. These include:

- Temperature fluctuations [4]
- Strong abundance gradients [5]
- Density inhomogeneities [6]
- Hydrogen-deficient clumps [2]
- X-ray irradiation of quasi-neutral material [7]
- κ-distributed electrons [8]

Abundance discrepancies in H II regions behave quantitatively differently from those in PNe, suggesting that (at least) two mechanisms are responsible [9].

The work in [10] noted that in several cases, the most extreme values of the abundance discrepancy occurred in PNe known to have formed through the ejection of a common envelope (CE; [11,12]), having a binary central star with such a short period that the orbital radius is less than the radius of the PN's Asymptotic Giant Branch (AGB) progenitor [13]. The work in [14] strengthened this connection by observing NGC 6778, known to have a binary central star with a period of 3.68 hours [15], and noted by [16] to show a very strong recombination line spectrum. New deep spectroscopic observations revealed an extreme abundance discrepancy as predicted from the binary nature of the central star, together with spatial patterns also seen in several other high-*adf* nebulae: the *adf* is not constant across the nebula, but rather strongly centrally peaked. In the case of NGC 6778, the value derived from the spatially-integrated spectrum is ∼18, while the spatially-resolved values peak at ∼40 close to the central star.

To investigate this connection further, we observed ∼40 PNe with known binary central stars, to measure their chemistry, predicting that we should find high abundance discrepancies in a significant fraction of them. The nebulae were observed in 2015–2016 using FORS2 at the VLT in Chile, with spectra covering wavelengths from 3600–9300 Å at a resolution of 1.5–3 Å.

2. Results

Spectra of sufficient depth to obtain recombination line abundances were obtained for eight objects in our sample. The observations of NGC 6778 were published in [14]. The seven additional objects included Hf 2-2, already known to have an extreme abundance discrepancy [2], which we included in our sample as a benchmark to verify our methodology (our results from the integrated spectrum are in good agreement with those of [2]) and to study spatially. Emission lines in the spectra were measured using the code ALFA [17], which operates autonomously, first fitting and subtracting a continuum from each spectrum and then optimising Gaussian fits to emission lines using a genetic algorithm. The code detected ∼100 lines in each spectrum. These line fluxes were then analysed using NEAT [18], which also fully autonomously carries out an empirical analysis, calculating temperatures and densities from traditional CEL diagnostics, as well as from hydrogen continuum jumps, helium emission line ratios and oxygen recombination line ratios. The code then determines ionic abundances using a three-zone scheme and corrects for unseen ionisation stages using the scheme of [19].

For four objects, their angular size and the depth of the spectra obtained permitted a spatially-resolved analysis. In these cases, we generally found that the *adf* was strongly centrally peaked; this was most clearly seen in Hf 2–2 and NGC 6326. In Fg 1, the *adf* showed central peaking, but the *adf* was also seen to be higher at the outer edge of the bright inner region of the nebula. NGC 6337 was thought to be a bipolar nebula viewed edge-on [20]; on-sky, it appears as a ring, and the highest values of the *adf* were seen at the inner edge of the ring. Figure 1 shows the variation of the *adf* along the slit for each of these objects.

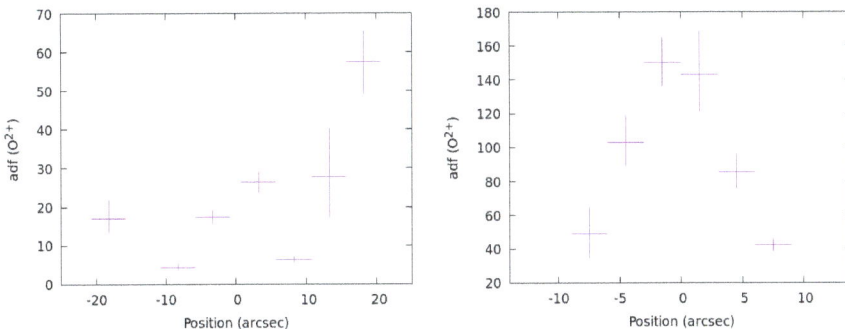

Figure 1. *Cont.*

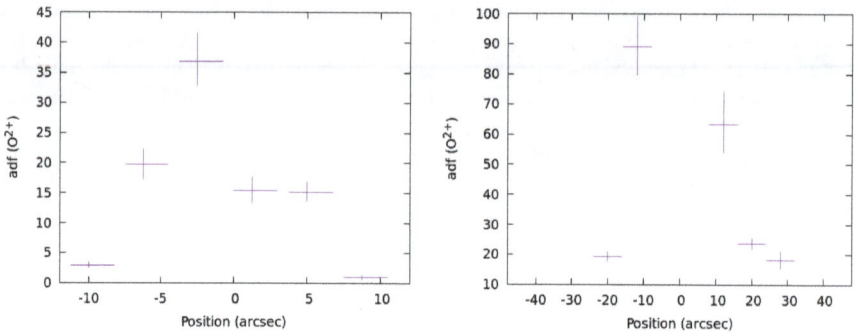

Figure 1. Spatially-resolved analyses of Fg 1 (**top left**), Hf 2-2 (**top right**), NGC 6326 (**bottom left**) and NGC 6337 (**bottom right**), showing the variation with slit position of the abundance discrepancy factor ($adf(O^{2+})$).

3. Discussion

Figure 2 shows the measured *adfs* of 207 objects in rank order, with H II regions highlighted in purple and PNe with close binary central stars shown in blue. This shows that the *adfs* of H II regions are mostly lower than the median value of 2.3, while those of post-CE PNe are typically much higher. Henceforth, we refer to abundance discrepancy factors of less than 5 as "normal", factors between 5 and 10 as "elevated" and those greater than 10 as "extreme".

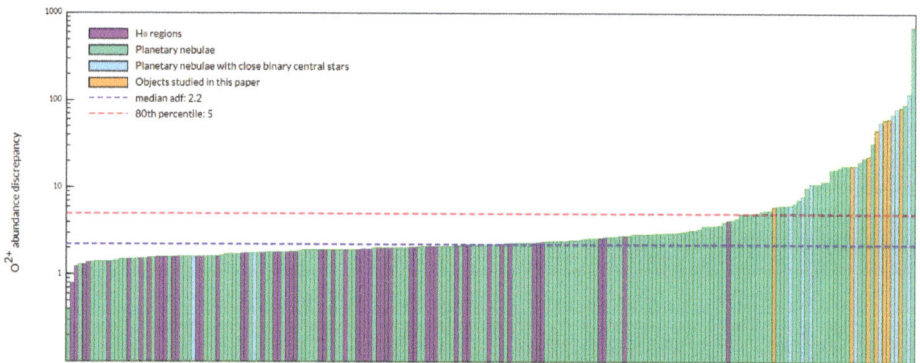

Figure 2. Two hundred and seven measurements of $adf(O^{2+})$ available in the literature as of September 2018, including the new measurements presented in this paper, shown in rank order. Objects with close binary central stars are highlighted in blue, H II regions in purple and the objects studied in the current work in orange. A full list of the individual objects and references used to compile this figure is available at https://www.nebulousresearch.org/adfs.

3.1. Relationship between adfs and Central Star Properties

3.1.1. The Binary Period

We have measured the abundance discrepancy for the first time in six post-common envelope PNe; the integrated spectra reveal five extreme and one elevated *adf*. Nine further post-CE nebulae with measured abundance discrepancy are found in the literature, of which two have normal *adfs*, two have elevated *adfs* and five have extreme *adfs*. Thus, the majority of post-CE nebulae have extreme abundance discrepancies, but a few have much lower values. In Table 1, we list some key observational parameters for all 15 objects.

Table 1. Properties of the 15 nebulae with close binary central stars and a measured *adf*.

Object	*adf*	Period (days)	N/H	O/H	Electron Density (cm^{-3})	Reference
A 46	120	0.47	6.77	7.93	1590	[10]
A 63	8	0.46	7.46	8.59	1560	[10]
Fg 1	46^{+10}_{-8}	1.2	7.84	8.26	500	this work
Hen 2-283	5.1 ± 0.5	1.15	8.56	8.75	3200	this work
Hen 2-155	6.3	0.148	7.85	8.21	1300	[21]
Hen 2-161	11	1.01?	8.06	8.34	1600	[21]
Hf 2-2	83 ± 15	0.4	7.60	7.92	700	[22]; this work
IC 4776	1.75	9	7.80	8.57	~20,000	[23]
MPA 1759	62 ± 8	0.5	7.52	8.25	740	this work
NGC 5189	1.6	4.04	8.60	8.77	1000	[24]
NGC 6326	23 ± 3	0.37	7.10	7.43	820	this work
NGC 6337	18 ± 2	0.173	8.31	8.40	500	this work
NGC 6778	18	0.153	8.61	8.45	600	[14]
Pe 1-9	60 ± 10	0.140	7.85	7.98	740	this work
Ou 5	56	0.36	7.58	8.40	560	[10]

We searched for a relationship between the period of the binary central star and the abundance discrepancy. A continuous relationship would suggest that the mechanism giving rise to extreme *adfs* operates regardless of the binary period, but its magnitude is determined by it. On the other hand, if a threshold period can be identified that divides objects with elevated *adfs* from those with normal *adfs*, then it would imply that the mechanism is only triggered for shorter period binaries. Figure 3 shows the relation between the two properties, and although the number of points remains quite small, there is clearly no simple relationship between orbital period and *adf*. Instead, it suggests a threshold period, as three groups of objects can be identified: those with periods of less than one day all have elevated or extreme *adfs*; the objects with periods longer than 1.2 days have normal *adfs*; and the several objects with periods of around 1.15 days have a wide range of *adfs*, including Fg 1 with an *adf* of ~46, Hen 2-283 with an *adf* of 5.1 and the Necklace, with no measured *adf*, but with no RLs detected in deep spectra by [25], and thus likely a normal value.

Detection biases mean that the observed period distribution of central star binaries is strongly concentrated to lower values, and thus, objects with periods of days rather than hours are quite rare [13]. The absence of high-*adf* objects with longer periods could arise by chance, given that only two longer period objects appear in our sample of 15 objects. However, the likelihood of the two longest period binaries having the two lowest *adfs* by chance alone if the two properties were uncorrelated is just under 1%. A likely third such object is MyCn 18, for which [26] recently reported a binary period of 18.15 days, while [27] measured an *adf* of 1.8. The probability of the three lowest *adfs* coinciding with the three longest periods by chance is much less than 1% and strengthens the hypothesis that only when the binary period is shorter than a threshold value of around 1.15 days will the PN exhibit an elevated or extreme *adf*. Further measurements of *adf* in objects with known binary orbital period are of course still necessary to better constrain this proposed relationship.

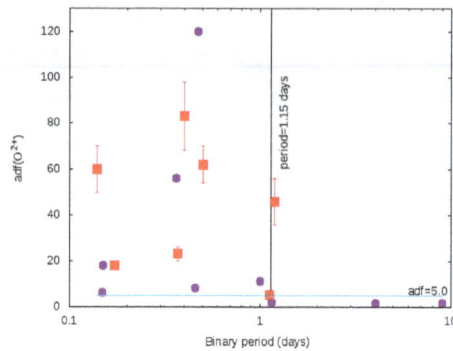

Figure 3. Abundance discrepancy for O^{2+} plotted against binary period for the 15 objects where both are known: nine literature values (purple dots) and seven from this study (red squares). Also plotted is a point for the Necklace, which has a period of 1.16 days and an unmeasured abundance discrepancy, but with an upper limit reported to be low and plotted here as a factor of 3.0. A horizontal line indicates an *adf* of 5.0, which we consider the dividing line between "normal" and "elevated", and a vertical line indicates the period of 1.15 days, which roughly divides objects with low and extreme *adf*s.

3.1.2. Stellar Abundances

The work in [28] suggested that there might be a relation between central star abundances and nebular abundance discrepancies, based on their discovery of the extreme abundance discrepancy of NGC 1501, which has a hydrogen-deficient central star, in common with several other then-recently identified extreme *adf* objects. However, they also noted that several extreme *adf* objects with H-rich central stars were known, concluding that no clear relationship existed. The work in [24] measured the *adf* in several objects with known H-deficient central stars, and did not find any elevated or extreme abundance discrepancies.

Given that extreme *adf*s can be reproduced by invoking cold hydrogen-deficient clumps embedded in a hot gas of normal composition, a source for these clumps needs to be identified. Two possibilities have been discussed; firstly, a very late thermal pulse (VLTP), in which a single star experiences a thermal pulse after having begun its descent of the white dwarf cooling track [29]; the second scenario is a nova-like eruption relying on a binary central star. Additional and more complex scenarios are possible: ref. [30] suggested that some combination of VLTP and nova in which the former triggered the latter could explain the properties of the hydrogen-deficient knot in Abell 58. Given the lack of observational constraints on such scenarios, we consider only these two relatively simple cases. The two scenarios make contrasting predictions for the central star abundances. The VLTP scenario would result in a hydrogen-deficient central star, and indeed, that scenario is commonly invoked specifically as a mechanism for creating such stars. Meanwhile, the nova-like scenario is as yet ill-defined. An eruption in which some hydrogen is neither burned nor ejected would be required to leave behind a hydrogen-rich post-nova object.

We compiled all available literature central star classifications, to compare *adf*s of nebulae with H-deficient central stars to those of nebulae with H-rich central stars (excluding nebulae with weak emission line stars (*wels*), which have a high likelihood of being due to either nebular contamination [31] or irradiation of the secondary [15]). The left-hand panel of Figure 4 shows a quantile-quantile (Q-Q) plot comparing the quantiles of the distributions of *adf*s, for nebulae with H-deficient central stars and those without. In such a plot, for two different datasets, if the two sets are drawn from the same underlying probability distribution, the points will lie close to the line of $y = x$. Populations drawn from differing probability distributions will diverge from the 1:1 relation. In this case, the points indeed mostly lie close to $y = x$. In the right-hand panel of the figure, we show the Q-Q plot for PNe with binary central stars against those without a known binary central star. The large deviations from the $y = x$ line indicate that the *adfs* of nebulae with binary central stars come from a strongly differing distribution to those without. This argues against a VLTP as the source of hydrogen-deficient material, while a nova-like outburst remains plausible.

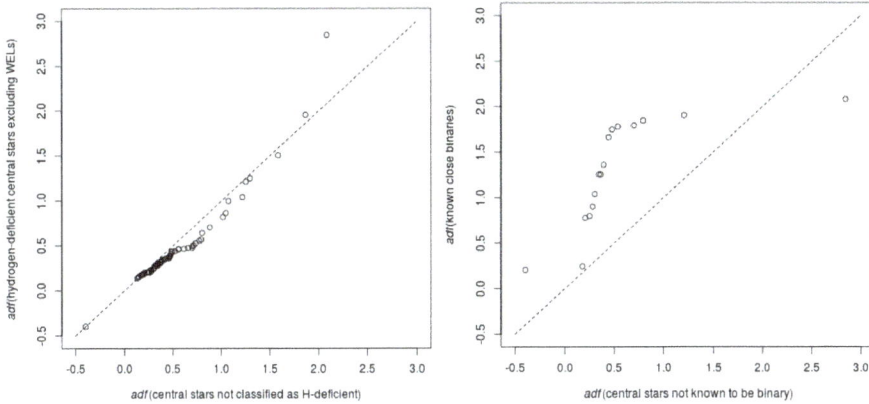

Figure 4. Q-Q plots for *adf* for (l) nebulae with hydrogen-deficient central stars against those without; (r) nebulae with close binary central stars against those without. When samples are drawn from the same underlying distribution, points in a Q-Q plot will lie close to $y = x$.

3.2. Relationship of adfs with Nebular Properties

Electron Density

As well as the proposed cut-off period of around 1.15 days separating elevated and extreme *adf* objects from normal *adf* objects, we have identified a relation between electron density and *adf*. In the seven objects studied in this work, all except Hen 2-283 have extreme abundance discrepancies, and while Hen 2-283 has an electron density of $\sim$3200 cm^{-3}, the other objects all have densities of <1000 cm^{-3}. To investigate this further, we compiled literature measurements of the electron density from lines of [O II], [S II], [Cl III] and [Ar IV], for all objects with a measured *adf*. Figure 5 shows the *adf* against electron density for each of the four diagnostics. These figures clearly show that the highest *adfs* only occur in the lowest density objects. Dashed lines indicate two bounds inside which almost all objects lie: a lower limit of *adf* = 1.3 and an upper limit of *adf* < 1.2×10^4 n$_e^{-0.8}$. The low densities associated with extreme *adfs* point to a low ionised mass, as found by [10] for the extreme-*adf* objects Ou 5 and Abell 48, and consistent also with the finding by [32] that post-CE nebulae have systematically lower ionised masses and surface brightnesses compared to the overall PN population.

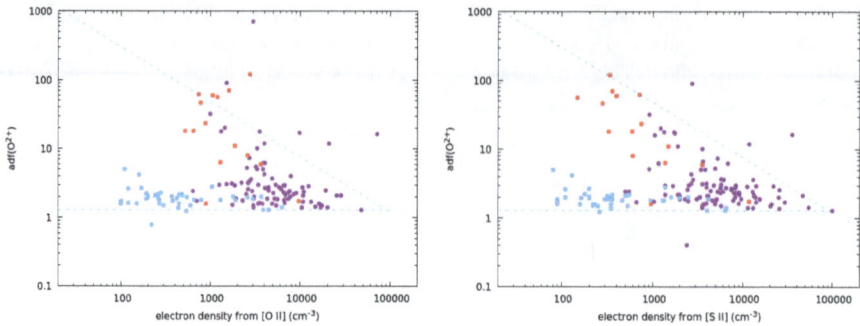

Figure 5. *Cont.*

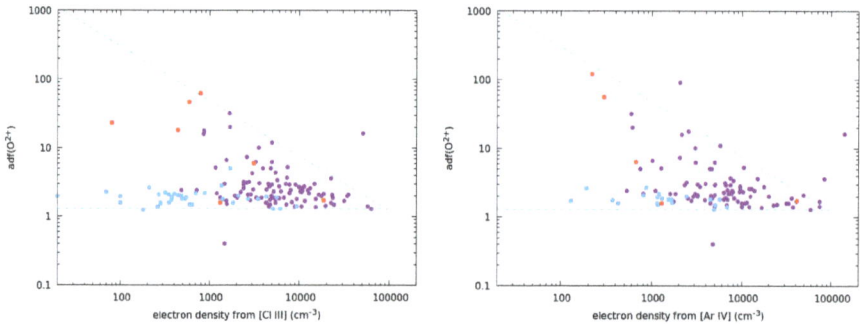

Figure 5. *adf* against electron density estimated from [O II] (**top left**), [S II] (**top right**), [Cl III] (**bottom left**) and [Ar IV] (**bottom right**) line ratios. Planetary nebulae with binary central stars are shown with red points; other PNe are shown with purple points; and H II regions are shown with light blue points. Dashed lines indicate the empirical limits inside which almost all objects are found (see the text for details).

4. Conclusions

The relationships between central star binarity, nebular density and *adf* show that common envelope evolution and nebular chemistry are strongly connected. We conclude that *adfs* provide a reliable way to infer the presence of a close binary central star; any object with an extreme *adf* must host a very short period binary. These objects will tend to have electron densities of $\ll 1000\,\text{cm}^3$. Meanwhile, if an object has a low *adf*, but a morphology indicative of binarity, we would predict that its binary period should be much longer than one day. One possible mechanism that could account for these observations is if the shortest period post-CE binaries experience fallback following the ejection of the common envelope, triggering an outburst of enriched material, which gives rise to extreme *adfs*. The implications of our results are discussed in greater detail in [33].

Author Contributions: R.W. led the analysis of the data, obtained through proposals for which D.J. was the Principal Investigator. All authors contributed to the interpretation of the results and the discussion.

Funding: This research received no external funding.

Acknowledgments: We are grateful to both referees for their careful reviews and helpful suggestions. R.W. was supported by European Research Grant SNDUST 694520. J.G.-R. acknowledges support from an Advanced Fellowship from the Severo Ochoa excellence program (SEV-2015-0548). We thank the organisers and participants of the APN7conference for a highly productive week of science and progress.

Conflicts of Interest: The authors declare no conflict of interest.

References

1.   Wyse, A.B. The Spectra of Ten Gaseous Nebulae. *Astrophys. J.* **1942**, *95*, 356. [CrossRef]
2.   Liu, X.W.; Storey, P.J.; Barlow, M.J.; Danziger, I.J.; Cohen, M.; Bryce, M. NGC 6153: A super-metal-rich planetary nebula? *Mon. Not. R. Astron. Soc.* **2000**, *312*, 585–628. [CrossRef]
3.   Peimbert, M.; Peimbert, A.; Delgado-Inglada, G. Nebular Spectroscopy: A Guide on Hii Regions and Planetary Nebulae. *Publ. Astron. Soc. Pac.* **2017**, *129*, 082001. [CrossRef]
4.   Peimbert, M. Temperature Determinations of H II Regions. *Astrophys. J.* **1967**, *150*, 825. [CrossRef]
5.   Torres-Peimbert, S.; Peimbert, M.; Pena, M. Planetary nebulae with a high degree of ionization—NGC 2242 and NGC 4361. *Astron. Astrophys.* **1990**, *233*, 540–552.
6.   Viegas, S.M.; Clegg, R.E.S. Density Condensations in Planetary Nebulae and the Electron Temperature. *Mon. Not. R. Astron. Soc.* **1994**, *271*, 993. [CrossRef]
7.   Ercolano, B. Can X-rays provide a solution to the abundance discrepancy problem in photoionized nebulae? *Mon. Not. R. Astron. Soc.* **2009**, *397*, L69–L73. [CrossRef]
8.   Nicholls, D.C.; Dopita, M.A.; Sutherland, R.S. Resolving the Electron Temperature Discrepancies in H II Regions and Planetary Nebulae: κ-distributed Electrons. *Astrophys. J.* **2012**, *752*, 148. [CrossRef]
9.   García-Rojas, J.; Esteban, C. On the Abundance Discrepancy Problem in H II Regions. *Astrophys. J.* **2007**, *670*, 457–470. [CrossRef]
10. Corradi, R.L.M.; García-Rojas, J.; Jones, D.; Rodríguez-Gil, P. Binarity and the Abundance Discrepancy Problem in Planetary Nebulae. *Astrophys. J.* **2015**, *803*, 99. [CrossRef]
11. Paczynski, B. Common Envelope Binaries. In *Structure and Evolution of Close Binary Systems*; Eggleton, P., Mitton, S., Whelan, J., Eds.; Cambridge University Press: Cambridge, UK, 1976; Volume 73, p. 75.
12. Ivanova, N.; Justham, S.; Chen, X.; De Marco, O.; Fryer, C.L.; Gaburov, E.; Ge, H.; Glebbeek, E.; Han, Z.; Li, X.D.; et al. Common envelope evolution: Where we stand and how we can move forward. *Astronomy Astrophys. Rev.* **2013**, *21*, 59. [CrossRef]
13. Jones, D.; Boffin, H.M.J. Binary stars as the key to understanding planetary nebulae. *Nat. Astronomy* **2017**, *1*, 0117. [CrossRef]
14. Jones, D.; Wesson, R.; García-Rojas, J.; Corradi, R.L.M.; Boffin, H.M.J. NGC 6778: Strengthening the link between extreme abundance discrepancy factors and central star binarity in planetary nebulae. *Mon. Not. R. Astron. Soc.* **2016**, *455*, 3263–3272. [CrossRef]
15. Miszalski, B.; Jones, D.; Rodríguez-Gil, P.; Boffin, H.M.J.; Corradi, R.L.M.; Santander-García, M. Discovery of close binary central stars in the planetary nebulae NGC 6326 and NGC 6778. *Astron. Astrophys.* **2011**, *531*, A158. [CrossRef]
16. Czyzak, S.J.; Aller, L.H. Spectrophotometric studies of nebulae. XXI. The remarkable planetary NGC 6778. *Astrophys. J.* **1973**, *181*, 817–823. [CrossRef]
17. Wesson, R. ALFA: An automated line fitting algorithm. *Mon. Not. R. Astron. Soc.* **2016**, *456*, 3774–3781. [CrossRef]
18. Wesson, R.; Stock, D.J.; Scicluna, P. Understanding and reducing statistical uncertainties in nebular abundance determinations. *Mon. Not. R. Astron. Soc.* **2012**, *422*, 3516–3526. [CrossRef]
19. Delgado-Inglada, G.; Morisset, C.; Stasińska, G. Ionization correction factors for planetary nebulae—I. Using optical spectra. *Mon. Not. R. Astron. Soc.* **2014**, *440*, 536–554. [CrossRef]
20. García-Díaz, M.T.; Clark, D.M.; López, J.A.; Steffen, W.; Richer, M.G. The Outflows and Three-Dimensional Structure of NGC 6337: A Planetary Nebula with a Close Binary Nucleus. *Astrophys. J.* **2009**, *699*, 1633–1638. [CrossRef]
21. Jones, D.; Boffin, H.M.J.; Rodríguez-Gil, P.; Wesson, R.; Corradi, R.L.M.; Miszalski, B.; Mohamed, S. The post-common envelope central stars of the planetary nebulae Henize 2-155 and Henize 2-161. *Astron. Astrophys.* **2015**, *580*, A19. [CrossRef]
22. Liu, M.C.; Leggett, S.K.; Golimowski, D.A.; Chiu, K.; Fan, X.; Geballe, T.R.; Schneider, D.P.; Brinkmann, J. SDSS J1534+1615AB: A Novel T Dwarf Binary Found with Keck Laser Guide Star Adaptive Optics and the Potential Role of Binarity in the L/T Transition. *Astrophys. J.* **2006**, *647*, 1393–1404. [CrossRef]

23. Sowicka, P.; Jones, D.; Corradi, R.L.M.; Wesson, R.; García-Rojas, J.; Santander-García, M.; Boffin, H.M.J.; Rodríguez-Gil, P. The planetary nebula IC 4776 and its post-common-envelope binary central star. *Mon. Not. R. Astron. Soc.* **2017**, *471*, 3529–3546. [CrossRef]
24. García-Rojas, J.; Peña, M.; Morisset, C.; Delgado-Inglada, G.; Mesa-Delgado, A.; Ruiz, M.T. Analysis of chemical abundances in planetary nebulae with [WC] central stars. II. Chemical abundances and the abundance discrepancy factor. *Astron. Astrophys.* **2013**, *558*, A122. [CrossRef]
25. Corradi, R.L.M.; Sabin, L.; Miszalski, B.; Rodríguez-Gil, P.; Santander-García, M.; Jones, D.; Drew, J.E.; Mampaso, A.; Barlow, M.J.; Rubio-Díez, M.M.; et al. The Necklace planetary nebula: Equatorial and polar outflows from a post-common-envelope system. In Proceedings of the Asymmetric Planetary Nebulae 5 Conference, Bowness-on-Windermere, UK, 20–25 June 2010.
26. Miszalski, B.; Manick, R.; Mikołajewska, J.; Van Winckel, H.; Iłkiewicz, K. SALT HRS Discovery of the Binary Nucleus of the Etched Hourglass Nebula MyCn 18. *arXiv* **2018**, *35*, arXiv:1805.07602.
27. Tsamis, Y.G.; Barlow, M.J.; Liu, X.W.; Storey, P.J.; Danziger, I.J. A deep survey of heavy element lines in planetary nebulae—II. Recombination-line abundances and evidence for cold plasma. *Mon. Not. R. Astron. Soc.* **2004**, *353*, 953–979. [CrossRef]
28. Ercolano, B.; Wesson, R.; Zhang, Y.; Barlow, M.J.; De Marco, O.; Rauch, T.; Liu, X.W. Observations and three-dimensional photoionization modelling of the Wolf-Rayet planetary nebula NGC 1501. *Mon. Not. R. Astron. Soc.* **2004**, *354*, 558–574. [CrossRef]
29. Iben, I., Jr.; Renzini, A. Asymptotic giant branch evolution and beyond. *Ann. Rev. Astronomy Astrophys.* **1983**, *21*, 271–342. [CrossRef]
30. Lau, H.H.B.; De Marco, O.; Liu, X.W. V605 Aquilae: A born-again star, a nova or both? *Mon. Not. R. Astron. Soc.* **2011**, *410*, 1870–1876. [CrossRef]
31. Basurah, H.M.; Ali, A.; Dopita, M.A.; Alsulami, R.; Amer, M.A.; Alruhaili, A. Problems for the WELS classification of planetary nebula central stars: Self-consistent nebular modelling of four candidates. *Mon. Not. R. Astron. Soc.* **2016**, *458*, 2694–2709. [CrossRef]
32. Frew, D.J.; Parker, Q.A.; Bojičić, I.S. The Hα surface brightness-radius relation: A robust statistical distance indicator for planetary nebulae. *Mon. Not. R. Astron. Soc.* **2016**, *455*, 1459–1488. [CrossRef]
33. Wesson, R.; Jones, D.; García-Rojas, J.; Boffin, H.M.J.; Corradi, R.L.M. Confirmation of the link between central star binarity and extreme abundance discrepancy factors in planetary nebulae. *Mon. Not. R. Astron. Soc.* **2018**, *480*, 4589–4613. [CrossRef]

Article

Analysis of Multiple Shell Planetary Nebulae Based on HST/WFPC2 Extended 2D Diagnostic Diagrams

Daniela Barría [1,*] and Stefan Kimeswenger [1,2]

[1]   Instituto de Astronomía, Universidad Católica del Norte, Av. Angamos 0610, 1240000 Antofagasta, Chile; skimeswenger@ucn.cl

[2]   Institut für Astro- und Teilchenphysik, Universität Innsbruck, Technikerstr. 25, 6020 Innsbruck, Austria

*   Correspondence: daniela.barria@ucn.cl; Tel.: +54-55-235-5480

Received: 27 June 2018; Accepted: 1 August 2018; Published: 3 August 2018

Abstract: The investigation of gaseous nebulae, emitting in forbidden lines, is often based extensively on diagnostic diagrams. The special physics of these lines often allows for disentangling with a few line ratios normally coupled thermodynamic parameters like electron temperature, density and properties of the photo-ionizing radiation field. Diagnostic diagrams are usually used for the investigation of planetary nebulae as a total. We investigated the extension of such integrated properties towards spatially resolved 2D diagnostics, using Hubble Space Telescope/Wide Field Planetary Camera 2 (HST/WFPC2) narrow band images. For this purpose, we also derived a method to isolate pure Hα emission from the [N II] contamination as normally suffering in the F656N HST/WFPC2 filter.

Keywords: planetary nebulae; stars: AGB and post-AGB; late stage stellar evolution

---

1. Introduction

According to the standard Interacting Stellar Winds (ISW) model, Planetary Nebula (PN singular, PNe plural) result from the interaction of a fast, hot and, thin wind developed at the post-AGB phase, which compresses and accelerates the dense material ejected during the final AGB phase [1]. While this model can explain the formation of main structures such as the shell and halo in PNe [2–4], it is not able to characterize the observed additional micro structures such as filaments, knots, clumpiness and low ionization outflows (see e.g., [5–7]). Certainly, radiative and dynamical processes taking place in the interaction of the stellar winds [8] might play a significant role in the origin and evolution of such structures, making clear that the mechanism(s) involved in the nebulae formation cannot be as straightforward as the ISW model predicts. In addition, some round/elliptical PNe show extra outer shells and/or halos (see e.g., [9–11]). These so-called Multiple Shell Planetary Nebulae (MSPNe) are then ideal candidates to study both macro and micro structures at PNe.

Mainly thanks to the great detail on HST images, remarkable low ionization structures such as FLIERs (Fast Low Ionization Emission Regions) or LISs (Low Ionization emission line Structures) have been observed in several MSPNe (see e.g., [12–14]). Compared to their neighboring gas, FLIERs and LISs are characterized to show a remarkable enhancement of low ionization species together to different kinematic properties. Some physical parameters such as electron temperature of FLIERs appears slightly but not so different to their surrounding nebular material (see e.g., [12,15–17]). A 3D photoionization modelling of a PN similar in structure to NGC 7009 shows that FLIERs may have chemical abundances similar to the main shell, but different density [18]. LISs, on the other hand, have shown to be less dense than their surroundings [19]. Thus, FLIERs/LISs 'exceptional' ionization properties seem to be associated to either shock regions or ionization fronts [15]. However, 'pure' shock models are not able to explain the entire observed properties of FLIERs [20]. By means of a series of numerical simulations, in [21] the authors predict the emission line spectrum of knots (low-ionization structures), which agrees approximately with observed spectra when shown in two-line ratio diagnostic diagrams. For different initial conditions, simulations predict for emission spectrum in a range from shock-excited spectra to spectra similar to those of gas in photoionized equilibrium. In [22] the authors studied near-IR H_2 emission from LISs at two PNe (one of them a MSPN). They found that the low ionization structures in these nebulae show similar traits to photodissociation regions.

Under this scenario and looking for a better interpretation of observed properties, we perform a detailed photometric analysis (using narrow band HST images) of three MSPNe. The line ratios log (Hα/[N II]) versus log (Hα/[S II]) has been used in the past to identify photoionized PNe from shocked nebulae, such as those found in supernova remnants or in H II regions. This scheme was first introduced by [23] and refined in [24]. An extension of this diagnostic diagram to two dimensions was first explored by [6]. Moreover, different diagnostic diagrams for spatially resolved PNe have been used in the last years to distinguish shock ionization from photoionization regions (see e.g., [19,21,25–27]). The high spatial resolution of HST images make them a perfect tool to investigate on line emission ratios at the entire surface of nebulae at each region individually. However, the HST/WFPC2 Hα images taken with the F656N filter might be strongly contaminated by [N II] emission. We present here a method to extract the nitrogen line contamination in the HST Hα images, by using the [N II] image taken with the F658N filter (Figure 1). Using these 'pure' Hα images together with [N II] and [S II] in two line ratios, we investigate the general ionization trends of the nebula versus that of the FLIERs and LISs regions in MSPNe.

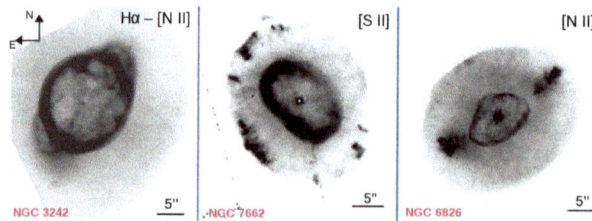

Figure 1. The HST images of some MSPNe. Note that the Hα image of NGC 3242 was already corrected for the contamination of the nitrogen line emission.

2. Methods and Results

The log (Hα/[N II]) versus log (Hα/[S II]) diagnostic diagrams normally were used for spectroscopic results of the nebulae as a total [23,24,28]. As to our knowledge, Reference [6] for the first time used this diagnostics to separate and classify different regions within the PN NGC 2438. There, the data along a long slit spectroscopy was used in detail, to derive a regression line, which then was applied to the entire 2D image from HST. We call these investigations thus thereafter Extended 2D Diagnostic Diagrams (E2DDs).

At this work and by means of HST/WFPC2 images (F656N, F658N, F673N and F502N band filters), we present our outcomes in the analysis of the nebular ionization trend using the E2DD at three MSPNe. We used pre-processed HST/WFPC2 level 2 images, which were flux-calibrated using the IRAF/STSDAS task *imcalc* and sky-subtracted by means of the *sky* task from XVISTA image processing software [29]. As we mentioned, however, the bandwidth of HST/WFPC2 Hα F656N filter include contamination from the [N II] 6583 Å and [N II] 6548 Å lines. Making use of the transmission curves for the F656N filter, we estimated a contamination in the F656N band of 39% for [N II] 6548 Å and 2% for [N II] 6583 Å. In a way to isolate for the Hα emission, we used the F658N filter image, which contains only [N II] 6583 Å, and thus considering for the known line ratio intensity R = ([N II] 6583 Å/[N II] 6548 Å) = 3, we were able to estimate for a total [N II] contribution in the F656N bandpass of 15%. Thus, cleaning the F656N images was possible using this fraction of the flux taken from the F658N image. Hereafter, line ratios for the E2DD were calculated using these 'pure' Hα data. Each image was then divided in 36 rectangular boxes covering the whole nebula. Each one of these was then sub-divided according to the main fragmentation structures observed in the nebula (see inserts in Figures 2 and 3). In this way, we were able to estimate for line ratios at different sectors and then following on the ionization tendency within each nebulae. Main results (E2DDs) are displayed in Figures 2 and 3. Different colors represent for different main sectors in the nebula. A linear fit to the data is displayed as a black line. Deviating points over a 2σ level were not considered at the fitting.

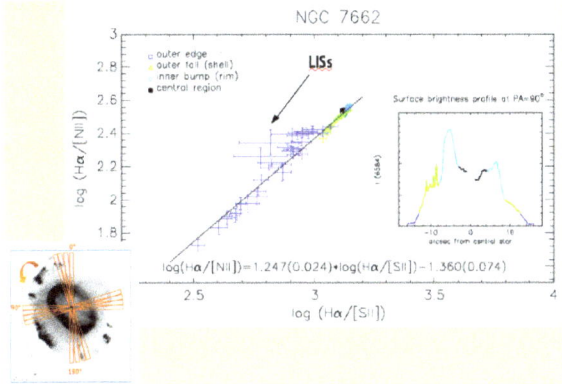

Figure 2. The E2DD derived from sectors of NGC 7662. The bottom-left insert illustrates the method how sectors are arranged in investigating the nebula. Different colors are related to different sectors within the nebula. The right inner plot shows (as an example) the extracted [N II] brightness profile at PA = 90° showing the main fragmentation structures.

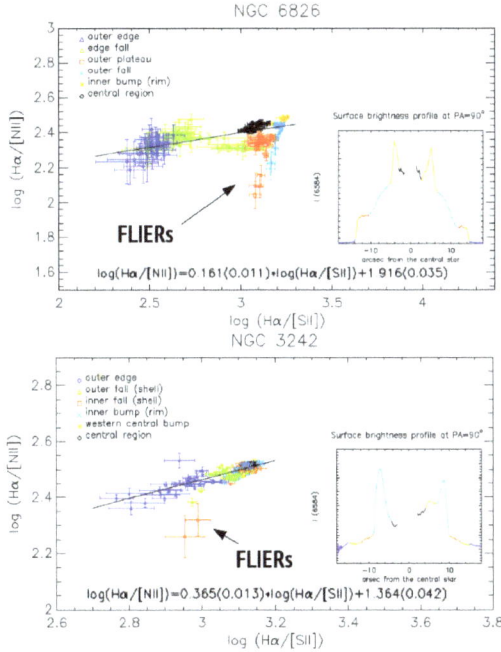

Figure 3. The E2DDs for of a nebula with very pronounced FLIERS and one with very weak ones.

3. Discussion and Conclusions

After careful inspection on the E2DD displayed in Figures 2 and 3, we observe that major deviations from the inside to outside excitation tendency seem to arrive only from regions where FLIERs or LISs are placed. From the outer edge (shell) at NGC 7662, the lower ionization contribution from LISs (compared to the surrounding nebular gas) slightly deviates from the main ionization tendency of the gas. In particular, this nebula exhibits the steepest ionization trend in our sample. On the other hand, showing the lowest inclination trend, NGC 6826 presents significant deviations from the general tendency emerging from the outer regions of the shell (so-called the outer plateau and outer fall at Figure 2). We notice at these regions of the nebula the presence of pronounced FLIERs (see the right-panel at Figure 1). Finally, the low ionization contribution from FLIERs at NGC 3242 becomes clear in its E2DD (Figure 3). Here, moderate deviating points emerge from the external regions of the shell where weak FLIERs are located.

Considering these outcomes, we conclude that no additional deviating data points beyond the ones related to FLIERs or LISs presence has been detected in the E2DD at this small MSPNe sample. If we assume that regions without the influence of FLIERs or LISs are dominated by photoionization as the main excitation mechanism, the influence of further processes could be affecting regions were these low ionization emission structures are present.

On the other hand, the three MSPNe investigated here exhibit similar properties like slight sub-solar abundances, high range of excitation, small expansion velocities and similar kinetic ages. Thus, we wonder about the strong differences observed at the inclination of the main ionization trend. Some open questions thus arise: do and, if so, how do the CSPNe parameters affect the inclination of the ionization trends? Moreover, is there any correlation between the properties of FLIERs (or LISs) and the ionization tendency of the surrounding gas? Why FLIERs/LISs can show different locations in the E2DD (below or above the main ionization trend)?

On the other hand, the presence of low ionization structures has been advertised as an indicator for central star binarity (see e.g., [30] and references therein). The fast rotating central star of NGC 6826 has been shown to be product of a binary evolution [31]. At this point, we wonder if this fact is somehow correlated to the lowest ionization tendency we observe in this nebula. As a future work, it becomes necessary to explore, using a bigger and more diversified sample of MSPNe, to study their E2DD ionization trends and its possible correlation to the stellar and/or nebular parameters.

Author Contributions: Conceptualization, D.B. and S.K.; Investigation, D.B.; Writing—Original Draft Preparation, D.B.; Writing—Review & Editing, D.B. and S.K.; Project Administration, D.B.; Funding Acquisition, S.K.

Funding: This research was funded by FONDO ALMA-Conicyt Programa de Astronomía/PCI 31150001.

Acknowledgments: Data used in this paper were based on observations made with the NASA/ESA Hubble Space Telescope, and obtained from the Hubble Legacy Archive, which is a collaboration between the Space Telescope Science Institute (STScI/NASA), the Space Telescope European Coordinating Facility (ST-ECF/ESA) and the Canadian Astronomy Data Centre (CADC/NRC/CSA). HST archival images were taken under proposal program ID 6117 (P.I. Bruce Balick). This research has made use of the SIMBAD database, operated at CDS, Strasbourg, France. We kindly thank the conference organizers for their help and Martin Guerrero and Quentin Parker for helpful discussion.

Conflicts of Interest: The authors declare no conflict of interest.

References

1. Kwok, S.; Purton, C.R.; Fitzgerald, P.M. On the origin of planetary nebulae. *Astrophys. J.* **1978**, *219*, L125–L127. [CrossRef]
2. Schönberner, D.; Steffen, M. Planetary Nebulae with Double Shells and Haloes: Insights from Hydrodynamical Simulations. *Rev. Mex. Astron. Astrofis.* **2002**, *12*, 144–145.
3. Perinotto, M.; Schönberner, D.; Steffen, M.; Calonaci, C. The evolution of planetary nebulae. I. A radiation-hydrodynamics parameter study. *Astron. Astrophys.* **2004**, *414*, 993–1015. [CrossRef]
4. Schönberner, D. The dynamical evolution of planetary nebulae. *J. Phys. Conf.* **2016**, *728*, 032001. [CrossRef]

5.   Matsuura, M.; Speck, A.K.; McHunu, B.M.; Tanaka, I.; Wright, N.J.; Smith, M.D.; Zijlstra, A.A.; Viti, S.; Wesson, R. A "Firework" of H_2 Knots in the Planetary Nebula NGC 7293 (The Helix Nebula). *Astrophys. J.* **2009**, *700*, 1067–1077. [CrossRef]

6.   Öttl, S.; Kimeswenger, S.; Zijlstra, A.A. Ionization structure of multiple-shell planetary nebulae. I. NGC 2438. *Astron. Astrophys.* **2014**, *565*, A87. [CrossRef]

7.   Fang, X.; Guerrero, M.A.; Miranda, L.F.; Riera, A.; Velázquez, P.F.; Raga, A.C. Hu 1-2: A metal-poor bipolar planetary nebula with fast collimated outflows. *Mon. Not. R. Astron. Soc.* **2015**, *452*, 2445–2462. [CrossRef]

8.   García-Segura, G.; López, J.A.; Steffen, W.; Meaburn, J.; Manchado, A. The Dynamical Evolution of Planetary Nebulae after the Fast Wind. *Astrophys. J. Lett.* **2006**, *646*, L61–L64. [CrossRef]

9.   Gómez-Muñoz, M.A.; Blanco Cárdenas, M.W.; Vázquez, R.; Zavala, S.; Guillén, P.F.; Ayala, S. Morpho-kinematics of the planetary nebula NGC 3242: An analysis beyond its multiple-shell structure. *Mon. Not. R. Astron. Soc.* **2015**, *453*, 4175–4184. [CrossRef]

10. Guerrero, M.A.; Manchado, A. On the Chemical Abundances of Multiple-Shell Planetary Nebulae with Halos. *Astrophys. J.* **1999**, *522*, 378–386. [CrossRef]

11. Corradi, R.L.M.; Steffen, M.; Schönberner, D.; Jacob, R. A hydrodynamical study of multiple-shell planetary nebulae. II. Measuring the post-shock velocities in the shells. *Astron. Astrophys.* **2007**, *474*, 529–539. [CrossRef]

12. Balick, B.; Alexander, J.; Hajian, A.R.; Terzian, Y.; Perinotto, M.; Patriarchi, P. FLIERs and Other Microstructures in Planetary Nebulae. IV. Images of Elliptical PNs from the Hubble Space Telescope. *Astron. J.* **1998**, *116*, 360–371. [CrossRef]

13. García-Díaz, M.T.; López, J.A.; Steffen, W.; Richer, M.G. A Detailed Morpho-kinematic Model of the Eskimo, NGC 2392: A Unifying View with the Cat's Eye and Saturn Planetary Nebulae. *Astrophys. J.* **2012**, *761*, 172. [CrossRef]

14. Gonçalves, D.R.; Corradi, R.L.M.; Mampaso, A.; Quireza, C. Do the Various Types of PNe Knots Differ in Terms of their Physical Properties? The case of NGC 7662. *Rev. Mex. Astron. Astrofis.* **2009**, *35*, 64–65.

15. Balick, B.; Rugers, M.; Terzian, Y.; Chengalur, J.N. Fast, low-ionization emission regions and other microstructures in planetary nebulae. *Astrophys. J.* **1993**, *411*, 778–793. [CrossRef]

16. Balick, B.; Perinotto, M.; Maccioni, A.; Terzian, Y.; Hajian, A. FLIERs and other microstructures in planetary nebulae, 2. *Astrophys. J.* **1994**, *424*, 800–816. [CrossRef]

17. Danehkar, A.; Parker, Q.A.; Steffen, W. Fast, Low-ionization Emission Regions of the Planetary Nebula M2-42. *Astron. J.* **2016**, *151*, 38. [CrossRef]

18. Gonçalves, D.R.; Ercolano, B.; Carnero, A.; Mampaso, A.; Corradi, R.L.M. On the nitrogen abundance of fast, low-ionization emission regions: the outer knots of the planetary nebula NGC 7009. *Mon. Not. R. Astron. Soc.* **2006**, *365*, 1039–1049. [CrossRef]

19. Akras, S.; Gonçalves, D.R. Low-ionization structures in planetary nebulae-I. Physical, kinematic and excitation properties. *Mon. Not. R. Astron. Soc.* **2016**, *455*, 930–961. [CrossRef]

20. Riera, A.; Raga, A.C. Physical conditions of the shocked regions in collimated outflows of planetary nebulae. In Proceedings of the Asymmetrical Planetary Nebulae IV, La Palma, Spain, 18–22 June 2007.

21. Raga, A.C.; Riera, A.; Mellema, G.; Esquivel, A.; Velázquez, P.F. Line ratios from shocked cloudlets in planetary nebulae. *Astron. Astrophys.* **2008**, *489*, 1141–1150. [CrossRef]

22. Akras, S.; Gonçalves, D.R.; Ramos-Larios, G. H_2 in low-ionization structures of planetary nebulae. *Mon. Not. R. Astron. Soc.* **2017**, *465*, 1289–1296. [CrossRef]

23. Garcia Lario, P.; Manchado, A.; Riera, A.; Mampaso, A.; Pottasch, S.R. IRAS 22568+6141—A new bipolar planetary nebula. *Astron. Astrophys.* **1991**, *249*, 223–232.

24. Magrini, L.; Perinotto, M.; Corradi, R.L.M.; Mampaso, A. Spectroscopy of planetary nebulae in M33. *Astron. Astrophys.* **2003**, *400*, 511–520. [CrossRef]

25. Gonçalves, D.R.; Mampaso, A.; Corradi, R.L.M.; Quireza, C. Low-ionization pairs of knots in planetary nebulae: Physical properties and excitation. *Mon. Not. R. Astron. Soc.* **2009**, *398*, 2166–2176. [CrossRef]

26. Ali, A.; Dopita, M.A. IFU Spectroscopy of Southern Planetary Nebulae V: Low-Ionisation Structures. *Publ. Astron. Soc. Aust.* **2017**, *34*, e036. [CrossRef]

27. Danehkar, A.; Karovska, M.; Maksym, W.P.; Montez, R., Jr. Mapping Excitation in the Inner Regions of the Planetary Nebula NGC 5189 Using HST WFC3 Imaging. *Astrophys. J.* **2018**, *852*, 87. [CrossRef]

28. Frew, D.J.; Parker, Q.A. Planetary Nebulae: Observational Properties, Mimics and Diagnostics. *Publ. Astron. Soc. Aust.* **2010**, *27*, 129–148. [CrossRef]

29. Barría, D.; Kimeswenger, S. HST/WFPC2 imaging analysis and Cloudy modelling of the multiple shell planetary nebulae NGC 3242, NGC 6826 and NGC 7662. *Mon. Not. R. Astron. Soc.* **2018**, *480*, 1626–1638 [CrossRef]

30. Jones, D.; Boffin, H.M.J. Binary stars as the key to understanding planetary nebulae. *Nat. Astron.* **2017**, *1*, 0117. [CrossRef]

31. De Marco, O.; Long, J.; Jacoby, G.H.; Hillwig, T.; Kronberger, M.; Howell, S.B.; Reindl, N.; Margheim, S. Identifying close binary central stars of PN with Kepler. *Mon. Not. R. Astron. Soc.* **2015**, *448*, 3587–3602. [CrossRef]

galaxies

MDPI

Article

X-ray Shaping of Planetary Nebulae

Martín A. Guerrero

Instituto de Astrofísica de Andalucía, IAA-CSIC, 18080 Granada, Spain; mar@iaa.es

Received: 30 July 2018; Accepted: 4 September 2018; Published: 11 September 2018

Abstract: The stellar winds of the central stars of planetary nebulae play an essential role in the shaping of planetary nebulae. In the interacting stellar winds model, the fast stellar wind injects energy and momentum, which are transferred to the nebular envelope through an X-ray-emitting hot bubble. Together with other physical processes, such as the ionization of the nebular envelope, the asymmetrical mass-loss in the asymptotic giant branch (AGB), and the action of collimated outflows and magnetic fields, the pressurized hot gas determines the expansion and evolution of planetary nebulae. *Chandra* and *XMM-Newton* have provided us with detailed information of this hot gas. Here in this talk I will review our current understanding of the effects of the fast stellar wind in the shaping and evolution of planetary nebulae and give some hints of the promising future of this research.

Keywords: planetary nebula; X-ray; stellar evolution

1. Introduction

Planetary nebulae (PNe) are the descendants of low- to intermediate-mass (0.8–$1.0\ M_\odot \leq M_i \leq$ 8–$10\ M_\odot$) stars. They are formed when such stars climb up to the tip of the asymptotic giant branch (AGB) and heavy mass loss through slow (~ 10 km·s^{-1}) and dense winds powered by radiation pressure on neutral dust grains eject the stellar envelope. As the hot stellar core is exposed, the sudden increase in ionizing flux and fast stellar wind (1000–4000 km·s^{-1}) [1] developed by the central star will ionize and sweep up the slow AGB wind to form a PN.

In the now classical interacting stellar winds (ISW) model [2,3], the expansion of PNe is powered by isotropic fast stellar winds. As these stellar winds encounter the slow AGB wind, a reverse-shock heats the stellar wind up to X-ray-emitting million-Kelvin temperatures, resulting in an onion-like structure similar to that proposed for Wolf–Rayet bubbles by Weaver et al. [4]. The extreme heating of this shock produces an over-pressurized hot bubble that works like a spherical piston, displacing undisturbed gas upstream into into a thin dense rim. Here I review our current understanding of the hot gas in PNe and describe some of the advances that can be achieved by future X-ray observatories.

2. Past and Current X-ray Observations of PNe

Hot bubbles have an X-ray limb-brightened morphology, with diffuse X-ray emission confined within the central cavity defined by the inner rim of PNe, as revealed in the first *Chandra* observations of PNe [5,6]. Since these early observations, *Chandra* and *XMM-Newton* have produced exquisite images and spectra of the diffuse X-ray emission from the hot bubbles of PNe [7–14].

All these results have been put in context and expanded by the analysis and new deep X-ray observations obtained in the framework of the *Chandra* Planetary Nebula Survey (ChanPlaNS) [15–17], which has surveyed the X-ray emission from ~50 PNe, all within 1.5 kpc, with a total exposure time over 1 Ms. These *Chandra* X-ray observations have made clear that the detections of diffuse X-ray emission is always confined within sharp closed innermost optical shells (i.e., the hot bubbles) [15,16]. So far, there is no evidence of X-ray emission associated with fast collimated outflows, but for the highly collimated ultrafast (>1000 km·s^{-1}) proto-PN Hen 3-1475 [18].

The emerging picture implies that a closed inner shell is necessary but not sufficient for the detection of hot X-ray-emitting gas. This is only present in young PNe, as testified by the high electron density ($N_e > 1000$ cm^{-3}) and small nebular radius (<0.15 pc) of PNe with diffuse X-ray emission [16]. Obviously, the rapidly changing stellar wind, whose mechanical power declines in a few thousand years, and the fast nebular expansion make the X-ray luminosity drop quickly [19]. As a general rule, diffuse X-ray emission is very unlikely among PNe with dynamical ages above 5000 years [14].

3. The Future of X-ray Studies of PNe

3.1. Future X-ray Observations of PNe

The current *Chandra* and *XMM-Newton* CCD spectroscopic observations of PNe have limited spectral resolution (typically $\approx$ 70 eV), which is degraded in cases of low count rate [20]. High-dispersion grating spectroscopic observations of the two X-ray brightest PNe, BD+30°3639 and NGC 6543, have provided interesting insights into the physical processes associated with the production of hot gas in PNe.

The *Chandra* X-ray Observatory's Low Energy Transmission Gratings and Advanced CCD Imaging Spectrometer (LETG/ACIS-S) were used to obtain a deep high-dispersion spectrum of BD+30°3639 [21]. The brightest lines in this spectrum are associated with highly ionized (H- and He-like) species of C, O, and Ne, such as O VIII λ18.97, C VI λ33.6, Ne IX $\lambda\lambda$13.45,13.55,13.7, and O VII $\lambda\lambda$21.60,21.80,22.10. The hot plasma has a range of temperatures between 1.7×10^6 K and 2.9×10^6 K, with chemical abundances consistent with the enhancement of carbon and neon.

As for NGC 6543, a 435 ks *XMM-Newton* Reflection Grating Spectrometer (RGS) exposure provided only a clear detection of the He-like triplet of O VII at 22 Å [22], although it must be noted that only 70 ks were useful because high background conditions affected most of these observations. The absence of the H-like O VIII points to lower T_X than in BD+30°3639, $kT \approx 0.147$ keV (or $T_X \approx 1.7 \times 10^6$ K). The N VII line at 24.78 Å was tentatively detected. The weakness of the N lines implies a low N/O ratio, favoring nebular abundances and high mixing. The lack of Ne and C lines indicates lower Ne/O and C/O than in BD+30°3639.

These high-dispersion grating X-ray spectroscopic observations are at the technical limit of *Chandra* and *XMM-Newton*. All other X-ray-emitting PNe are fainter or have lower surface brightness than BD+30°3639 and NGC 6543. We should not expect a breakthrough in our understanding of the production of hot gas in PNe and their effects in their expansion and evolution until the upcoming *ATHENA*, the Advanced Telescope for High-Energy Astrophysics. *ATHENA* is the second large-class ESA mission (L2), which is expected to be launched by 2028. *ATHENA* will have two main instruments, the Wide Field Imager (WFI) and the X-ray Integral Field Unit (XIFU). The former will have unprecedented sensitivity to diffuse X-ray emission, whereas the latter will be capable of providing high-dispersion spectra of sources as faint as PNe. This is illustrated in Figure 1. A single 10 ks WFI exposure can very easily determine whether the abundances of the X-ray-emitting plasma in NGC 6543 are nebular or stellar. Furthermore, a 20 ks XIFU exposure can even look into the details of the plasma physics to accurately determine the physical conditions and chemical abundances.

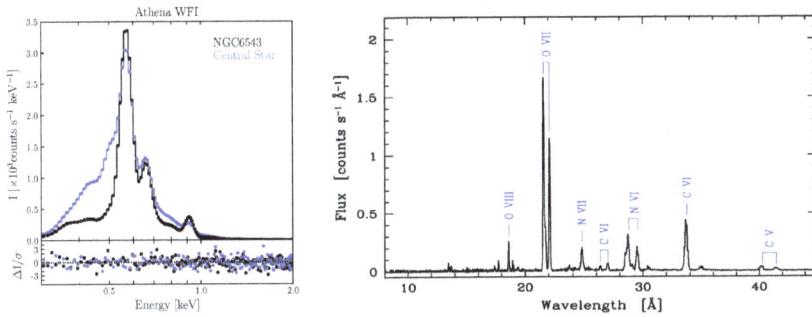

Figure 1. (left) Simulated *ATHENA* (Advanced Telescope for High-Energy Astrophysics) Wide Field Imager (WFI) 10 ks spectrum of NGC 6543 modeled using nebular (black) or stellar (blue) abundances. **(right)** Simulated *ATHENA* X-ray Integral Field Unit (XIFU) 20 ks spectrum of NGC 6543.

3.2. Effects of the Stellar and Nebular Evolution

The stellar evolution determines the mechanical luminosity of the stellar wind (i.e., the energy and momentum injected into the hot bubble). The nebular evolution, particularly the volume of the hot bubble, determines the energy density inside it. The most complete 1-D models accounting for the effects of the stellar and nebular evolution are those by Steffen et al. [19]. Their predictions compare reasonably well with detailed X-ray observations of PNe [14].

More complete 2-D models have the possibility to include effects of turbulent mixing. Previous models [23,24] have been superseded by the recent 2-D hydro-dynamical simulations presented by Toalá & Arthur [25]. These new models show the relevance of turbulent mixing and heat conduction on the evolution of the hot gas in PNe [20]. Rayleigh–Taylor instabilities at the interface are shown to be responsible for shadowing instabilities, which have notable effects in the mid-IR (dust) and near-IR H_2 (molecular) morphologies [26]. Future multi-dimensional hydro-dynamical models will also need to account for faster and brighter CSPN evolutionary tracks [27].

The end of the pressure-dominated hot bubble occurs when the stellar wind mechanical power drops below a limit where it is no longer able to push the nebular rim. The rim thickness then increases and, at some moment in the late evolution, it would even be able to backfill the central cavity [28,29]. However, it is difficult to compare theoretical predictions of this effect with observations given the different distances (and spatial scales) of PNe, projection effects, and small-scale nebular features [30].

Alternatively, the hot bubble can get pinched. Then, it would be unable to retain the hot gas, as shown in most PNe with open bipolar lobes or pinched hot inner rims [7]. Similar processes have been investigated in Wolf–Rayet bubbles, given their large angular size. The nebula S 308 is one of these cases, where the highly pressurized hot gas is producing a blowout feature on the otherwise round nebular morphology [31,32].

3.3. Effects of Asymmetries in the AGB Envelope or Fast Wind

The effects of asymmetric mass loss on the hot bubble have received little attention up to now. The symmetric fast stellar wind can impinge on an asymmetrical distribution of the nebular envelope, due either to asymmetrical mass loss during the AGB or to a symmetric AGB wind shaped by collimated outflows [33].

If the fast stellar wind is much faster than the AGB wind, the hot bubble is expected to become isobaric and push away the asymmetric envelope. The shape of the PN becomes self-similar for stellar winds with constant (unrealistic) properties [34]. More detailed simulations show that the action of a symmetric fast stellar wind on an asymmetric AGB envelope may result in elliptical PNe, whereas spherical PNe are unlikely to evolve from bipolar proto-PNe or elliptical PNe [35].

Alternatively, the fast stellar wind might be collimated. The interaction of a fast collimated fast winds (CFW) with a spherical AGB wind has been investigated by Lee & Sahai [36] and Akashi & Soker [37]. A 1000 km·s^{-1} CFW would be able to produce a bow-shock driven shell, where the wind itself would be surrounded by a hot X-ray-emitting cocoon.

3.4. Effects of Magnetic Fields

Heat conduction is largely influenced by magnetic fields, suppressing conduction normal to the field. Even a weak stellar magnetic field, $B_\star \leq 1$ G, can result in asymmetric thermal conduction in colliding stellar winds, because the reduced mixing along the equatorial direction implies an additional pumping (more pressure) along the main axis [38,39].

However, these results do not account for radiative and photo-ionization effects, which tend to make isobaric the pressure inside the hot bubble. Magneto-radiative-hydrodynamical simulations are required to solve this issue.

3.5. Ignored Physics

The limitations of the current X-ray observations (and theoretical models) of hot bubble in PNe do not allow us to investigate a series of physical processes with potential implications on the evolution of PNe. For instance, the hot bubble is assumed to be isobaric. We have seen above that magnetic fields may change this situation, but the interaction of the hot bubble with the AGN envelope is certainly complex and can produce short-lived non-isobaric regions. The 2-D radiative hydro-dynamical simulations of Toalá & Arthur [20] predict the presence of pockets of plasma with varying physical conditions at the interface between the hot bubble and the nebular rim. Slow-moving cold clumps may even provide pick-up ions, that is, suprathermal particles that may lower the post-shock velocity of the fast wind and reduce the temperature of the X-ray-emitting plasma in the hot bubble [40].

Charge-exchange reactions (CXE) may be important between ions of the stellar winds and the nebular envelope. These have been suggested to play an important role in the interactions of the post-born-again and present fast stellar winds with H-poor ejecta in the born-again PNe A 30 and A 78. In this particular case, soft thermal emission from H-poor knots ablated by the stellar wind, which is mass-loaded to raise its density and damp its velocity, produces the extremely soft spectrum mostly consistent with a single C VI line emission [41,42]. The evaporation of clumps of cold material, which may survive inside the hot bubble and photo-evaporate [43,44], may produce similar effects.

As important can be the recombination of carbon ions from the hot bubble with cool e$^-$. These can cross the contact discontinuity into the cold nebula to produce the continuum emission excess attributed to recombination lines of C VI in BD+30°3639 [45].

4. Summary

The hot gas resulting from the wind–wind interaction is one main PN shaping agent (in addition to collimated outflows, ionization, azimuthal density gradients, etc.) Hot gas mostly affects the early post-AGB evolution. Once the stellar wind power declines, the hot bubble pressure is not sufficient to maintain the nebular expansion. Rims depressurize very quickly, within 5000 years from the PN formation, or even faster if the hot bubble gets pinched.

Hot gas reveals complex physical processes: RRC, CEX, local non-equilibrium, weak magnetic fields, etc. Delicate interactions can produce subtle effects, such as shadowing instabilities, cold clumps evaporation, etc. The community needs to get ready for the next generation of X-ray telescopes, with all its observational and theoretical homework done.

Funding: This research was funded by Ministerio de Economía, Industria y Competitividad, Gobierno de España, grant number: AYA2014-57280-P.

Acknowledgments: The author acknowledges the support of the grant AYA2014-57280-P, co-funded with FEDER funds. He also appreciates the invitation of the Asymmetrical Planetary Nebulae VII Scientific Organizing Committee for the invited talk, which is reported here.

Conflicts of Interest: The author declares no conflict of interest.

References

1. Guerrero, M.A.; De Marco, O. Analysis of far-UV data of central stars of planetary nebulae: Occurrence and variability of stellar winds. *Astron. Astrophys.* **2013**, *553*, A126. [CrossRef]

2. Kwok, S. Effects of stellar mass loss on the formation of planetary nebulae. In *Planetary Nebulae; Proceedings of the Symposium*; Dordrecht, D., Ed.; Reidel Publishing Co.: London, UK, 1983; pp. 293–302.

3. Frank, A.; Mellema, G. A radiation-gasdynamical method for numerical simulations of ionized nebulae: Radiation-gasdynamics of PNe I. *Astron. Astrophys.* **1994**, *289*, 937–945.

4. Weaver, R.; McCray, R.; Castor, J.; Shapiro, P.; Moore, R. Interstellar bubbles. II—Structure and evolution. *Astrophys. J. Lett.* **1977**, *218*, 377–395. [CrossRef]

5. Kastner, J.H.; Soker, N.; Vrtilek, S.D.; Dgani, R. Chandra X-ray Observatory Detection of Extended X-ray Emission from the Planetary Nebula BD+30°3639. *Astrophys. J. Lett.* **2000**, *545*, L57–L59. [CrossRef]

6. Chu, Y.-H.; Guerrero, M.A.; Gruendl, R.A.; Williams, R.M.; Kaler, J.B. Chandra Reveals the X-ray Glint in the Cat's Eye. *Astrophys. J. Lett.* **2001**, *553*, L69–L72. [CrossRef]

7. Gruendl, R.A.; Guerrero, M.A.; Chu, Y.-H.; Williams, R.M. XMM-Newton Observations of the Bipolar Planetary Nebulae NGC 2346 and NGC 7026. *Astrophys. J. Lett.* **2006**, *653*, 339–344. [CrossRef]

8. Guerrero, M.A.; Chu, Y.-H.; Gruendl, R.A.; Meixner, M. XMM-Newton detection of hot gas in the Eskimo Nebula: Shocked stellar wind or collimated outflows? *Astron. Astrophys.* **2005**, *430*, L69–L72. [CrossRef]

9. Guerrero, M.A.; Gruendl, R.A.; Chu, Y.-H. Diffuse X-ray emission from the planetary nebula NGC 7009. *Astron. Astrophys.* **2002**, *387*, L1–L5. [CrossRef]

10. Kastner, J.H.; Vrtilek, S.D.; Soker, N. Discovery of Extended X-ray Emission from the Planetary Nebula NGC 7027 by the Chandra X-ray Observatory. *Astrophys. J. Lett.* **2001**, *550*, L189–L192. [CrossRef]

11. Kastner, J.H.; Montez, R., Jr.; Balick, B.; De Marco, O. Serendipitous Chandra X-ray Detection of a Hot Bubble within the Planetary Nebula NGC 5315. *Astrophys. J. Lett.* **2008**, *672*, 957–961. [CrossRef]

12. Montez, R., Jr.; Kastner, J.H.; De Marco, O.; Soker, N. X-ray Imaging of Planetary Nebulae with Wolf-Rayet-type Central Stars: Detection of the Hot Bubble in NGC 40. *Astrophys. J. Lett.* **2005**, *635*, 381–385. [CrossRef]

13. Montez, R., Jr.; Kastner, J.H.; Balick, B.; Frank, A. Serendipitous XMM-Newton Detection of X-ray Emission from the Bipolar Planetary Nebula Hb 5. *Astrophys. J. Lett.* **2009**, *694*, 1481–1484. [CrossRef]

14. Ruiz, N.; Chu, Y.H.; Gruendl, R.A.; Guerrero, M.A.; Jacob, R.; Steffen, M. Detection of Diffuse X-ray Emission from Planetary Nebulae with Nebular O VI. *Astrophys. J. Lett.* **2013**, *767*, A35. [CrossRef]

15. Kastner, J.H.; Montez, R., Jr.; Balick, B.; Frew, D.J.; Miszalski, B.; Sahai, R.; Blackman, E.; Chu, Y.H.; De Marco, O.; Frank, A.; et al. The Chandra X-ray Survey of Planetary Nebulae (ChanPlaNS): Probing Binarity, Magnetic Fields, and Wind Collisions. *Astrophys. J. Lett.* **2012**, *144*, A58. [CrossRef]

16. Freeman, M.; Montez, R., Jr.; Kastner, J.H.; Balick, B.; Frew, D.J.; Jones, D.; Miszalski, B.; Sahai, R.; Blackman, E.; Chu, Y.H.; et al. The Chandra Planetary Nebula Survey (ChanPlaNS). II. X-ray Emission from Compact Planetary Nebulae. *Astrophys. J. Lett.* **2014**, *794*, A99. [CrossRef]

17. Montez, R., Jr.; Kastner, J.H.; Balick, B.; Behar, E.; Blackman, E.; Bujarrabal, V.; Chu, Y.H.; Corradi, R.L.; De Marco, O.; Frank, A.; et al. The Chandra Planetary Nebula Survey (ChanPlaNS). III. X-ray Emission from the Central Stars of Planetary Nebulae. *Astrophys. J. Lett.* **2015**, *800*, A8. [CrossRef]

18. Sahai, R.; Kastner, J.H.; Frank, A.; Morris, M.; Blackman, E.G. X-ray Emission from the Pre-planetary Nebula Henize 3-1475. *Astrophys. J. Lett.* **2003**, *599*, L87–L90. [CrossRef]

19. Steffen, M.; Schönberner, D.; Warmuth, A. The evolution of planetary nebulae. V. The diffuse X-ray emission. *Astron. Astrophys.* **2008**, *489*, 173–194. [CrossRef]

20. Toalá, J.A.; Arthur, S.J. Formation and X-ray emission from hot bubbles in planetary nebulae—II. Hot bubble X-ray emission. *Mon. Not. R. Astron. Soc.* **2016**, *463*, 4438–4458. [CrossRef]

21. Yu, Y.S.; Nordon, R.; Kastner, J.H.; Houck, J.; Behar, E.; Soker, N. The X-ray Spectrum of a Planetary Nebula at High Resolution: Chandra Gratings Spectroscopy of BD+30°3639. *Astrophys. J. Lett.* **2009**, *690*, 440–452. [CrossRef]

22. Guerrero, M.A.; Toalá, J.A.; Chu, Y.-H.; Gruendl, R.A. XMM-Newton RGS observations of the Cat's Eye Nebula. *Astron. Astrophys.* **2015**, *574*, A1. [CrossRef]

23. Mellema, G.; Frank, A. Radiation gasdynamics of planetary nebulae—V. Hot bubble and slow wind dynamics. *Mon. Not. R. Astron. Soc.* **1995**, *273*, 401–410. [CrossRef]

24. Stute, M.; Sahai, R. X-ray Emission from Planetary Nebulae. I. Spherically Symmetric Numerical Simulations. *Astrophys. J. Lett.* **2006**, *651*, 882–897. [CrossRef]

25. Toalá, J.A.; Arthur, S.J. Formation and X-ray emission from hot bubbles in planetary nebulae—I. Hot bubble formation. *Mon. Not. R. Astron. Soc.* **2014**, *443*, 3486–3505. [CrossRef]

26. Fang, X.; Zhang, Y.; Kwok, S.; Hsia, C.H.; Chau, W.; Ramos-Larios, G.; Guerrero, M.A. Extended Structures of Planetary Nebulae Detected in H_2 Emission. *Astrophys. J. Lett.* **2018**, *859*, A92. [CrossRef]

27. Miller Bertolami, M.M. New models for the evolution of post-asymptotic giant branch stars and central stars of planetary nebulae. *Astron. Astrophys.* **2016**, *588*, A25. [CrossRef]

28. Soker, N. Backflow in post-asymptotic giant branch stars. *Mon. Not. R. Astron. Soc.* **2001**, *328*, 1081–1084. [CrossRef]

29. Chen, Z.; Frank, A.; Blackman, E.G.; Nordhaus, J. The creation of AGB fallback shells. *Mon. Not. R. Astron. Soc.* **2016**, *457*, 3219–3224. [CrossRef]

30. Ruiz, N.; Guerrero, M.A.; Chu, Y.-H.; Gruendl, R.A. Physical Structure of the Planetary Nebula NGC 3242 from the Hot Bubble to the Nebular Envelope. *Astron. J.* **2011**, *142*, A91. [CrossRef]

31. Chu, Y.-H.; Guerrero, M.A.; Gruendl, R.A.; García-Segura, G.; Wendker, H.J. Hot Gas in the Circumstellar Bubble S308. *Astrophys. J. Lett.* **2003**, *599*, 1189–1195. [CrossRef]

32. Toalá, J.A.; Guerrero, M.A.; Chu, Y.H.; Gruendl, R.A.; Arthur, S.J.; Smith, R.C.; Snowden, S.L. X-ray Emission from the Wolf-Rayet Bubble S 308. *Astrophys. J. Lett.* **2013**, *755*, A77. [CrossRef]

33. Sahai, R.; Trauger, J.T. Multipolar Bubbles and Jets in Low-Excitation Planetary Nebulae: Toward a New Understanding of the Formation and Shaping of Planetary Nebulae. *Astron. J.* **1998**, *116*, 1357–1366. [CrossRef]

34. Dwarkadas, V.V.; Balick, B. The Morphology of Planetary Nebulae: Simulations with Time-evolving Winds. *Astrophys. J. Lett.* **1998**, *497*, 267–275. [CrossRef]

35. Huarte-Espinosa, M.; Frank, A.; Balick, B.; Blackman, E.G.; De Marco, O.; Kastner, J.H.; Sahai, R. From bipolar to elliptical: Simulating the morphological evolution of planetary nebulae. *Mon. Not. R. Astron. Soc.* **2012**, *424*, 2055–2068. [CrossRef]

36. Lee, C.-F.; Sahai, R. Shaping Proto-Planetary and Young Planetary Nebulae with Collimated Fast Winds. *Astrophys. J. Lett.* **2003**, *586*, 319–337. [CrossRef]

37. Akashi, M.; Soker, N. Shaping planetary nebulae by light jets. *Mon. Not. R. Astron. Soc.* **2008**, *391*, 1063–1074. [CrossRef]

38. Soker, N. Heat conduction fronts in planetary nebulae. *Astron. J.* **1994**, *107*, 276–279. [CrossRef]

39. Zhekov, SA.; Myasnikov, A.V. Colliding Stellar Winds: "Asymmetric" Thermal Conduction. *Astrophys. J. Lett.* **2000**, *543*, L53–L56. [CrossRef]

40. Soker, N.; Rahin, R.; Behar, E.; Kastner, J.H. Comparing Shocks in Planetary Nebulae with the Solar Wind Termination Shock. *Astrophys. J. Lett.* **2010**, *725*, 1910–1917. [CrossRef]

41. Guerrero, M.A.; Ruiz, N.; Hamann, W.-R.; Chu, Y.H.; Todt, H.; Schönberner, D.; Oskinova, L.; Gruendl, R.A.; Steffen, M.; Blair, W.P.; et al. Rebirth of X-ray Emission from the Born-again Planetary Nebula A 30. *Astrophys. J. Lett.* **2012**, *755*, A129. [CrossRef]

42. Toalá, J.A.; Guerrero, M.A.; Todt, H.; Hamann, W.R.; Chu, Y.H.; Gruendl, R.A.; Schönberner, D.; Oskinova, L.M.; Marquez-Lugo, R.A.; Fang, X.; et al. The Born-again Planetary Nebula A 78: An X-ray Twin of A 30. *Astrophys. J. Lett.* **2015**, *799*, A67. [CrossRef]

43. López-Martín, L.; Raga, A.C.; Mellema, G.; Henney, W.J.; Cantó, J. Photoevaporating Flows from the Cometary Knots in the Helix Nebula (NGC 7293). *Astrophys. J. Lett.* **2001**, *548*, 288–295. [CrossRef]

44. Mellema, G.; Raga, A.C.; Cantó, J.; Lundqvist, P.; Balick, B.; Steffen, W.; Noriega-Crespo, A. Photo-evaporation of clumps in planetary nebulae. *Astron. Astrophys.* **1998**, *331*, 335–346.

45. Nordon, R.; Behar, E.; Soker, N.; Kastner, J.H.; Yu, Y.S. Narrow Radiative Recombination Continua: A Signature of Ions Crossing the Contact Discontinuity of Astrophysical Shocks. *Astrophys. J. Lett.* **2009**, *695*, 834–843. [CrossRef]

galaxies

MDPI

Article

Simulations of the Formation and X-ray Emission from Hot Bubbles in Planetary Nebulae

Jesus A. Toalá * and S. Jane Arthur

Instituto de Radioastronomía y Astrofísica (IRyA), UNAM Campus Morelia, 58090 Morelia, Michoacan, Mexico; j.arthur@irya.unam.mx
* Correspondence: j.toala@irya.unam.mx

Received: 26 June 2018; Accepted: 23 July 2018; Published: 30 July 2018

Abstract: High-quality X-ray observations of planetary nebulae (PNe) have demonstrated that the X-ray-emitting gas in their hot bubbles have temperatures in the small range $T_X = (1 - 3) \times 10^6$ K. However, according to theoretical expectations, adiabatically-shocked wind-blown bubbles should have temperatures up to two orders of magnitude higher. Numerical simulations show that instabilities at the interface between the hot bubble and the nebular material form clumps and filaments that generate an intermediate-temperature turbulent mixing layer. We describe the X-ray properties resulting from simulations of PNe in our Galaxy and the Magellanic Clouds.

Keywords: planetary nebulae; mass-loss; stellar evolution; X-rays

1. Introduction

One of the major probes of the interacting stellar winds model of planetary nebulae (PNe) formation scenario is the detection of hot bubbles. *XMM-Newton* and *Chandra* X-ray satellites have unveiled in great detail the distribution of the X-ray-emitting gas and its physical properties in PNe (e.g., [1–3]). The accumulation of the early X-ray work led to the design of the *Chandra* Planetary Nebulae Survey (CHANPLANS). This consists of a series of *Chandra* Large Programs and archival data to study the volume-limited sample of PNe within 1.5 kpc of the Sun [4]. The diffuse X-ray emission is mainly detected in the 0.3–2.0 keV energy range and is mostly associated with compact ($\lesssim$0.15 pc) and young ($\lesssim$5000 year) PNe.

Similar to wind-blown bubbles around massive stars, the X-ray temperatures obtained by means of spectral fitting from hot bubbles in PNe are in the $T_X = (1 - 3) \times 10^6$ K range [5,6], which is not in accordance to what is expected from analytical predictions. The temperature of an adiabatically-shocked wind-blown bubble is a function of the terminal wind velocity (V_∞) from the central star and can be estimated to be

$$T = 2.3 \times 10^7 \mu \left(\frac{V_\infty}{1000 \text{ km s}^{-1}} \right)^2 \text{ [K]}, \tag{1}$$

where μ is the mean molecular weight. The winds from central stars of PNe can reach $V_\infty \gtrsim 500 - 4000$ km s^{-1} [7]. That is, the hot bubble's temperature should be at least an order of magnitude higher than values obtained from observations. To resolve this discrepancy, several mechanisms have been proposed, with thermal conduction being the leading idea. Thermal conduction diffuses heat into the surrounding photoionised ($T \approx 10^4$) nebular material [8]. The inner surface of the nebular material evaporates into the hot bubble reducing the temperature and increasing its density (see [9] and references therein). A strong argument against the idea of thermal conductivity is that the presence of even a very small magnetic field would inhibit this mechanism.

As a result of the interaction of the current fast wind with material previously ejected from the asymptotic giant branch (AGB) phase, the interface becomes unstable [10]. Hydrodynamical ablation and photoevaporation of the dense clumps and filaments leads to turbulent mixing of cooler material into the hot gas producing favorable conditions for soft X-ray emission. Here, we discuss our most recent results on the X-ray emission from PNe in our Galaxy and in the Magellanic Clouds [11–13].

2. Methods

We performed high-resolution 2D axisymmetric radiation–hydrodynamic simulations of the formation of hot bubbles in PNe for different initial stellar masses (1, 1.5, 2, and 2.5 $M_\odot$) with and without thermal conduction. We used stellar evolution models from the AGB [14] to the post-AGB phase [15]. The stellar wind parameters for the post-AGB phase were computed using the WM-Basic code [16].

The first step for studying the X-ray properties from our models is to calculate the differential emission measure (DEM). This has been defined as $DEM(T_b) = \sum_k n_e^2 \Delta V_k$, where n_e is the electron number density in cell k, ΔV_k is the cell volume, and the sum is performed over cells with gas temperature T_b. Examples of DEMs obtained from our simulations are presented in Figure 1.

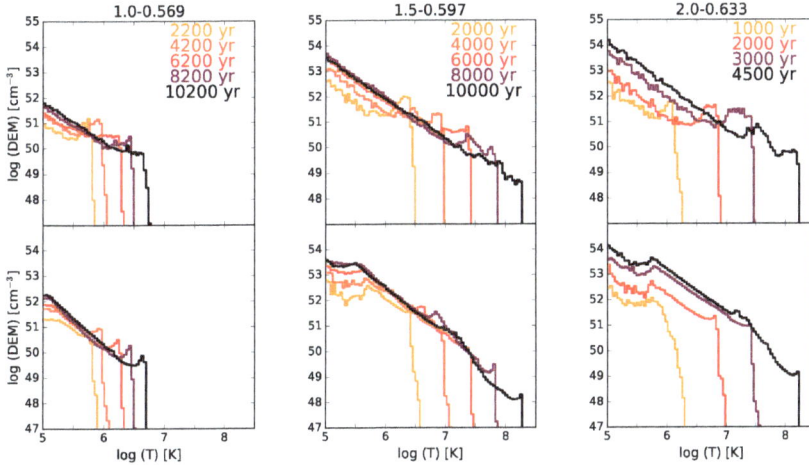

Figure 1. Evolution with time of the differential emission measure (DEM) of hot bubbles for the stellar models with initial masses of: 1 $M_\odot$ (**left**); 1.5 $M_\odot$ (**middle**); and 2.0 $M_\odot$ (**right**). Upper and lower panels correspond to cases without and with thermal conduction, respectively. Times are marked with different colors. These panels have been taken from [12].

Using the extensively tested CHIANTI database version 7.1.3 [17], we calculated the emission coefficient $\epsilon(T)$ corresponding to the soft X-ray energy band (0.3–2.0 keV). Figure 2 (left) shows the emission coefficient for PNe chemical abundances in our Galaxy and in the Large and Small Magellanic Clouds (LMC and SMC, respectively) (see Appendix in [13] for details). Note that the different $\epsilon(T)$ curves show two peaks: the main one peaks around $\log_{10}(T) = 6.1 - 6.3$ depending on the abundance set. The secondary peak at $\log_{10}(T) \sim 7.5$ is only present for the hydrogen rich cases.

We used the emission coefficient in combination with the DEM computed for each simulation to estimate emission-coefficient-weighted averaged temperatures as

$$T_A = \frac{\int \epsilon(T) DEM(T) T dT}{\int \epsilon(T) DEM(T) dT}. \tag{2}$$

Figure 2 (middle) presents the results of applying this equation in combination with the emission coefficient shown in Figure 2 (left) to the results obtained from the numerical simulations with initial mass of 1.5 $M_\odot$.

Finally, CHIANTI was also used to calculate the detailed synthetic spectrum of each simulation at different times for all the abundance sets used here. Each spectrum was integrated in the soft X-ray band (0.3–2.0 keV) to calculate its corresponding X-ray luminosity (L_X). Figure 2 (right) presents the evolution with time of L_X for all abundance sets corresponding to simulations with initial stellar mass of 1.5 $M_\odot$.

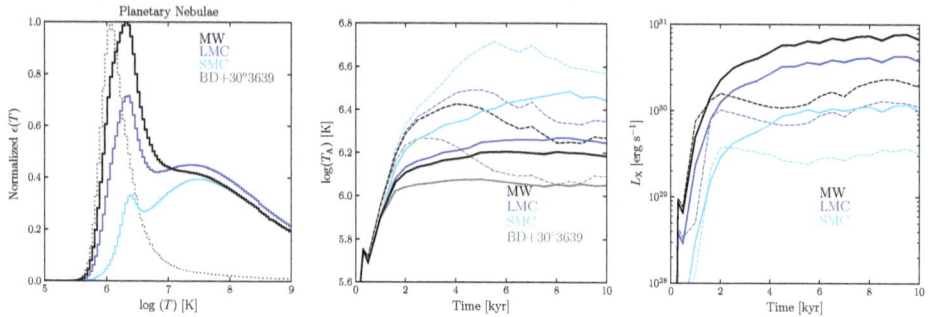

Figure 2. (**Left**) Emission coefficient for the 0.3–2.0 keV energy range for different PNe abundance sets; (**Middle**) evolution of the emission-coefficient-weighted mean temperature of simulated PNe for a 1.5 $M_\odot$ progenitor star for different abundance sets; and (**Right**) evolution of the X-ray luminosity for the same models presented in the middle panel. In the middle and right panels, solid (dashed) lines correspond to simulations with (without) thermal conduction. See [13] for details.

3. Results

All of our models present a clumpy interface between the hot bubble and the outer nebular material formed as a result of a diversity of hydrodynamical and cooling instabilities (see Figures 1 and 2 in [12]). *Hubble Space Telescope* observations of inner bubbles in X-ray-emitting PNe reveal clumps and filaments as part of the interface layers between the hot bubbles and the outer nebular material (e.g., [5,18]). This supports the idea that the presence of clumps and filaments is related to the production of X-ray emission in these objects.

Whether thermal conduction is included in our simulations or not, the mean temperature obtained from our simulations is similar to the values obtained to single-temperature fits to the observations. Whilst models with thermal conduction converge to a constant temperature value, models without conduction reach an early peak in their mean temperature due to the increasing stellar wind velocity. When the turbulent mixing layer becomes important, the mean temperature tends to a lower constant value. Our results show that even in the presence of a magnetic field, soft X-ray emission can be produced as a result of mixing by instabilities.

It is important to note that the emission coefficient used to compute the mean temperature of the hot bubble acts as a efficient filter and, depending on the abundances for which it is computed, it selects the contribution from gas at particular temperatures. The position of the narrow peak in the $\epsilon(T)$ curves (Figure 2, left) and the relative heights of the peak and the broad bump change with metallicity. These differences are reflected in the mean temperature shown in Figure 2 (middle). For example, we note that the $\epsilon(T)$ computed for the carbon rich hot gas in the PN BD+30°3639 exhibits a single dominant temperature peak at $\log_{10}(T) \sim 6.1$. This is the reason its predicted mean temperature for models with and without thermal conduction are the lowest.

We expect PNe in the lower metallicity environments to have higher X-ray-emitting temperatures but lower X-ray luminosities. However, there is still no evidence of diffuse X-ray emission from PNe in external galaxies.

Author Contributions: Most of the results presented here were developed as part of J.A.T.'s PhD thesis and correspond to original work. J.A.T. and S.J.A. performed the simulations and analysis presented here. S.J.A. developed the radiation–hydrodynamic code used for running the simulations presented in this paper.

Funding: J.A.T. and S.J.A. were funded by UNAM DGAPA PAPIIT projects IA100318 and IN112816, respectively.

Conflicts of Interest: The authors declare no conflict of interest.

References

1. Kastner, J.H.; Soker, N.; Vrtilek, S.D.; Dgani, R. Chandra X-ray Observatory detection of Extended X-ray Emission from the planetary nebula BD+30°3639. *Astrophys. J.* **2000**, *545*, L57. [CrossRef]
2. Chu, Y.-H.; Guerrero, M.A.; Gruendl, R.A.; Williams, R.M.; Kaler, J.B. Chandra reveals the X-ray Glint in the Cat's Eye Nebula. *Astrophys. J.* **2001**, *553*, L69. [CrossRef]
3. Guerrero, M.A.; Gruendl, R.A.; Chu, Y.-H. Diffuse X-ray emission from the planetary nebula NGC 7009. *Astron. Astrophys.* **2002**, *387*, L1. [CrossRef]
4. Kastner, J.H.; Montez, R., Jr.; Balick, B.; Sahai, R.; Blackman, E.; Chu, Y.-H.; de Marco, O.; Frank, A. The Chandra X-ray Survey of Planetary Nebulae (CHANPLANS): Probing Binarity, Magnetic Fields, and Wind Collisions. *Astrophys. J.* **2012**, *144*, 58. [CrossRef]
5. Ruiz, N.; Chu, Y.-H.; Gruendl, R.A.; Jacob, R.; Schönberner, D.; Steffen, M. Detection of Diffuse X-ray Emission from Planetary Nebulae with Nebular O IV. *Astrophys. J.* **2013**, *767*, 35. [CrossRef]
6. Toalá, J.A.; Marston, A.P.; Guerrero, M.A.; Chu, Y.-H.; Gruendl, R.A. Hot Gas in the Wolf-Rayet Nebula NGC 3199. *Astrophys. J.* **2017**, *846*, 76. [CrossRef]
7. Guerrero, M.A.; De Marco, O. Analysis of far-UV data of central stars of planetary nebulae: Occurrence and variability of stellar winds. *Astron. Astrophys.* **2013**, *553*, A126. [CrossRef]
8. Soker, N. Heat Conduction fronts in Planetary nebulae. *Astrophys. J.* **1994**, *107*, 276–279. [CrossRef]
9. Steffen, M.; Schönberner, D.; Warmuth, A. The evolution of planetary nebulae. V. The diffuse X-ray emission. *Astron. Astrophys.* **2008**, *489*, 173–194. [CrossRef]
10. Stute, M.; Sahai, R. X-ray Emission from Planetay Nebulae. I. Spherically symmetric Numerical Simulations. *Astrophys. J.* **2006**, *651*, 882. [CrossRef]
11. Toalá, J.A.; Arthur, S.J. Formation and X-ray emission from hot bubbles in planetary nebulae. I. Hot bubble formation. *Mon. Not. R. Astron. Soc.* **2014**, *443*, 3486–3505. [CrossRef]
12. Toalá, J.A.; Arthur, S.J. Formation and X-ray emission from hot bubbles in planetary nebulae. II. Hot bubble X-ray emission. *Mon. Not. R. Astron. Soc.* **2016**, *463*, 4438–4458. [CrossRef]
13. Toalá, J.A.; Arthur, S.J. On the X-ray temperature of hot gas in diffuse nebulae. *Astrophys. Galax.* **2018**, *478*, 1218. [CrossRef]
14. Vassiliadis, E.; Wood, P.R. Evolution of low- and intermediate-mass stars to the end of the asymptotic giant branch with mass loss. *Astrophys. J.* **1993**, *413*, 641–657. [CrossRef]
15. Vassiliadis, E.; Wood, P.R. Post-asymptotic giant branch evolution of low- to intermediate-mass stars. *Astrophys. J. Suppl. Ser.* **1994**, *92*, 125–144. [CrossRef]
16. Pauldrach, A.W.A.; Vanbeveren, D.; Hoffmann, T.L. Radiation-driven winds of hot luminous stars. XVI. Expanding atmospheres of massive and very massive stars and the evolution of dense stellar clusters. *Astron. Astrophys.* **2012**, *538*, A75. [CrossRef]
17. Landi, E.; Young, P.R.; Dere, K.P.; Del Zanna, G.; Mason, H. E. CHIANTI–An atomic Database for Emission Lines. XIII. Soft X-ray Improvements and Other Changes. *Astrophys. J.* **2013**, *763*, 86. [CrossRef]
18. Freeman, M.; Montez, R., Jr.; Kastner, J.H.; Frew, D.J.; Jones, D.; Miszalski, B.; Sahai, R.; Blackman, E.; Chu, Y.-H. The The Chandra Planetary nebulae Survey (CHANPLANS). II. X-ray Emission from Compact Planetary Nebulae. *Astrophys. J.* **2014**, *794*, 99. [CrossRef]

galaxies

MDPI

Article

Radio Continuum Spectra of Planetary Nebulae

Marcin Hajduk [1,*], Peter A. M. van Hoof [2], Karolina Śniadkowska [1], Andrzej Krankowski [1], Leszek Błaszkiewicz [2,3], Bartosz Dąbrowski [1] and Albert A. Zijlstra [4]

[1] Space Radio-Diagnostics Research Centre, University of Warmia and Mazury in Olsztyn, Prawocheńskiego 9, 10-720 Olsztyn, Poland; karolina.sniadkowska@uwm.edu.pl (K.Ś.); kand@uwm.edu.pl (A.K.); bartosz.dabrowski@uwm.edu.pl (B.D.)
[2] Royal Observatory of Belgium, Ringlaan 3, B-1180 Brussels, Belgium; p.vanhoof@oma.be (P.A.M.v.H.); leszekb@matman.uwm.edu.pl (L.B.)
[3] Faculty of Mathematics and Computer Sciences, University of Warmia and Mazury in Olsztyn, Słoneczna 54, 10-720 Olsztyn, Poland
[4] Department of Astronomy and Astrophysics, The University of Manchester, Manchester M13 9PL, UK; a.zijlstra@manchester.ac.uk
* Correspondence: marcin.hajduk@uwm.edu.pl

Received: 4 December 2018; Accepted: 21 December 2018; Published: 27 December 2018

Abstract: Radio continuum emission of planetary nebulae is a rich source of information about their structure and physical parameters. Although radio emission is well studied, planetary nebulae show higher spectral indices than expected for homogeneous sphere. A few competing models exist in the literature to explain this discrepancy. We propose that it is related to non-spherical morphology of most of planetary nebulae.

Keywords: planetary nebulae; AGB and post-AGB; interstellar medium; radio continuum; winds; outflows

1. Introduction

Planetary nebulae (PNe) mark the last phase of evolution of stars with the initial mass in the range of $\sim$1–8 $M_\odot$ prior to descending to white dwarf cooling track. PNe shells have been studied in radio frequencies for over seven decades now. Continuum radio emission in PNe originates predominantly from thermal free-free emission of electrons [1,2]. Short variability timescale and negative spectral index suggested non-thermal origin of radio emission for a few young PNe [3].

A lot of PNe show 5 GHz/1.4 GHz spectral index higher than a model of a homogeneous spherical shell. Reference [4] introduced a model in which the nebular image consists of regions having two different values of optical thickness to fit the observations. On the other hand, Reference [5] claimed that at least 10–20% of PN shells are associated with strong radial density gradients, which also produces higher 5 GHz/1.4 GHz flux ratio with respect to homogeneous model.

Reference [6] showed that the strong radial density gradient could not explain the observed 5 GHz/1.4 GHz spectral index. They proposed that the observed spectral index is related to morphology, or results from temperature variations within a nebula. Here, we further investigate the observed spectral index in PNe.

2. Methodology

The radio flux S_ν from a source with a uniform surface brightness distribution (hereinafter referred to as uniform model) covering a solid angle Ω is

$$S_\nu = \frac{2\nu^2 k T_e}{c^2}(1 - e^{-\tau_\nu})\Omega \tag{1}$$

and the optical thickness of the nebula τ_ν

$$\tau_\nu = 5.44 \times 10^{-2} T_e^{-1.5} \nu^{-2} g_{ff}(\nu, T_e) \int n(H^+) n_e dS. \qquad (2)$$

Reference [7] where T_e is the electron temperature, $n(H^+)$ and n_e are proton and electron density, respectively. The Gaunt factor, g_{ff}, is given by [8]. The emission measure $EM = \int n_e\, n(H^+)\, dS$ is integrated along the line of sight.

EM determines the turnover frequency ν of the spectrum for which $\tau_\nu \sim 1$. The optically thin spectrum has a spectral index of -0.1, after applying an analytical approximation for $g_{ff} = 8.235 \times 10^{-2} T_e^{-1.35} \nu^{-2.1}$, and is proportional to EM. The optically thick spectrum has a spectral index of 2 and is proportional to T_e. Additionally, Ω scales the absolute flux in the whole range of the spectrum.

Brightness temperature at 5 GHz can be defined as

$$T_b = 73.87 F_{5GHz}/\Theta^2, \qquad (3)$$

where Θ is the angular diameter in arcsec and F_{5GHz} radio flux at 5 GHz in mJy. Brightness temperature is a function of optical thickness for uniform surface brightness emission:

$$T_b = (1 - e^{-\tau_{5GHz}}). \qquad (4)$$

We studied the observed 5 GHz/1.4 GHz spectral index as a function of T_b using the same dataset as in [6], combining fluxes measured by single-dish instruments and interferometers. The fluxes from single-dish instruments showed different results from interferometer measurements [4]. However, this applied to large PNe in some data sets and did not affect the results significantly [6].

We used optical diameters from [9]. Optical $H\alpha$ and radio free-free emission should represent the same region. In the case of non-spherical PNe, we adopted the geometric mean of the two axes, which we used to compute the brightness temperature at 5 GHz. Radio diameters were available for about half of PNe, observed with interferometers. However, these measurements are unreliable for small PNe [4]. For large PNe, radio diameters are consistent with optical diameters. We used T_e from optical spectra [10].

The spectral index rises steeply at $log(T_b) \geq$ 2–2.5 (Figure 1). For an optically thick source, T_b approaches electron temperature and the spectral index approaches 2. The upper right corner of the Figure 1 is populated by young and dense PNe, whereas evolved PNe have low T_b and populate the lower left corner. For most PNe, the spectral index is higher than predicted by the uniform model for the same T_b (Figure 1).

In the next step, we modeled spectra of individual PNe using derived brightness temperatures and available radio continuum fluxes. The uniform model gives very similar results to a model of homogeneous sphere (see, e.g., Figure 1), so, for simplicity, we used the former one, though it is less realistic. The only free parameter was EM. All the spectra were consistent with free–free emission mechanism. However, the uniform model cannot fit the turnover frequency observed for lowest frequencies in the majority of PNe. To reproduce a non-negligible optical thickness at the lowest frequencies, a higher value of EM is required in most PNe. However, higher EM results in exceeding the observed flux in the optically thin part of the spectrum.

In order to scale down the flux, non-uniform brightness distribution is needed, with part of the nebula with higher EM dominating the total radio flux, or lower T_e in the outermost parts of a PN. However, this latter possibility seems to be ruled out by spatially resolved optical spectroscopy [11].

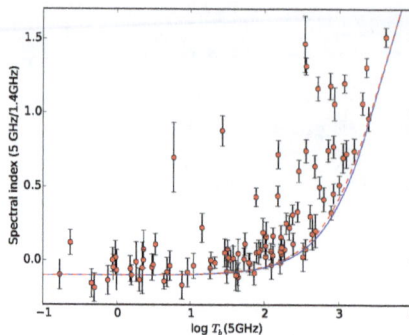

Figure 1. The 5 GHz to 1.4 GHz spectral index as a function of brightness temperature at 5 GHz for uniform (blue line) and spherical models (red dashed line) for $T_e = 10{,}000$ K.

We introduced a $(1 - \eta^2)$ factor in Equation (1) so that we could increase *EM* (and thus optical thickness) to achieve a good fit of the turnover frequency and simultaneously scale down the absolute flux:

$$S_v = \frac{2v^2 k T_e}{c^2}(1 - e^{-\tau_v})\Omega(1 - \eta^2), \tag{5}$$

where the fitted parameters are η and *EM*. In this model, $1 - \eta^2$ is the ratio of the emitting area to area of zero surface brightness within the PN. Thus, only part of the solid angle $\Omega(1 - \eta^2)$ contributes to the total flux of the PN, while $\Omega\eta^2$ has zero contribution. The emitting area is smaller but has higher optical thickness compared to a uniform model. A new model fits well for both turnover frequency and absolute flux. Reference [6] presents fits for individual PNe.

The η parameter depends on the surface brightness distribution of a PN. PNe with $\eta \approx 0$ can be fit equally well with a uniform or spherical model. However, almost all PNe require $\eta > 0$. $\eta \approx 1$ indicate that a small part of the PN with high *EM* dominates the radio flux. This cannot be simply the central part of the PN closest to the central star. Otherwise, the measured radio diameter would be smaller and represents this central region. Thus, radio emission can be distributed in a bright rim (e.g., cylindrical shell projected along its axis), or there are small scale blobs distributed over the entire solid angle of the PN.

We searched for a somewhat more realistic model than described by Equation (5). We considered a case of strong limb brightening in a projected spherical shell. However, the optical thickness of the limb brightened rim still appeared too low to reproduce the turnover frequency, even for a very thin shell.

In order to explain the observed large limb brightening in an optical range, Reference [12] applied a model of a prolate ellipsoidal shell (PES). Reference [13] showed that this model can successfully explain variety of morphologies. This tempted us to explore this model for our purposes. In this case dense, the limb brightened region with high *EM* could make a major contribution to the radio flux.

The inner and outer shell are defined as ellipsoids with minor and major axes of (a, b) and $(a + t, b + t)$. n_e^2 is inversely proportional to the inner radius in each direction. Due to the large parameter space (inclination angle i, a/b ratio, t, n_e), we did not use the PES model to fit individual PNe, but to fit spectral indices and brightness temperatures. The PES model explains the observed spectral indices of PNe better than the spherical model, but, for some PNe, it would require extreme parameter values (Figure 2).

The PNe with high brightness temperature are usually compact. The diameters are less reliable. Uncertainty in the determination of Ω propagates to T_b. In order to reproduce the observed spectral index—brightess temperature diagram and diameters should be systematically underestimated. This would decrease T_b and shift PNe leftward from the model.

We checked if the position of the PNe in the diagram correlates with morphological type. We used morphological types from [14–16]. The location of various morphological types appears to be correlated with position in a spectral index—T_b diagram.

Multipolar PNe are only seen among relatively high brightness temperatures, thus they are represented only among young PNe. Bipolar PNe are more likely to show the spectral index excess from other types of PNe. Perhaps, bipolar PNe show more complicated brightness distribution—the radio flux may be dominated by compact core (which becomes optically thick), and the lobes give less contribution to the total flux but are still detected in radio images (otherwise, observed radio diameters would be much lower from optical diameters, as they would only refer to the core, which is not the case). Unexpectedly, the model also failed to fit some PNe classified as round. This suggests that they have non-spherical morphologies producing a projected circular appearance. Confirmation would require morpho-kinematical analysis.

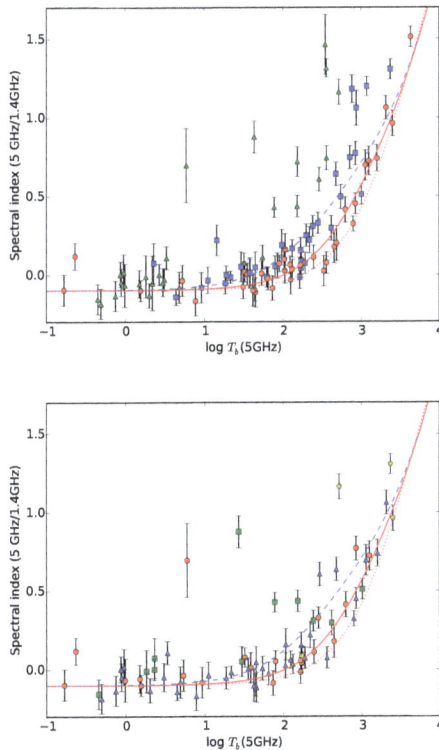

Figure 2. The 5 GHz to 1.4 GHz spectral index as a function of brightness temperature at 5 GHz. The solid red line and blue dashed line denote a PES model with $a/b = 3$ and $t/b = 0.1$ and $t/b = 0.01$, respectively. The dotted line marks the spherical model. In the top figure, PNe are split into three groups with respect to the η parameter: $\eta < 0.5$ (red circles), $0.5 < \eta < 0.9$ (blue squares), and $\eta > 0.9$ (green triangles). In the bottom figure, PNe are split according to morphological type: spherical (red circles), elliptical (blue triangles), bipolar (green squares), and multipolar PNe (yellow pentagons) (bottom).

3. Discussion

We confirm that uniform and spherical models cannot fit the observed radio spectra and brightness temperatures of most of the PNe. Spherical, homogeneous PNe appear to be very rare, even among PNe having circular appearances. The PES model can explain the observed spectral indices of a some of PNe. However, it cannot reproduce most of the PNe with $\eta > 0.7$ or bipolar PNe. For these PNe, most likely, a small part of the nebula with high EM contributes most of the radio flux.

4. Conclusions

Radio continuum fluxes and diameters are not sufficient to definitely solve the problem. Fitting observed radio brightness distribution of individual PNe, especially the ones characterized with large excess of the spectral index with respect to the uniform model, would certainly reduce ambiguities.

Author Contributions: M.H. performed the analysis and wrote the paper. P.A.M.v.H. and A.Z. commented on the paper and contributed to the data interpretation. K.Ś., A.K., L.B., B.D. and A.Z. commented on the paper.

Funding: We gratefully acknowledge financial support from National Science Centre, Poland, Grant No. 2016/23/B/ST9/01653.

Conflicts of Interest: The authors declare no conflict of interest.

References

1. Pazderska, B.M.; Gawroński, M.P.; Feiler, R.; Birkinshaw, M.; Browne, I.W.A.; Davis, R.; Kus, A.J.; Lancaster, K.; Lowe, S.R.; Pazderski, E.; et al. Survey of planetary nebulae at 30 GHz with OCRA-p. *Astron. Astrophys.* **2009**, *498*, 463. [CrossRef]

2. Chhetri, R.; Ekers, R.D.; Kimball, A.; Miszalski, B.; Cohen, M.; Manick, R. High radio frequency sample of bright planetary nebulae in the Southern hemisphere detected in the AT20G survey. *Mon. Not. R. Astron. Soc.* **2007**, *382*, 1607. [CrossRef]

3. Suárez, O.; Gómez, J.F.; Bendjoya, P.; Miranda, L.F.; Guerrero, M.A.; Uscanga, L.; Green, J.A.; Rizzo, J.R.; Ramos-Larios, G. Time-variable Non-thermal Emission in the Planetary Nebula IRAS 15103-5754. *Astrophys. J.* **2015**, *806*, 105. [CrossRef]

4. Siódmiak, N.; Tylenda, R. An analysis of the observed radio emission from planetary nebulae. *Astron. Astrophys.* **2001**, *373*, 1032–1042. [CrossRef]

5. Phillips, J.P. Density gradients in Galactic planetary nebulae. *Mon. Not. R. Astron. Soc.* **2007**, *378*, 231. [CrossRef]

6. Hajduk, M.; van Hoof, P.A.; Śniadkowska, K.; Krankowski, A.; Błaszkiewicz, L.; Dąbrowski, B.; Zijlstra, A.A. Radio observations of planetary nebulae: No evidence for strong radial density gradients. *Mon. Not. R. Astron. Soc.* **2018**, *479*, 4931.

7. Olnon, F.M. Thermal bremsstrahlung radiospectra for inhomogeneous objects, with an application to MWC 349. *Astron. Astrophys.* **1975**, *39*, 217.

8. Van Hoof, P.A.M.; Williams, R.J.R.; Volk, K.; Chatzikos, M.; Ferland, G.J.; Lykins, M.; Porter, R.L.; Wang, Y. Accurate determination of the free-free Gaunt factor—I. Non-relativistic Gaunt factors. *Mon. Not. R. Astron. Soc.* **2014**, *444*, 420. [CrossRef]

9. Frew, D.J.; Parker, Q.A.; Bojičić, I.S. The Hα surface brightness-radius relation: A robust statistical distance indicator for planetary nebulae. *Mon. Not. R. Astron. Soc.* **2016**, *455*, 1459. [CrossRef]

10. Cahn, J.H.; Kaler, J.B.; Stanghellini, L. A catalogue of absolute fluxes and distances of planetary nebulae. *Astron. Astrophys. Suppl. Ser.* **1992**, *94*, 399.

11. Sandin, C.; Schönberner, D.; Roth, M.M.; Steffen, M.; Böhm, P.; Monreal-Ibero, A. Spatially resolved spectroscopy of planetary nebulae and their halos. I. Five galactic disk objects. *Astron. Astrophys.* **2004**, *486*, 545. [CrossRef]

12. Masson, C.R. On the structure of ionization-bounded planetary nebulae. *Astrophys. J.* **1990**, *348*, 580. [CrossRef]

13. Aaquist, O.B.; Kwok, S. Radio Morphologies of Planetary Nebulae. *Astrophys. J.* **1996**, *462*, 813.

14. Phillips, J.P. The relation between elemental abundances and morphology in planetary nebulae. *Mon. Not. R. Astron. Soc.* **2003**, *340*, 883. [CrossRef]

15. Stanghellini, L.; Villaver, E.; Manchado, A.; Guerrero, M.A. The Correlations between Planetary Nebula Morphology and Central Star Evolution: Analysis of the Northern Galactic Sample. *Astrophys. J.* **2002**, *576*, 285. [CrossRef]

16. Sahai, R.; Morris, M.R.; Villar, G.G. Young Planetary Nebulae: Hubble Space Telescope Imaging and a New Morphological Classification System. *Astron. J.* **2011**, *141*, 134.

logo

MDPI

Article

The Formation of Fullerenes in Planetary Nebulae

Jan Cami [1,2,*], Els Peeters [1,2], Jeronimo Bernard-Salas [3,4], Greg Doppmann [5] and
James De Buizer [6]

1   Department of Physics and Astronomy and Centre for Planetary Science and Exploration (CPSX),
    The University of Western Ontario, London, ON N6A 3K7, Canada; epeeters@uwo.ca
2   SETI Institute, 189 Bernardo Ave, Suite 100, Mountain View, CA 94043, USA
3   Robert Hooke Building, Department of Physical Sciences, The Open University,
    Milton Keynes MK7 6AA, UK; Jeronimo.Bernard-Salas@open.ac.uk
4   ACRI-ST, 260 Route du Pin Montard, 06904 Sophia-Antipolis, France
5   W. M. Keck Observatory, 65-1120 Mamalahoa Highway, Kamuela, HI 96743, USA;
    gdoppmann@keck.hawaii.edu
6   Stratospheric Observatory for Infrared Astronomy-USRA, NASA Ames Research Center, MS N232-12,
    Moffett Field, CA 94035, USA; jdebuizer@sofia.usra.edu
*   Correspondence: jcami@uwo.ca; Tel.: +1-519-661-2111 (ext. 80978)

Received: 31 July 2018; Accepted: 18 September 2018; Published: 21 September 2018

Abstract: In the last decade, fullerenes have been detected in a variety of astrophysical environments, with the majority being found in planetary nebulae. Laboratory experiments have provided us with insights into the conditions and pathways that can lead to fullerene formation, but it is not clear precisely what led to the formation of astrophysical fullerenes in planetary nebulae. We review some of the available evidence, and propose a mechanism where fullerene formation in planetary nebulae is the result of a two-step process where carbonaceous dust is first formed under unusual conditions; then, the fullerenes form when this dust is being destroyed.

Keywords: planetary nebulae; fullerenes

1. Introduction

When Kroto et al. [1] conducted a series of experiments to simulate the chemistry occurring in the surroundings of carbon-rich evolved stars, they discovered a new and particularly stable carbonaceous molecule: Buckminsterfullerene, C_{60}. We now know that C_{60} is the most stable (and best known) member of an entire class of large, cage-like carbonaceous molecules. Given the stability of the molecule and the nature of the simulation experiments, Kroto et al. [1] immediately concluded that C_{60} was most likely widespread and abundant in space, and as soon as spectroscopic data were available, astronomers searched for its telltale signature in interstellar and circumstellar environments see [2].

We can now confirm that C_{60} is indeed widespread and abundant in space. Since the first unambiguous detection of all IR active vibrational modes of C_{60} in the Spitzer-IRS spectrum of the planetary nebula (PN) Tc 1 at 7.0, 8.5, 17.4, and 18.9 μm [3], the same spectral features have been found in a variety of evolved star environments (see Section 3), as well as in Reflection Nebulae RNe; see e.g., [4,5], the diffuse ISM [6], and young stellar objects and Herbig Ae/Be stars [7]. Recently, laboratory experiments and astronomical observations have confirmed the identification of two strong (and 3 weaker) diffuse interstellar bands DIBs; see [8] as due to electronic transitions of the C_{60}^+ cation see e.g., [9–12], and references therein. It is estimated that, on average, C_{60} locks up 10^{-4}–10^{-3} of the cosmic carbon [9,13,14], a considerable abundance for a single species!

While it is thus established that C_{60} is abundantly present in interstellar and circumstellar environments, the life cycle of the C_{60} molecule itself is not crystal clear. Here, we review what we know about the conditions that lead to the formation of fullerenes and how this can be reconciled with the body of available observational evidence to construct a coherent formation mechanism for fullerenes in evolved star environments.

2. The Conditions Required to form Fullerenes

Laboratory experiments (as well as theoretical calculations) on the condensation of carbonaceous gas have provided important information about the carbon chemistry that can occur in circumstellar environments. The key parameter that determines the outcome is the temperature (see Figure 1 for a schematic overview). At low temperatures, the condensation products are a variety of (small) molecules (e.g., C_2H_2, HCN...); any dust that condenses out will be primarily in the form of amorphous carbon [15]. Above ∼1000 K, the nature of the possible chemical reactions changes, and now benzene can form through a series of chemical reactions and from this species, a whole family of polycyclic aromatic hydrocarbon (PAH) molecules [16,17]. These PAHs are then the condensation nuclei for the formation of soot; the soot is formed quickly, and is graphitic in nature [17]. Finally, similar experiments at much higher temperatures (above ∼3500 K) result in the formation of fullerene molecules in addition to a soot that is fullerenic in nature [17]. Thus, different temperatures result in distinctly different products: no fullerenes are formed at the PAH-favoured temperatures, and similarly no PAHs appear in the high temperature experiments. This is partly due to hydrogen atoms that inhibit the formation of fullerenes at the lower temperatures; when similar experiments are carried out in H-poor conditions, fullerenes do form at these lower temperatures [18].

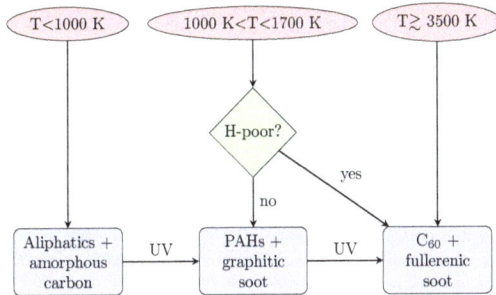

Figure 1. Formation routes for carbonaceous materials based on [15–18].

Irradiation of these carbonaceous compounds by UV photons can then significantly change their nature. It has long been known that UV irradiation of amorphous carbon leads to aromatization in the compounds. More recently, theoretical calculations [13,19] and laboratory experiments [20] have shown that large PAHs are first stripped of their H atoms by UV photons; the resulting pure graphene flake then curls up and shrinks by emission of C_2 units until it reaches the most stable configuration—C_{60}. It is thus conceivable that a significant amount of carbonaceous dust can be transformed by these processes over their lifetime—from aliphatic to aromatic to fullerenic material.

3. Observations of Fullerenes in PNe

While the IR C_{60} bands have been detected in various environments, the majority of detections originates from evolved stars (including post-AGB and pre-PN objects, but primarily in PNe in the Milky Way and the Magellanic Clouds; see e.g., [21–23] and references therein). All C_{60}-PNe have very similar properties and IR spectra. Indeed, they are all fairly young, low-excitation objects, the IR spectra of which are characterized by a rising, featureless continuum on which several emission features are superposed: The C_{60} bands, a strong 11–13 μm emission plateau (possibly due to SiC), a 6–9 μm emission plateau, and a strong 30 μm feature [22,23]. Thus, the conditions that are conducive to the formation and excitation of C_{60} also result in the appearance of other dust components. Notably, PAH emission in these objects is very weak or absent [23]. When comparing optical and near-IR colors of these objects to other evolved stars, the fullerene sources are the least reddened objects (i.e., the direct lines of sight to the central stars do not contain much dust that can cause reddening) and cluster together in near-IR color-color diagrams, while other evolved objects have a large range in colors [24].

Spatial studies of the different emission components can give clues to the conditions and origin of the different components. Narrow-band images of Tc 1 clearly show different distributions of the ionized gas, the dust responsible for the continuum emission, and the fullerenes (see Figure 2). They reveal that the fullerenes emit in a ring $\sim 5''$ away from the central star, at the interface between the ionized region and the surrounding Photo-Dissociation Region (PDR). There is no clear evidence for PAHs in this object, but if a very weak 11.3 µm bump is to be associated with PAHs, then those are all located in the top part of the central horse-shoe shaped region associated with the continuum dust emission. In that case, the PAHs and fullerenes would clearly be spatially separated, with the fullerenes at a location where the radiation field should be weaker than at the location of the PAHs.

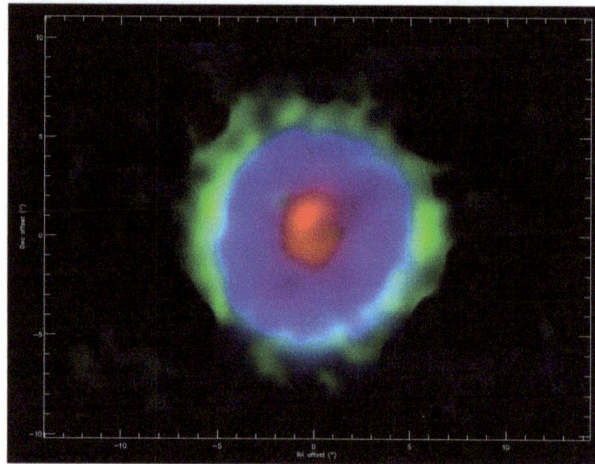

Figure 2. Decomposition of Gemini/T-ReCS narrow-band infrared images of Tc 1 into the different components (Cami et al., in prep): the continuum dust emission (red), the ionized gas (blue), and the fullerene emission (green).

Finally, an important observational result is that the C_{60} detection rate in PNe is fairly low: Only 3–5% of the Galactic PNe observed by Spitzer-IRS at high resolution exhibit the C_{60} bands [23]; the fraction of C_{60} detections in the Magellanic Clouds is slightly higher [22]. These low rates imply that the C_{60} molecules either form in a small fraction of objects (possibly related to requiring specific formation and/or excitation conditions), or alternatively that pure C_{60} has only a short lifetime in these environments, possibly related to chemically altering the molecular structure that would significantly affect the spectroscopic signatures.

4. Discussion

The circumstellar environments of evolved stars represent what happened to an atomic gas (the stellar envelope) that cooled after ejection due to mass loss processes. When considering the formation of molecules and dust in these environments, we thus always start from bottom up processes. In most cases (for typical condensation temperatures), we would expect the end products in C-rich environments to be PAHs and graphitic dust. Only when the temperatures are high or the environment is H-poor would fullerenes and fullerenic dust result.

The most straightforward scenario for the formation of fullerenes is thus one where the C_{60}-PNe represent objects where—for some reason—the condensation happens at higher temperatures or under H-poor conditions. This must have happened close to the central star (otherwise temperatures and densities would be too low), and consequently the condensation products moved outward since its formation—at least for Tc 1 where we see the C_{60} emission far from the central star. Since the formation of C_{60} also results in large amounts of fullerenic soot, while no dust is seen at the location of the C_{60} emission, dust destruction must have occurred as well. This would then explain why the colors of the C_{60}-PNe are so similar and appear unreddened. Perhaps the C_{60} molecules are the only species to survive this process, but it is conceivable that the destruction of fullerenic dust actually results in the formation of C_{60} molecules, e.g., by "liberating" C_{60} molecules from the dust grains. This dust destruction process must then be related to the transition of a post-AGB object to a PN since we only see the C_{60} bands in young PNe. Shock waves related to a fast wind overtaking a slow wind, or the development of an ionization front appear to be the most likely direct causes. The formation of C_{60} in PNe would thus involve a two-step process: Condensation (in an earlier phase, likely the AGB) under unusual conditions to produce C_{60} and fullerenic dust, followed by the destruction of the dust (that possibly creates more C_{60}). Such a scenario can explain why the C_{60}-PNe have the observational properties they have.

There is of course a fundamentally different formation route for C_{60} that we should also consider. Given that the central stars of PNe are hot stars that produce copious amounts of UV photons, photo-processing of carbonaceous materials seems like a plausible way to produce fullerenes, and indeed, it has been suggested that the formation of C_{60} is the result of photo-processing of hydrogenated amorphous carbon (HAC) dust grains or similar carbon clusters or PAHs see e.g., [21,25]. However, if this process can efficiently convert carbonaceous dust to fullerenes in the low-excitation C_{60}-PNe, the same process should be even more efficient in the more mature objects where timescales for processing have been longer, and UV photons are more abundant. Similarly, much of the dust in Tc 1 would have been converted to fullerenes as well, since it is located five times closer to the source of UV photons than the fullerenes themselves. In other words, if photo-processing of HACs (or even PAHs) would be the general process for fullerene formation in young PNe, there should only be very few PNe left that have circumstellar PAHs or in fact any carbonaceous material at all other than fullerenes.

The only remaining question is then why the dust condensation conditions were different in the C_{60}-PNe, and this may be tied to the evolutionary history of the central stars. For about half of the C_{60}-PNe, it has been established that their central stars are of [WC] type see e.g., [23]; certainly for such objects, it is plausible that condensation could happen in H-poor environments that would be more suited to create fullerenes and fullerenic soot than other objects. It is not clear what the status of the other C_{60}-PNe is in that sense, and we should also point out that not all PNe with [WC] stars display the fullerenes. Future observations that allow to measure e.g., the carbon abundance would be crucial to determine the underlying cause for the different condensation conditions.

Author Contributions: Conceptualization, J.C., E.P. and J.B.-S.; Data curation, J.D.B.; Formal analysis, J.C.; Funding acquisition, J.C., E.P. and G.D.; Methodology, J.C.; Software, J.C.; Visualization, J.C.; Writing—original draft, J.C.; Writing—review & editing, J.C., E.P., J.B.-S. and G.D.

Funding: Natural Sciences and Engineering Research Council of Canada: RGPIN-2016-06047.

Acknowledgments: J.C. and E.P. acknowledge support from an NSERC DG.

Conflicts of Interest: The authors declare no conflict of interest.

References

1. Kroto, H.W.; Heath, J.R.; Obrien, S.C.; Curl, R.F.; Smalley, R.E. C_{60}: Buckminsterfullerene. *Nature* **1985**, *318*, 162–163. [CrossRef]
2. Herbig, G.H. The Search for Interstellar C_{60}. *Astrophys. J.* **2000**, *542*, 334–343. [CrossRef]
3. Cami, J.; Bernard-Salas, J.; Peeters, E.; Malek, S.E. Detection of C_{60} and C_{70} in a Young Planetary Nebula. *Science* **2010**, *329*, 1180–1182. [CrossRef] [PubMed]
4. Sellgren, K.; Werner, M.W.; Ingalls, J.G.; Smith, J.D.T.; Carleton, T.M.; Joblin, C. C_{60} in Reflection Nebulae. *Astrophys. J. Lett.* **2010**, *722*, L54–L57. [CrossRef]
5. Peeters, E.; Tielens, A.G.G.M.; Allamandola, L.J.; Wolfire, M.G. The 15–20 µm Emission in the Reflection Nebula NGC 2023. *Astrophys. J.* **2012**, *747*, 44. [CrossRef]
6. Berné, O.; Cox, N.L.J.; Mulas, G.; Joblin, C. Detection of buckminsterfullerene emission in the diffuse interstellar medium. *Astron. Astrophys.* **2017**, *605*, L1. [CrossRef] [PubMed]
7. Roberts, K.R.G.; Smith, K.T.; Sarre, P.J. Detection of C_{60} in embedded young stellar objects, a Herbig Ae/Be star and an unusual post-asymptotic giant branch star. *Mon. Not. R. Astron. Soc.* **2012**, *421*, 3277–3285. [CrossRef]
8. Foing, B.H.; Ehrenfreund, P. Detection of Two Interstellar Absorption Bands Coincident with Spectral Features of C_{60}^{+}. *Nature* **1994**, *369*, 296. [CrossRef]
9. Campbell, E.K.; Holz, M.; Gerlich, D.; Maier, J.P. Laboratory confirmation of C_{60}^{+} as the carrier of two diffuse interstellar bands. *Nature* **2015**, *523*, 322–323. [CrossRef] [PubMed]
10. Walker, G.A.H.; Bohlender, D.A.; Maier, J.P.; Campbell, E.K. Identification of More Interstellar C_{60}^{+} Bands. *Astrophys. J. Lett.* **2015**, *812*, L8. [CrossRef]
11. Walker, G.A.H.; Campbell, E.K.; Maier, J.P.; Bohlender, D. The 9577 and 9632 Å Diffuse Interstellar Bands: C_{60}^{+} as Carrier. *Astrophys. J.* **2017**, *843*, 56. [CrossRef]
12. Cordiner, M.A.; Cox, N.L.J.; Lallement, R.; Najarro, F.; Cami, J.; Gull, T.R.; Foing, B.H.; Linnartz, H.; Lindler, D.J.; Proffitt, C.R.; et al. Searching for Interstellar C_{60}^{+} Using a New Method for High Signal-to-noise HST/STIS Spectroscopy. *Astrophys. J. Lett.* **2017**, *843*, L2. [CrossRef]
13. Berné, O.; Tielens, A.G.G.M. Formation of buckminsterfullerene (C60) in interstellar space. *Proc. Natl. Acad. Sci. USA* **2012**, *109*, 401–406. [CrossRef] [PubMed]
14. Omont, A. Interstellar fullerene compounds and diffuse interstellar bands. *Astron. Astrophys.* **2016**, *590*, A52. [CrossRef]
15. Gail, H.P.; Sedlmayr, E. Formation of crystalline and amorphous carbon grains. *Astron. Astrophys.* **1984**, *132*, 163–167.
16. Frenklach, M.; Feigelson, E.D. Formation of polycyclic aromatic hydrocarbons in circumstellar envelopes. *Astrophys. J.* **1989**, *341*, 372–384. [CrossRef]
17. Jäger, C.; Huisken, F.; Mutschke, H.; Jansa, I.L.; Henning, T. Formation of Polycyclic Aromatic Hydrocarbons and Carbonaceous Solids in Gas-Phase Condensation Experiments. *Astrophys. J.* **2009**, *696*, 706–712. [CrossRef]
18. Wang, X.K.; Lin, X.W.; Mesleh, M.; Jarrold, M.F.; Dravid, V.P.; Ketterson, J.B.; Chang, R.P.H. The effect of hydrogen on the formation of carbon nanotubes and fullerenes. *J. Mater. Res.* **1995**, *10*, 1977–1983. [CrossRef]
19. Berné, O.; Montillaud, J.; Joblin, C. Top-down formation of fullerenes in the interstellar medium. *Astron. Astrophys.* **2015**, *577*, A133. [CrossRef] [PubMed]
20. Zhen, J.; Castellanos, P.; Paardekooper, D.M.; Linnartz, H.; Tielens, A.G.G.M. Laboratory Formation of Fullerenes from PAHs: Top-down Interstellar Chemistry. *Astrophys. J. Lett.* **2014**, *797*, L30. [CrossRef]
21. García-Hernández, D.A.; Manchado, A.; García-Lario, P.; Stanghellini, L.; Villaver, E.; Shaw, R.A.; Szczerba, R.; Perea-Calderón, J.V. Formation of Fullerenes in H-containing Planetary Nebulae. *Astrophys. J. Lett.* **2010**, *724*, L39–L43. [CrossRef]
22. García-Hernández, D.A.; Iglesias-Groth, S.; Acosta-Pulido, J.A.; Manchado, A.; García-Lario, P.; Stanghellini, L.; Villaver, E.; Shaw, R.A.; Cataldo, F. The Formation of Fullerenes: Clues from New C_{60}, C_{70}, and (Possible) Planar C_{24} Detections in Magellanic Cloud Planetary Nebulae. *Astrophys. J. Lett.* **2011**, *737*, L30. [CrossRef]
23. Otsuka, M.; Kemper, F.; Cami, J.; Peeters, E.; Bernard-Salas, J. Physical properties of fullerene-containing Galactic planetary nebulae. *Mon. Not. R. Astron. Soc.* **2014**, *437*, 2577–2593. [CrossRef]

24. Sloan, G.C.; Lagadec, E.; Zijlstra, A.A.; Kraemer, K.E.; Weis, A.P.; Matsuura, M.; Volk, K.; Peeters, E.; Duley, W.W.; Cami, J.; et al. Carbon-rich Dust Past the Asymptotic Giant Branch: Aliphatics, Aromatics, and Fullerenes in the Magellanic Clouds. *Astrophys. J.* **2014**, *791*, 28. [CrossRef]
25. Micelotta, E.R.; Jones, A.P.; Cami, J.; Peeters, E.; Bernard-Salas, J.; Fanchini, G. The Formation of Cosmic Fullerenes from Arophatic Clusters. *Astrophys. J.* **2012**, *761*, 35. [CrossRef]

Article

The Astrochemistry Implications of Quantum Chemical Normal Modes Vibrational Analysis

SeyedAbdolreza Sadjadi * and Quentin Andrew Parker

Laboratory for Space Research and the Department of Physics, Faculty of Science, The University of Hong Kong, Pokfulam Road, Hong Kong, China; quentinp@hku.hk
* Correspondence: ssadjadi@hku.hk; Tel.: +852-3962-1438

Received: 8 October 2018; Accepted: 20 November 2018; Published: 23 November 2018

Abstract: Understanding the molecular vibrations underlying each of the unknown infrared emission (UIE) bands (such as those found at 3.3, 3.4, 3.5, 6.2, 6.9, 7.7, 11.3, 15.8, 16.4, 18.9 µm) observed in or towards astronomical objects is a vital link to uncover the molecular identity of their carriers. This is usually done by customary classifications of normal-mode frequencies such as stretching, deformation, rocking, wagging, skeletal mode, etc. A large literature on this subject exists and since 1952 ambiguities in classifications of normal modes via this empirical approach were pointed out by Morino and Kuchitsu New ways of interpretation and analyzing vibrational spectra were sought within the theoretical framework of quantum chemistry. Many of these methods cannot easily be applied to the large, complex molecular systems which are one of the key research interests of astrochemistry. In considering this demand, a simple and new method of analyzing and classifying the normal mode vibrational motions of molecular systems was introduced. This approach is a fully quantitative method of analysis of normal-mode displacement vector matrices and classification of the characteristic frequencies (fundamentals) underlying the observed IR bands. Outcomes of applying such an approach show some overlap with customary empirical classifications, usually at short wavelengths. It provides a quantitative breakdown of a complex vibration (at longer wavelengths) into the contributed fragments such as their aromatic or aliphatic components. In addition, in molecular systems outside the classical models of chemical bonds and structures where the empirical approach cannot be applied, this quantitative method enables an interpretation of vibrational motion(s) underlying the IR bands. As a result, further modifications in the structures (modeling) and the generation of the IR spectra (simulating) of the UIE carriers, initiated by proposing a PAH model, can be implemented in an efficient way. Here fresh results on the vibrational origin of the spectacular UIE bands based on astrochemistry molecular models, explored through the lens of the quantitative method applied to thousands of different vibrational motion matrices are discussed. These results are important in the context of protoplanetary nebulae and planetary nebulae where various molecular species have been uncovered despite their harsh environments.

Keywords: astrochemistry; planetary nebulae; UIE bands; normal modes; displacement vectors

---

1. Introduction

Detection in or towards various astronomical objects of broad emission infrared bands at certain wavelengths, i.e., 3.3, 3.4, 3.5, 6.2, 6.9, 7.7, 8.7, 11.3, 15.8, 16.4 & 18.9 µm led to the hypothesis that organic molecules are the carriers of these observed bands. These are a diverse range of organic molecular compounds from polyaromatic hydrocarbons(PAHs) [1,2], hydrogenated PAHs [3], nitrogenated PAHs [4], nanodiamonds [5], oil fragments [6], quenched carbonaceous composites (QCC) [7] and mixed aromatic-aliphatic organic nanoparticles(MAON) [8]. The main purpose of this work is to briefly discuss some new results of our on-going theoretical efforts and to explore the molecular vibrational origin of these mysterious UIE bands [9] through the lens of each of the proposed astrochemistry models above.

Most of the assignments of molecular vibrational motions and their responsible fragments or bonds are still based on an empirical approach which can potentially lead to incorrect interpretation of the molecular structure(s) of unknown carrier(s) of IR spectra [10–12]. Our theoretical approach aims to provide complementary information to other methods of simulating fundamental, combination, overtone [13–16] and ro-vibrational [17,18] bands of the Infrared spectra so generated.

An example of an astrochemistry application of our methodology is our recent work on the vibrational origins of the 3.28 and 3.3 μm components of the 3.3 μm feature known in the framework of the current PAH model [19]. The customary classification and assignment of the first component at 3.28 μm to the so-called *bay* and the second at 3.3 μm, to the *non-bay* -C-H stretching modes in PAH molecules (for example there are two bay C-H bonds, each from one ring, in the phenanthrene molecule) cannot adequately explain the observed wavelength difference and flux ratio of these two UIE features. This is due to the large and complex coupling between the stretching modes of aromatic -C-H bondings (bay and non-bay) [19].

The primary empirical tool developed for interpreting and simulating the frequency and intensity of molecular fundamental infrared (IR) bands is the model of normal-mode vibration. Reference [10] noted are no rules to exactly define the customary (empirical) classifications of normal-mode frequencies as a result of bond stretching, deformation, rocking, wagging, skeleton mode, etc. The key role of having a well-defined system for assigning and classifying the vibrational motions in the interpretation of observed IR spectra was recently discussed by Tao et al. [20]. Their main argument is that normal modes are delocalized over all constituent atoms of a molecule. Hence, without any definite rules, assigning a vibration to a particular bond(s) or fragment(s) is a difficult and ambiguous task.

Although new mathematical methods to resolve this problem were introduced into quantum chemical models since 1998 [11,12] these are not suitable methods to apply for analysis of vibrational motions of large molecules with complex structures [12], especially the types of molecules of particular interest in astrochemistry. The recent generalized subsystem vibrational analysis, developed by Tao et al. [20] might be a good framework for our purposes but the reliability of this model needs to be confirmed in different classes of molecules. Clearly not all the possible 10^{33} to 10^{180} molecular species within the structural space of organic chemistry can be explored theoretically. Vibrational analysis and interpretation of IR spectra hence plays a key role in choosing the right bondings and fragments to modify in the astrochemistry molecular models and to inform a better understanding of the origin of the mysterious UIE bands.

In this regard, we have developed and introduced a simple approach for vibrational analysis based on displacement vector analysis of normal modes [9]. These are N × 3 matrices, where N is the number of atomic centers with a total number of 3N-6 normal modes for each molecule. We applied our methods to numerous examples of vibrational motions in simple and complex organic compounds composed of different types of chemical bonds, structures and sizes [9,19,21–23]. Our methodology provides consistent results compared to the more customary (empirical) approach and in a fully quantitative fashion while additionally revealing important details of vibrational motions that are not recovered by the more traditional techniques. Another example is our detailed and comprehensive vibrational motion analysis of the complex correlation of a band's peak wavelength position of the out of plane bending mode vibration of aromatic -C-H bondings (OOPs) with the exposed edges (neighbor -C-H bondings) in the structure of PAH molecules [23]. Such correlation was originally suggested as due to a monotonic increase in wavelength position by increasing the exposed edge [24].

2. Brief Summary of Our Previous Work

Here we present a brief summary of our latest findings and understanding of the types of vibrational motions underlying UIE bands. This is based on our application of displacement vector analysis of vibrational normal modes to thousands of Raman and IR active modes of PAHs [19] with different aliphatic and olefinic side chains [9,19] and MAON type molecules [21]. In the customary empirical approach, the vibrational origin of an IR band is assigned to the vibration with strongest intensity. Unfortunately, the type of vibration and the responsible fragment(s)/bonds cannot be correctly assigned in this way [10–12,20]. Neglecting the rest of the vibrational motions, including the Raman modes, will lead to incomplete or, in the case of long wavelength bands, an incorrect interpretation of the origin of an IR band. Below is an itemized summary of our current findings. Thereafter our new results are presented for the first time.

- In hydrocarbons within the realm of classical molecular structure only C-H stretching vibrations occur. This holds for any combination of different aromatic (sp^2), olefinic (sp^2) and aliphatic (sp^3) groups within a molecule. Considering the low amplitude motions in our analysis we did not find the signatures of other types of vibrations in this spectral region.
- For the above types of molecules the aromatic (sp^2) C-H vibrations do not show any common wavelength coverage with aliphatic (sp^3) stretching modes (non-overlap wavelength range). The vibrational motions of these two types of C-H bonds are reported to be uncoupled.
- Olefinic (sp^2) C-H stretching shows a common wavelength coverage with aromatic (sp^2) C-H vibrations (overlapped wavelength ranges). However, vibrational motions of these two bond types occur independently without any couplings.
- Olefinic (sp^2) C-H stretching motion couples partly with aliphatic (sp^3) C-H vibrations.
- These vibrational characteristics of olefinic C-H stretching motions can glue the wavelength range of aliphatic (sp^3) and aromatic (sp^2) C-H stretching vibrations. In other words, they can make an indirect coupling between these two uncoupled and non-overlapped vibrational motions. This effect can be considered as one of the origins of the formation of a plateau in this spectral region.
- The symmetric and asymmetric C-H stretching motions of methyl and methylene groups are highly coupled with common wavelength coverage of the resultant features. Thus, vibrational motions of these two functional groups in this wavelength region are difficult to discriminate against under different UIE features.
- In neutral honeycomb PAH molecules, the stretching vibrations of C-H bonds in the bay and non-bay positions show considerable coupling in addition to their common wavelength coverage.

3. New Results from This Work

The results presented here are from new quantum chemical calculations performed within density functional theory using mathematical tools to solve the Schrodinger equation for molecular systems.(B3LYP and BHandHLYP functionals, in combination with a PC1 basis set family). This model delivers an average error of 0.12–0.13 µm (within the wavelength range of 2 to 20 µm) in reproducing the fundamental bands of laboratory gas-phase IR spectra for the different classes of organic molecules [9]. The effects of anharmonicity in vibrations and rotational-vibrational couplings on final IR bands wavelength positions are considered by applying the Laury et al. [25] scheme of double-scale factors for harmonic normal-mode calculations.

3.1. The 3 µm Region

Models: PAH, PAHs with Side Groups, Aliphatic Hydrocarbons and MAONs

Although this part of the emission spectra seems to be well explored in terms of the types of vibrational motions in the framework of cited astrochemistry models, plateau formation, the flux ratio of bands used to estimate the aliphatic/aromatic contents of UIE carriers, the long wavelength UIE features (such as the 3.51 µm peak assigned to nano-diamond structures) and the relationship of these features to the other UIEs at different IR wavelengths have remained unsolved issues. A comprehensive vibrational analysis and classification of normal modes will play a key role in addressing these issues. Specifically, we found that by keeping the molecular structure of organic compounds within the framework of classical Lewis model of bonding and structure the classifications and assignments of normal modes are not simple and definite as implied from the customary (empirical) approach. Perhaps classifications, assignments, and interpretation of vibrational motions underlying IR bands within the empirical approach would be ambiguous for non-classical molecular structures such as an amorphous type complex hydrocarbon molecule shown in Figure 1. Here a modern and advanced molecular chemical bonding and structure theory such as the quantum theory of atoms in molecules should be recalled to assist the interpretation [26–28].

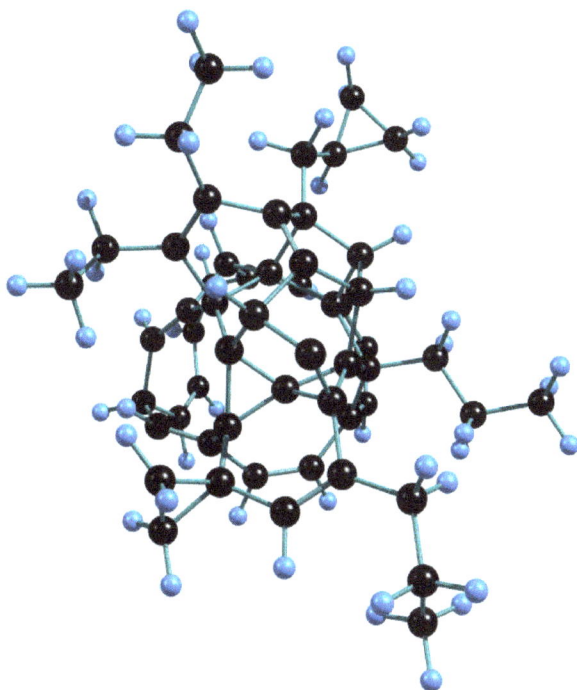

Figure 1. Local minimum geometry of an amorphous type complex hydrocarbon molecule ($C_{55}H_{52}$) at B3LYP/PC1 model. The molecule is composed of non-classical core and classical side groups.

3.2. The 6–10 μm Region

Model Used: PAH

Figure 2 summarizes the results of our new, quantitative vibrational analysis followed by statistical manipulations within the wavelength step size of 0.1 μm, performed on all normal modes of a set of 70 neutral honeycomb PAH molecules with the size of 6 to 136 carbon atoms [19,23]. Based on this new mathematical analysis we find that:

- All normal-mode vibrations (IR and Raman active modes) within this wavelength range occur in the symmetry plane of the PAH molecule. They are all in-plane modes. We did not detect any low amplitude out of plane modes through our quantitative analysis.
- All vibrations are reported as highly coupled modes. This includes the 6.2 μm feature described as an uncoupled C-C stretching band (zone (a) in Figure 2). Only the wavelength range of 8.6 to 8.9 μm is found to contain pure aromatic C-H in-plane bending modes (zone (c) in Figure 2).

These pictures of vibrations, derived and assigned purely from theory, are in very good agreement with the results of experimental spectroscopy reported by Jobelin et al. [29] from IR spectra of numbers of PAH molecules. *This agreement with experimental work demonstrates the reliability of our quantitative vibrational analysis methodology which is a key point to exploring the origin of UIE bands at longer wavelengths.*

- In Figure 2 we have labeled and described these vibrations as coupled C-C stretching and C-H in-plane modes. These are commonly used labels. These types of motions also occur at longer wavelengths in PAH molecules. We explain later a possible way of discriminating such vibrations and assigning them to intrinsic fragments. It should be mentioned that the comprehensive vibrational analysis of this 6–10 μm region and the 3 μm part of the spectra has led us to suggest that the olefinic carrier of the 6 μm UIE feature observed in the famous protoplanetary red rectangle nebulae HD44179 [22]. This feature was previously assigned to CO molecule.

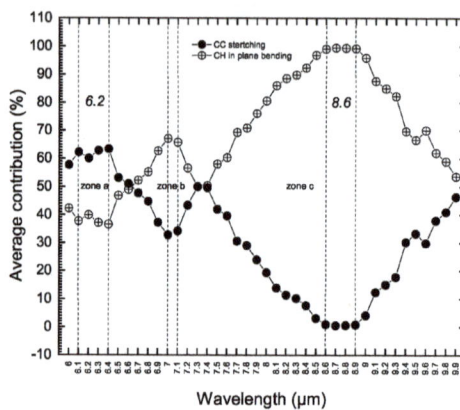

Figure 2. The newly calculated changes in average contributions of aromatic C-C stretching and C-H in-plane bending modes in neutral honeycomb PAH molecules.

3.3. The 11–15 μm Region

Mode Usesl: PAH

For neutral planar honeycomb PAH molecules, this wavelength range contains two major types of motions. They are either in-plane or out of molecular plane vibrations. We did not detect any vibrational motions composed of both types of these vibrations. Each of these two major types of vibrations has their own subdivision vibrations, where the out of plane bending mode of aromatic C-H bondings is one of the well-known categories. The origin of each UIE feature in this wavelength range is complex and cannot be assigned by one label. By filtering all coupled vibrational motions, it is found that the pure uncoupled C-H out of plane bending modes vibrate at different frequencies, independent of their classical exposed edge origin [23].

3.4. Skeletal Modes

Model Used: MAONs

One of the ambiguous concepts in vibrational analysis is the concept of the skeletal mode (or skeleton mode). Usually, this refers to the vibrational motion of a part of the molecules composed of elements heavier than hydrogen, such as the C-C stretching mode in PAH molecules [2]. Such a picture is also satisfied by the other types of vibrational motions over both short and long wavelength ranges in different classes of organic molecules. This concept is becoming more confusing by introducing heteroatoms such as nitrogen, oxygen, sulfur, etc... into the structure of organic molecules.

Here we provide an alternative way to define skeletal mode by considering all atoms not only heavy ones. We chose the MAON model [8], because of its 3D structural diversity. The results of vibrational analysis of 17,257 normal modes (IR and Raman active modes) of 56 MAON type molecules are plotted in Figure 3 from 2 to 30 μm. All MAONs were composed of benzene, methyl, and methylene classical functional groups. The statistical analysis has been carried out within the wavelength interval of 0.1 μm. The simulated IR signature of these MAONs can be found in [21]. Such a statistical plot (Figure 3) can be obtained for any class of organic molecule by applying our methodology [9].

In Figure 3, the smallest number of total atoms that contributes to the vibrations obtained within a 0.1 μm interval is plotted against wavelength. On average this number shows a sudden increase from 14.4 μm towards longer wavelengths. This enables us to split the IR spectra of MAONs into two major regions (Figure 3). We suggest that the long wavelength section is labeled as the skeletal mode section. Here the vibrational motions are delocalized in nature. The short wavelengths section which contains localized vibration　　　　　　　　　　　　　　　　　　　　　　　　　　ding modes.

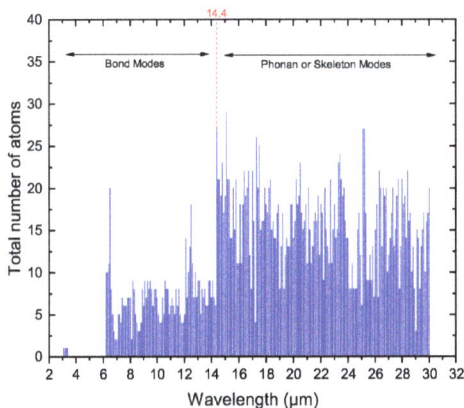

Figure 3. A new plot presented for the first time of statistical analysis done on the results of displacement vector matrix analysis of 17,257 normal modes (all Raman and IR active modes) in 56 different mixed aromatic-aliphatic organic nanoparticles (MAON) type molecules. Statistical analysis performed within the wavelength interval of 0.1 μm from 2 to 30 μm. The abrupt change in the total number of participant atoms occurs at a wavelength of ∼14.4 μm. Two regimes of vibrational motions are distinguishable as bond(bonding) and skeletal(skeleton) modes in MAON molecules.

4. Conclusions

Despite the harsh conditions, the rich chemistry of proto- and planetary nebulae is seen in their strong UIE bands. Here, new quantitative and comprehensive results of displacement vector analysis of normal modes are reported for PAH molecules in the range of (6–10 μm)—see Figure 2. The C-C stretching, C-H bending coupling diagram (Figure 2), provides theoretical explanations for laboratory experimental observations as well as describing UIE band origin in a quantitative way in light of the PAH model. The analysis of >17,000 vibrational normal modes of MAONs with different numbers of atoms and structures within the range of (2–30 μm) are presented for the first time (Figure 3). This shows a boundary at 14.4 μm between two regimes of complex molecular motions and provides clues to the bond(s) and fragment(s) responsible for different IR features. With such valuable quantitative information, modeling the structure of unknown UIE carries and further computations on their spectroscopic properties are performed effectively. With the correct description of the molecular vibrations underlying the UIE bands, we can get closer to explaining the mechanism of the formation of another spectacular feature of UIE bands, i.e., the plateaus at 8, 12, 17 and 21 μm. It should be emphasized that this type of molecular vibrational motion analysis is not restricted, like customary empirical approaches, to the molecules with classical bond types, fragments, and functional groups. The complex molecular vibrations in Figure 1 with non-classical bond types and structure (and unlimited numbers of such molecules), can be analyzed. This will push the modeling of UIE carriers to include new species with exotic properties.

Hence, computational quantum chemistry provides vital theoretical resources to discover the molecular carriers of these mysterious bands. This is currently performed by modeling the spectroscopic properties of large numbers of complex organic molecules. The calculated data is then processed via vibrational analysis for interpretation of the origin of UIE features. This provides valuable information linking the spectra to molecular structure. This assists further modifications of the proposed astrochemistry molecular models. Although the link between molecular structure and IR features is a complex relationship and sometimes confusing, our present results of vibrational analysis on thousands of normal modes on different classes of organic molecules can contribute significantly in addressing the vibrational origin of UIE bands and their unknown molecular carriers.

Author Contributions: S.S. undertook all the mathematical modelling based on the existing collaboration and wrote the bulk of the manuscript. Q.A.P. as supervisor provided some oversight, critique, discussion, vetting and re-writing of certain aspects of the paper.

Acknowledgments: We thank Sun Kwok for his previous leadership of the LSR Astrochemistry program. We are also grateful to colleagues Yong Zhang and Chih Hao Hsia for useful discussions. The computations were performed using the computing facilities provided by the University of Hong Kong.

Conflicts of Interest: The authors declare no conflict of interest.

References

1. Duley, W.W.; Williams, D.A. The infrared spectrum of interstellar dust: Surface functional groups on carbon. *Mon. Not. R. Astron. Soc.* **1981**, *196*, 269–274. [CrossRef]
2. Allamandola, L.J.; Tielens, A.G.G.M.; Barker, J.R. Polycyclic aromatic hydrocarbons and the unidentified infrared emission bands—Auto exhaust along the Milky Way. *Astrophys. J.* **1985**, *290*, L25–L28. [CrossRef]
3. Wagner, D.R.; Kim, H.S.; Saykally, R.J. Peripherally Hydrogenated Neutral Polycyclic Aromatic Hydrocarbons as Carriers of the 3 Micron Interstellar Infrared Emission Complex: Results from Single-Photon Infrared Emission Spectroscopy. *Astrophys. J.* **2000**, *545*, 854–860. [CrossRef] [PubMed]
4. Hudgins, D.M.; Allamandola, L.J. Polycyclic Aromatic Hydrocarbons and Infrared Astrophysics: The State of the PAH Model and a Possible Tracer of Nitrogen in Carbon-Rich Dust. *Astrophys. Dust ASP Conf. Ser.* **2004**, *309*, 665–668. [CrossRef]
5. Duley, W.W.; Grishko, V.I. Evolution of Carbon Dust in Aromatic Infrared Emission Sources: Formation of Nanodiamonds. *Astrophys. J. Lett.* **2001**, *554*, L209–L212. [CrossRef]
6. Cataldo, F.; Keheyan, Y.; Heymann, D. A new model for the interpretation of the unidentified infrared bands (UIBS) of the diffuse interstellar medium and of the protoplanetary nebulae. *Int. J. Astrobiol.* **2002**, *1*, 79–86. [CrossRef]
7. Wada, S.; Tokunaga, A.T. Carbonaceous onino-like particles: A possible component of the interstellar medium. In *Natural Fullerences and Related Structures of Elemental Carbon*; Rietmeijer, F.J.M., Ed.; Springer: Dordrecht, The Netherlands, 2006; pp. 31–52, ISBN 978-1-4020-4135-8.
8. Kwok, S.; Zhang, Y. Mixed aromatic-aliphatic organic nanoparticles as carriers of unidentified infrared emission features. *Nature* **2011**, *479*, 80–83. [CrossRef] [PubMed]
9. Sadjadi, S.; Zhang, Y.; Kwok, S. A Theoretical Study on the Vibrational Spectra of Polycyclic Aromatic Hydrocarbon Molecules with Aliphatic Sidegroups. *Astrophys. J.* **2015**, *801*, 34. [CrossRef]
10. Morino, Y.; Kuchitsu, K. A Note on the Classification of Normal Vibrations of Molecules. *J. Chem. Phys.* **1952**, *20*, 1809–1810. [CrossRef]
11. Konkoli, Z.; Cremer, D. A new way of analyzing vibrational spectra. I. Derivation of adiabatic internal modes. *Int. J. Quantum Chem.* **1998**, *67*, 1–9. [CrossRef]

12. Konkoli, Z.; Larsson, J.A.; Cremer, D. A New Way of Analyzing Vibrational Spectra. IV. Application and Testing of Adiabatic Modes within the Concept of the Characterization of Normal Modes. *Int. J. Quantum Chem.* **1998**, *67*, 41–55. [CrossRef]
13. Mackie, C.J.; Candian, A.; Huang, X.; Maltseva, E.; Petrignani, A.; Oomens, J.; Buma, W.J.; Lee, T.J.; Tielens, A.G.G.M. The anharmonic quartic force field infrared spectra of three aromatic hydrocarbons: Naphthalene, anthracene and tetracene. *J. Chem. Phys.* **2015**, *143*, 224314(1)–224314(15). [CrossRef] [PubMed]
14. Hanson-Heine, M.W.D.; George, M.W.; Besley, N.A. Investigating the Calculation of Anharmonic Vibrational Frequencies Using Force Fields Derived from Density Functional Theory. *J. Phys. Chem. A* **2012**, *116*, 4417–4425. [CrossRef] [PubMed]
15. Miani, A.; Cané, E.; Palmieri, P.; Trombetti, A. Experimental and theoretical anharmonicity for benzene using density functional theory. *J. Chem. Phys.* **2000**, *112*, 248–259. [CrossRef]
16. Guntram, R.; Tomica, H. A combined variational and perturbational study on the vibrational spectrum of P2F4. *Chem. Phys.* **2008**, *346*, 160–166. [CrossRef]
17. Dunham, J.L. The energy levels of rotating vibrator. *Phys. Rev.* **1932**, *41*, 721–731. [CrossRef]
18. Pavlyuchkoa, A.I.; Yurchenko, S.N.; Tennysonb, J. Hybrid variational–perturbation method for calculating ro-vibrational energy levels of polyatomic molecules. *Mol. Phys.* **2015**, *113*, 1559–1575. [CrossRef]
19. Sadjadi, S.; Zhang, Y.; Kwok, S. On the Origin of the 3.3 μm Unidentified Infrared Emission Feature. *Astrophys. J.* **2017**, *845*, 123. [CrossRef]
20. Tao, Y.; Tian, C.; Verma, N.; Zou, W.; Wang, C.; Cremer , D.; Kraka, E. Recovering Intrinsic Fragmental Vibrations Using the Generalized Subsytem Vibrational Analysis. *J. Chem. Theory Comput.* **2018**, *14*, 2558–2569. [CrossRef] [PubMed]
21. Sadjadi, S.; Kwok, S.; Zhang, Y. Theoretical infrared spectra of MAON molecules. *J. Phys. Conf. Ser.* **2016**, *728*, 062003. [CrossRef]
22. Hsia, C.-H.; Sadjadi, S.; Zhang, Y.; Kwok, S. The 6 μm Feature as a Tracer of Aliphatic Components of Interstellar Carbonaceous Grains. *Astrophys. J.* **2016**, *832*, 213. [CrossRef]
23. Sadjadi, S.; Zhang, Y.; Kwok, S. On the Origin of the 11.3 Micron Unidentified Infrared Emission Feature. *Astrophys. J.* **2015**, *807*, 95. [CrossRef]
24. Hudgins, D.M.; Allamandola, L.J. Interstellar PAH Emission in the 11-14 Micron Region: New Insightsfrom Laboratory Data and a Tracer of Ionized PAHs. *Astrophys. J. Lett.* **1999**, *516*, L41–L44. [CrossRef]
25. Laury, M.L.; Carlson, M.J.; Wilson, A.K. Vibrational frequency scale factors for density functional theory and the polarization consistent basis sets. *J. Comput. Chem.* **2012**, *33*, 2380–2387. [CrossRef] [PubMed]
26. Richard, F. Bader. In *Atoms in Molecules A Quantum Theory*; Oxford University: New York, NY, USA, 1990; ISBN 0-19-855168-1
27. Gillespie, R.J.; Popelier, P.L.A. *Chemical Bonding and Molecular Geometry*; Oxford University: New York, NY, USA, 2001; ISBN 0-19-510496-X
28. Keith, T.A. Atomic Response Properties. In *The Quantum Theory of Atoms in Molecules, From Solid State to DNA and Drug Design*; Wiley-VCH: Weinheim, Germany, 2007; ISBN 978-3-527-30748-7
29. Joblin, C.; Boissel, P.; Leger, A.; D'Hendecourt, L.; Defourneau, D. Infrared spectroscopy of gas-phase PAH molecules. I. Role of the physical environment. *Astron. Astrophys.* **1994**, *281*, 926–936.

galaxies

MDPI

Article

Spectroscopy of Planetary Nebulae with *Herschel*: A Beginners Guide

Katrina Exter [1,2]

[1]   Instituut voor Sterrenkunde, KU Leuven, B3001 Leuven, Belgium; katrinaexter@gmail.com
[2]   Herschel Science Centre, ESAC, Camino Bajo del Castillo sn/n, Villafranca del Castillo,
      Villanueva de la Cañada, 28692 Madrid, Spain

Received: 28 May 2018; Accepted: 13 July 2018; Published: 17 July 2018

Abstract: A brief overview of the *Herschel* Space Telescope PACS and SPIRE spectrographs is given, pointing out aspects of working with the data products that should be considered by anyone using them. Some preliminary results of Planetary Nebulae (PNe) taken from the *Herschel* Planetary Nebula Survey (HerPlaNs) programme are then used to demonstrate what can be done with spectroscopy observations made with PACS. The take-home message is that using the full 3D information that PACS spectroscopy observations give will greatly aid in the interpretation of PNe.

Keywords: infra-red; planetary nebulae; integral field spectroscopy

1. Introduction

ESA's *Herschel* Space Telescope was operational between 2009 and 2013. It gathered images and spectra in the far-IR with three instruments:

- HIFI: High resolution spectroscopy between 490 and 1900 GHz; single aperture observations but also some mapping.
- SPIRE and PACS photometry: Covering 3 bands each, namely 70, 100, and 160 µm (PACS) and 250, 350, and 500 µm (SPIRE), both using a scan-mapping over the requested field.
- SPIRE and PACS spectroscopy: SPIRE was a Fourier transform spectrometer (FTS) operating over 190–670 µm, and PACS an integral field unit (IFU) operating over 50–200 µm, in both cases gathering sets of discrete spectra or performing spectral mapping.

The Herschel Science Archive (HSA) provides access to all *Herschel* observations. Any download of the standard products (standard product generation, SPG) includes the raw, partially-processed, and the fully-reduced data, and associated calibration data. In addition to these data, Highly-Processed Data Products (HPDPs) can also be obtained via the Herschel Science Archive (HSA): the HPDPs are superior-quality or complementary products to the standard ones. Finally, all large programmes also provide their processing results as User-Provided Data Products.

All documentation related to *Herschel* data and processing these data in HIPE or elsewhere can be found on the Instrument Overview pages of the Herschel Science Centre (https://www.cosmos.esa. int/web/herschel/home). Of particular interest will be the documentation on the calibration of the three instruments and the corrections that users are recommended to carry out on data downloaded from the HSA.

2. PACS and SPIRE

Since no planetary nebulae programmes used the HIFI instrument, we only discuss PACS and SPIRE here. Both instruments included a photometer and a spectrometer, and it is the spectrometers that we concentrate on.

The PACS spectrograph was an integral field spectrograph. It had several observing modes—pointed, mapping, tiling, full spectral range or on selected spectral lines—and two fundamentally different methods of gathering the background spectra (the "sky" to be subtracted from the source). The main output of PACS were cubes (3D datasets with two spatial axes and a spectral axis): four types of cubes were created by the standard pipeline, and which are provided in any observation depends on the type of pointing mode that was used for the observation:

- Rebinned cubes have the native footprint of the IFU; a slightly irregular 5×5 grid of $9.4''$ spaxels. These are provided for all observations.
- Interpolated cubes: A mosaicking and spatial resampling of the rebinned cubes, creating mosaic cubes with regular spatial grid of $3''$ spatial pixels. These cubes are provided for all undersampled mapping and for pointed observations.
- Projected cubes: A mosaicking and spatial resampling of the rebinned cubes, creating mosaic cubes with regular spatial grid with $0.5''$ (pointed) or up to $3''$ (mapping) spatial pixels. These cubes are provided for all Nyquist and oversampled mapping and pointed observations.
- Drizzled cubes: A mosaicking and spatial resampling of the rebinned cubes, which also have a regular spatial grid with spatial pixels of a size optimised to the wavelength (i.e., to the beam). These cubes are provided only for short wavelength range oversampled mapping observations.

The FWHM for the beam for spectroscopy observations ranges from $9''$ to $14''$ from blue to red, but the native spaxels are all $9.4''$ in size. This means that all pointed observations and some mapping observations are *spatially undersampled*. Special mapping observing modes ("Nyquist" and "oversampled") were offered to improve the spatial sampling and at the same time to allow a larger field to be covered. Hence, the type of cube to use for science depends on the observing mode the data were taken with.

PACS data downloaded from the HSA require no extra pipeline processing, however, for certain types of observation and targets, extra calibration steps are necessary: for point sources, for semi-extended sources, and for larger sources which have a steep surface brightness distribution. These correction are variously described in user notes that can be obtained from the previously-mentioned webpages.

The SPIRE FTS consisted of rings of circular detectors, arranged in a set of nested circles, where each detector gathered the spectrum of that patch of sky. The observing modes were the following:

- Sparse mode observing had the spatial sampling of the native footprint, i.e., producing sets of single spectra for each detector in each ring as output.
- Mapping modes executed a jiggle pattern to fill the gaps and improve the spatial sampling, producing cubes as output.
- Raster mode was used to observe a large field, with cubes as an output.

If no background ("sky") could be observed within the FoV of the rings, then a separate sky observation was recommended.

Data downloaded from the HSA are fully processed and on the whole no more pipeline work is necessary. In most cases, the data are also science ready, however, for some combinations of observing mode and target, the SPIRE data may require additional post-pipeline considerations (e.g., choosing between the point- and extended-source calibration schemes, and semi-extended source calibrations). In addition, it is important to understand the difference between working with the spectra with the native Sinc spectral line profile and those apodised to have Gaussian profiles. These cases are explained in the instrument overview pages, which can be reached from previously-mentioned webpages.

3. PNe with the HerPlaNs Programme

Planetary nebulae are good targets for PACS and SPIRE observations, since both had observing modes that allowed for the collection of spatially-resolved spectra, and between the two one could cover the range 50 to 670 μm. There are over 200 separate observations of PNe with spectroscopy in the HSA, and most of these PNe will also have photometry observations.

HerPlaNS was a *Herschel* programme to observe 11 planetary nebula with PACS and SPIRE, and it is the data from this programme that we discuss here. The goals of HerPlaNS were:

- To map the cold dust in the nebulae/haloes: how much mass is contained in the cold haloes, and how extended are they? The haloes are important in the study of the total mass of nebulae, and hence of the precursor star: this has historically been measured mainly from the ionised nebular gas, which however underestimates the total mass. It is also interesting to look at how the halo material, which was expelled prior to the PN formation, was expelled from the star: how does the halo's shape compare to that of the present nebula?
- To do a full mapping of the energetics for a few objects. With this, one can create a photoionisation model of the nebula and look at how the conditions vary over its extent.
- To map the physical conditions over the nebulae. PNe generally have very complex structures, with various symmetry axes, embedded small-scale structures such as globules and knots, dust mixed with gas, multiple velocities in the outflows, and interaction with the ISM, to name a few. Ionised nebulae exist in many types of sources (from stars to galaxies), but studying these via PNe benefits from their relatively high brightness and their closeness to us (allowing for a good spatial resolution). To gain a full picture of the physical, as well as photoionisation, state of any PNe, multi-wavelength observations are necessary. By combining the *Herschel* data with those at other wavelengths, HerPlaNS aims to build 3D photoionisation models for the target PNe.
- To obtain *Herschel* photometry and spectroscopy data, to allow a combined mapping of the spatial distribution and the physical conditions of the dust (photometry) and gas (spectroscopy) of the nebulae.

The HerPlaNS observations include full spectral range coverage with photometry and spectroscopy from both instruments, and spectral mapping in selected spectral lines for some objects with PACS and SPIRE. A first HerPlaNS overview and demonstration paper has been published by Ueta et al. [1], and a presentation of the HerPlaNS PACS and SPIRE imaging data will soon be published (Asano et al.) and of the spectroscopy data by Exter et al. Other papers highlighting various spectral aspects of the HerPlaNS PNe can be found in Aleman [2,3] and Otsuka et al. [4].

In this work, we present the data analysis done that will lead to further scientific investigations of the HerPlaNS PNe. As our PACS spectral data are cubes, the spectral lines can be fit and turned into maps of intensity. Our SPIRE data are partly cubes and partly sets of discrete spectra. Where PACS and SPIRE photometry data both exist, a comparison of the two can yield information about the gas and the dust content of the nebulae.

The Preliminary Results: PACS Emission-Line Maps

The HerPlaNS PACS spectroscopy observations consisted of mapping observations in selected emission lines for some PNe, and pointed observations covering the full spectral range for all the PNe. The first useful data product to create are therefore emission-line maps for all targets. We used the projected or interpolated cubes for the mapping observations, and the interpolated or rebinned cubes for the pointed observations. The maps were made by fitting a Gaussian profile to the emission line and a low-order polynomial to the continuum, over the entire cube, and then summing up the area of the fitted Gaussian. For the bright lines, where the slightly non-Gaussian nature of the PACS instrumental profile is more obvious (in the form of low-level broadened wings), we instead integrated under the emission line after having subtracted a low-order fit to the continuum: this recovers a few per cent of the total flux that the fitting misses. In both cases, the resulting maps are in units of integrated flux (W/m^2) in each pixel, with the pixels being the same size as the spaxels of the cube used. Some example maps are shown in Figures 1–3. In these figures, the IR images trace the thermal dust continuum, while the PACS spectral maps trace the gas. This spatial resolution in the PACS photometer images is ∼6″, with pixels of 3″, and in the spectrometer images is ∼9″, with pixels of 3″. Optical images are also shown for each PN, for comparison.

NGC 2392

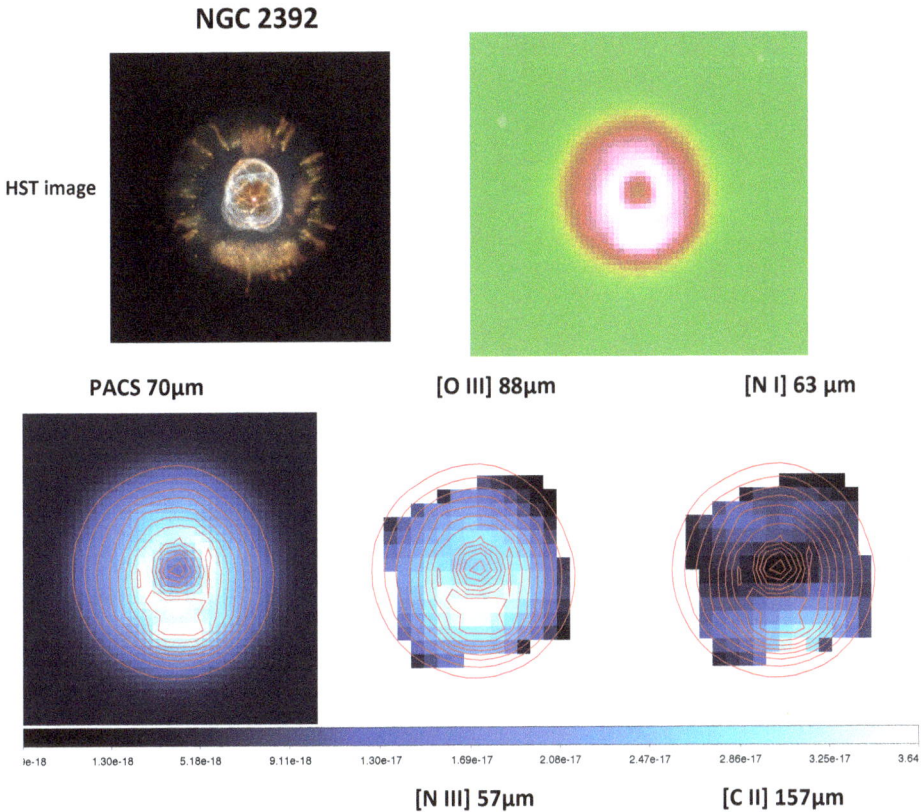

Figure 1. Images of selected PNe of the HerPlaNs programme. In all figures, the top shows an optical image for comparison to the PACS IR image at 70 μm, these being approximately similarly scaled (N up, E left). The IR image is shown again on the bottom with contours to help the eye pick out the shape. These contours are then overplotted on two spectral images, which were chosen to demonstrate the morphology of the nebula in the indicated ions. The mapped ion identifications are printed above the spectral images, and those ions with a similar appearance are printed below. NGC 2392. (**Top**) An optical image (HST, WFPC2) and the PACS 70 μm image. (**Bottom**) The PACS photometer image and two spectral images.

NGC 2392 is shown in Figure 1. The IR image at 70 μm is of about the same size as the optical image. The PACS image shows the dust in the nebula, while the HST image (WFPC2, taken from the HST's Hubblesite) is a composite of emission lines (Hα, He II, [O III], and [N II]) and hence mostly gaseous emission. Two general morphologies in the PACS emission lines are seen: one seen in [N I] and [C II] and the other in [N III] and [O III]. While both sets of ions peak in the south (where the optical and IR photometry do also), the region of highest surface brightness for the multiply ionised ions lies inside that of the singly ionised ions/atomic lines: at a first guess, this indicates a gradient in temperature.

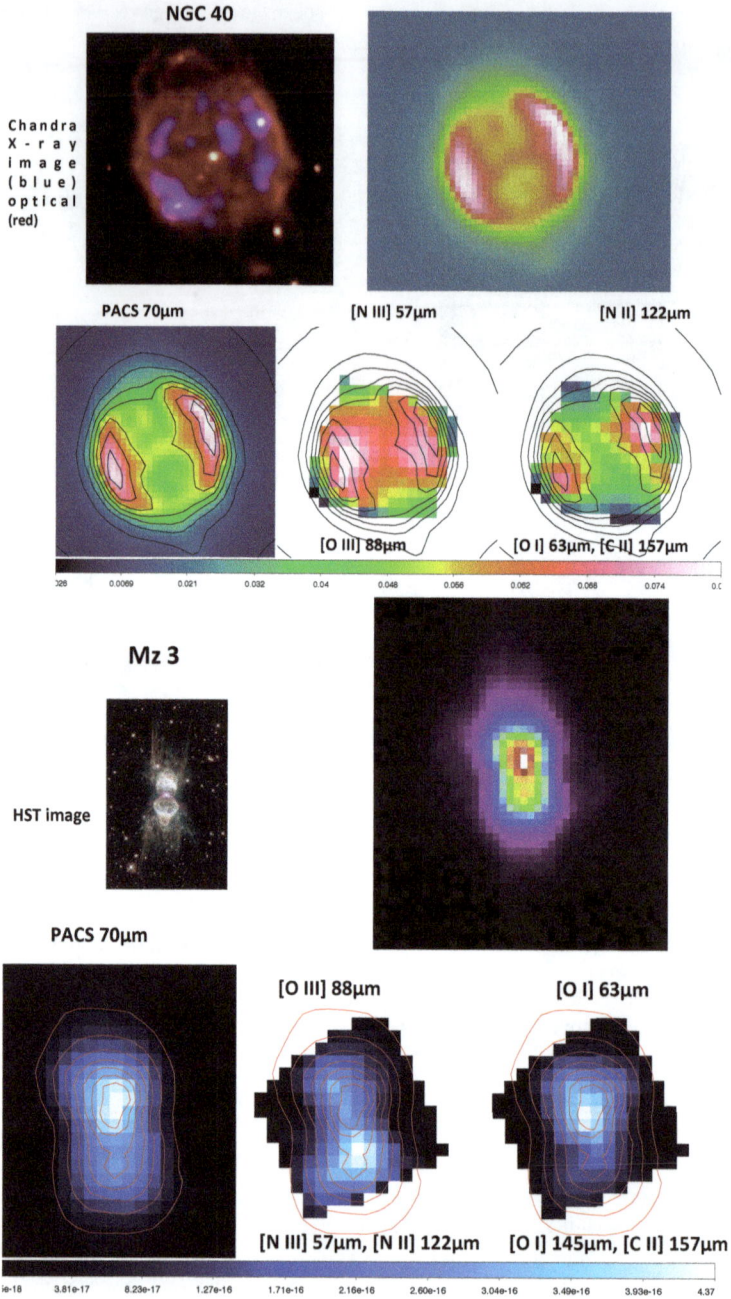

Figure 2. (**Top**) NGC 40, with an X-ray/optical image composite and the PACS 70 μm image, and the IR image with the two spectral images on the bottom. (**Bottom**) Mz 3, optical image (HST, WFPC2) and the PACS 70 μm image at the top, and the IR image with the two spectral images on the bottom.

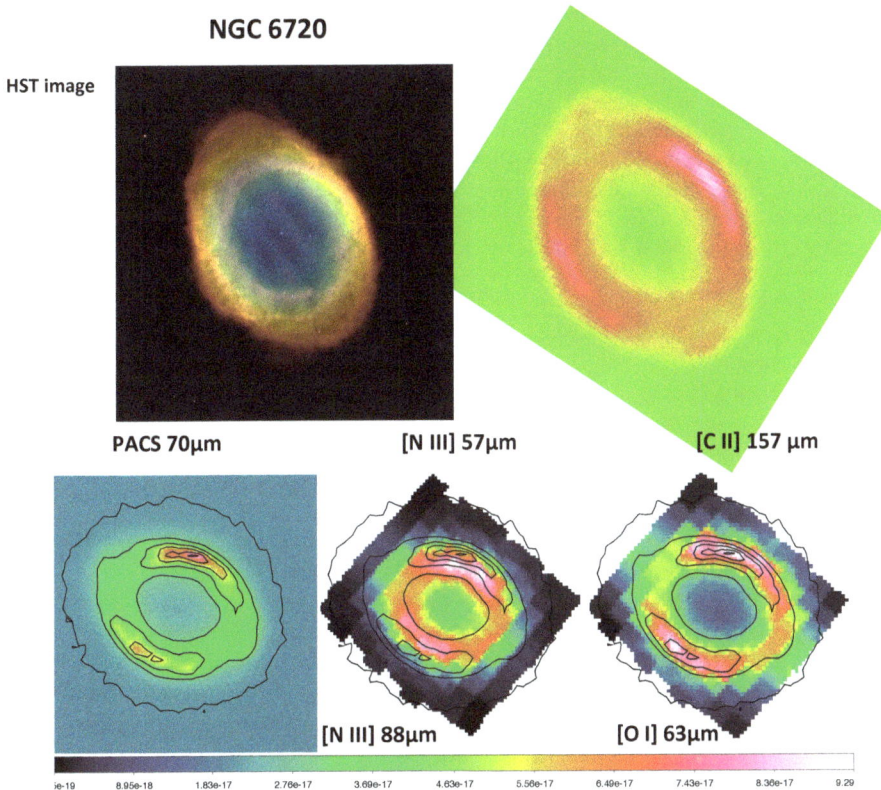

Figure 3. NGC 6720, optical image (HST, WFPC2) and the PACS 70 μm image at the top, and the IR image with the two spectral images on the bottom.

NGC 40 is shown in Figure 2. This nebula is described has a cylindrical or barrel-like shape in the IR. The X-ray/optical image composite (taken from the Chandra "photo album" website) and IR images in Figure 2 shows a similar E–W edge brightening. The X-ray image traces hot gas, suggested to result from the fast wind of the central star (Kastner et al. [5]), and it appears that this is delimited by the brightest parts of the dust in the nebula (i.e., the brightest parts of the PACS image). The IR emission lines also peak in these portions of the nebula. The peak of the multiply ionised ions ([N III] and [O III]) is again differently-located to the singly-ionised ions/atomic line, however here they could be said to be slightly circularly rotated with respect to each other.

MZ 3 is shown in Figure 2. The optical image (WFPC2, taken from the HST's Hubblesite) is a composite of emission lines (Hα, He II, [S II], and [N II]) and hence mostly gaseous emission. The PACS photometer image (which is dust emission) shows a similar shape—the centrally-bright part of the IR image is about the same size as the optical image—but additional fainter emission can be clearly seen outside of this region (note that even PACS image is nicely resolved, and so this difference is not due to the different instrument resolutions). The surface brightness peak in the 70 μm image is to the north, while the gas-tracing spectral lines peak in the centre (singly ionised ions/atomic lines) and south (multiply-ionised ions). This object certainly merits a closer inspection. Aleman et al. [3] reported on the detection of Hydrogen recombination laser lines in the PACS and SPIRE spectra of Mz 3.

NGC 6720 is the final PN shown here, in Figure 3. This nebula has a very similar ring-like appearance in the optical (WFPC2, taken from the HST's Hubblesite; is a composite of He II, [O III], and [N II])) and PACS images. This ring-like appearance is the result of the projection of bipolar lobes pointing almost along the observer's axis (e.g., O'Dell et al. [6]). The dust (measured from the IR photometry images) and gas (measured from the IR emission-line images) peak in the same location in the ring. It is clear that the surface brightness gradient is different between the multiply- and singly-ionised ions in the ring, pointing to a globally smooth variation in the physical conditions. NGC 6720 was included in the *Herschel* MESS programme (Groenewegen, et al. [7]), and PACS photometric data were used to study the knots in the ring (van Hoof et al. [8]).

4. Conclusions

Overall, we see a high degree of similarity between the the PACS photometer and spectrometer maps. However, the devil lies in the details: different ions peak at different locations in the nebulae, and this clearly will lead to information about the physical conditions and the structure of the nebulae. With such clear spatial variations, it is obvious that you do need to study the PNe in two dimensions.

A more careful comparison of the spectral maps to each other and to the photometer maps requires the additional step of matching the beam sizes at the different wavelengths, i.e., convolving the data to the same beam size.

We also need to consider the flux calibration more carefully. PACS cubes (and the spectral maps made from the cubes) are calibrated assuming the source is flat and extended, and for any other source morphology the flux calibration is not quite correct (this is documented on the PACS Overview page on the Herschel website). An attempt can be made to adjust the calibration to account for the unique source shape and observing (pointing) pattern of any target/observation. A script to do just this is provided on the HSC PACS pages and this is something that should be attempted for these sources.

Funding: This research made use of NASA's Astrophysics Data System, and the SIMBAD database operated at CDS, Strasbourg, France. This work is based on observations made with the *Herschel* Space Observatory, a European Space Agency (ESA) Cornerstone Mission with significant participation by NASA. This research received no external funding.

Conflicts of Interest: The author declares no conflict of interest.

References

1. Ueta, T.; Ladjal, D.; Exter, K.; Otsuka, M.; Szczerba, R.; Siódmiak, N.; Aleman, I.; van Hoof, P.A.; Kastner, J.H.; Montez, R.; et al. The Herschel Planetary Nebula Survey (HerPlaNS)-I. Data overview and analysis demonstration with NGC 6781. *Astron. Astrophys.* **2014**, *565*, A36. [CrossRef]
2. Aleman, I.; Ueta, T.; Ladjal, D.; Exter, K.M.; Kastner, J.H.; Montez, R.; Tielens, A.G.; Chu, Y.H.; Izumiura, H.; McDonald, I.; et al. Herschel Planetary Nebula Survey (HerPlaNS). First detection of OH+ in planetary nebulae. *Astron. Astrophys.* **2014**, *566*, A79. [CrossRef]
3. Aleman, I.; Exter, K.; Ueta., T.; Walton, S.; Tielens, A.G.; Zijlstra, A.; Montez, R., Jr.; Abraham, Z.; Otsuka, M.; Beaklini, P.P.; et al. Herschel Planetary Nebula Survey (HerPlaNS): Hydrogen Recombination Laser Lines in Mz 3. *Mon. Not. R. Astron. Soc.* **2018**, *47*, 4499–4510. [CrossRef]
4. Otsuka, M.; Ueta, T.; van Hoof, P.A.; Sahai, R.; Aleman, I.; Zijlstra, A.A.; Chu, Y.H.; Villaver, E.; Leal-Ferreira, M.L.; Kastner, J.; et al. The Herschel Planetary Nebula Survey (HerPlaNS): A Comprehensive Dusty Photoionization Model of NGC6781. *Astrophys. J. Suppl. Ser.* **2017**, *231*, 22. [CrossRef] [PubMed]
5. Kastner, J.H.; Montez, R.; De Marco, O.; Soker, N. X-ray Imaging of Planetary Nebulae with Wolf-Rayet-type Central Stars. *Bull. Am. Astron. Soc.* **2005**, *37*, 437.
6. O'Dell, C.R.; Ferland, G.J.; Henney, W.J.; Peimbert, M. Studies of NGC 6720 with Calibrated HST/WFC3 emission-Line Filter Images. I. Structure and Evolution. *Astron. J.* **2013**, *145*, 92. [CrossRef]

7.   Groenewegen, M.A.; Waelkens, C.; Barlow, M.J.; Kerschbaum, F.; Garcia-Lario, P.; Cernicharo, J.; Blommaert, J.A.; Bouwman, J.; Cohen, M.; Cox, N.; et al. MESS (Mass-loss of Evolved StarS), a Herschel key program. *Astron. Astrophys.* **2011**, *526*, A162. [CrossRef]

8.   Van Hoof, P.A.M.; Barlow, M.J.; Van de Steene, G.C.; Exter, K.M.; Wesson, R.; Ottensamer, R.; Lim, T.L.; Sibthorpe, B.; Matsuura, M.; Ueta, T.; et al. Herschel observations of PNe in the MESS key program. In Proceedings of the IAU Symposium 283: Planetary Nebulae: An Eye to the Future, Canary Islands, Spain, 25–29 July 2012.

galaxies

MDPI

Article

ALMA's Acute View of pPNe: Through the Magnifying Glass... and What We Found There

Carmen Sánchez Contreras [1,*], **Javier Alcolea** [2], **Valentín Bujarrabal** [3]
and Arancha Castro-Carrizo [4]

[1]   Centro de Astrobiología (CSIC-INTA), Camino Bajo del Castillo s/n, Urb. Villafranca del Castillo, 28691 Villanueva de la Cañada, Spain

[2]   Observatorio Astronómico de Madrid, Observatorio Astronómico Nacional (IGN), Alfonso XII No 3, 28014 Madrid, Spain; j.alcolea@oan.es

[3]   Centro de Investigaciones de Ciencias Geográficas y Astronomía, Observatorio Astronómico Nacional (IGN), Ap 112, 28803 Alcalá de Henares, Spain; vbujarrabal@oan.es

[4]   Institut de Radioastronomie Millimetrique, 300 rue de la Piscine, 38406 Saint Martin d'Heres, France; ccarrizo@iram.fr

*   Correspondence: csanchez@cab.inta-csic.es; Tel.: +34-918-131-205

Received: 28 June 2018; Accepted: 20 August 2018; Published: 4 September 2018

Abstract: We present recent Atacama Large Millimeter/submillimeter Array (ALMA)-based studies of circumstellar envelopes (CSEs) around Asymptotic Giant Branch (AGB) stars and pre-Planetary Nebulae (pPNe). In only a few years of operation, ALMA is revolutionising the field of AGB-to-PN research by providing unprecedentedly detailed information on the complex nebular architecture (at large but also on small scales down to a few $\sim$10 AU from the centre), dynamics and chemistry of the outflows/envelopes of low-to-intermediate mass stars in their late stages of the evolution. Here, we focus on continuum and molecular line mapping studies with high angular resolution and sensitivity of some objects that are key to understanding the complex PN-shaping process. In particular, we offer (i) a brief summary of ALMA observations of rotating disks in post-AGB objects and (ii) report on ALMA observations of OH 231.8+4.2 providing the most detailed and accurate description of the global nebular structure and kinematics of this iconic object to date.

Keywords: AGB and post-AGB stars; circumstellar matter; winds and outflows; mass-loss; jets

1. Introduction

Due to its unique capabilities, the Atacama Large Millimeter/submillimeter Array (ALMA) has an immense potential to make great advances, and to answer major questions, in the field of Asymptotic Giant Branch (AGB)-to-Planetary Nebulae (PNe) evolution. The first ALMA images of an AGB star, R Sculptoris [1], showed a spiral structure inscribed in the circumstellar envelope (CSE), and, for the first time, astronomers could get full three-dimensional information about this outrageous structure, created by a hidden companion star orbiting the red giant. These ALMA maps were the prelude for future discoveries and marked the beginning of a new era in AGB/post-AGB/pre-Planetary Nebulae (pPNe)/PNe research.

At the time of giving this presentation, i.e., during the Asymmetric Planetary Nebulae (APN) VII meeting, ALMA was in cycle five, offering already excellent angular resolution over a broad range of frequencies (from $\sim$100 to $\sim$870 GHz). At 1 mm the angular resolution was below 20 milliarcsec, which represents an improvement of more than a factor of 10 with respect to other mm/sub-mm-interferometers. This resolution enables us (i) to obtain a close up view of the central regions of pPNe and to look for relevant structures that are directly related to, or take part in, the wind collimation process and (ii) to gather a very detailed characterization of large-scale structures (and to discover new ones) that can help us to build up the nebular shaping history of these objects. In this presentation, we show some examples of these two types of studies. In Section 2, a very brief summary of ALMA observations of rotating disks in post-AGB objects is offered. In Section 3, we present (a small part of) our ALMA observations of OH 231.8+4.2, a key object to understand PN-shaping.

2. A Quest for Rotating Disks in Post-AGB Objects

Rotating disks are invoked by most wind collimation theories, e.g., [2–4], and are suspected to be present in a particular class of post-AGB stars with near-IR emission excess (see [5] for a review). The near-IR emission excess indicates hot dust located in a stable disk-like structure close to the star (at $\approx 10^{14}$–10^{15} cm). In contrast to most pPNe, these objects show narrow CO emission profiles consistent with rotating disks [6]. The prototype of this class of post-AGB objects is the Red Rectangle, which was the first source in which a Keplerian disk was directly confirmed and spatially resolved with interferometric CO emission observations [7]. Modelling of these data enabled a detailed characterization of the structure and kinematics of the disk, as well as an estimate of the mass of the central binary system of $\sim 1.5\,M_\odot$.

More recently, several molecular transitions in the Red Rectangle have been mapped with ALMA with better angular resolution and sensitivity [8,9]. This has enabled not only an improved characterization of the disk, but has also revealed the presence of a tenuous X-shaped wind that emanates from the disk at the low velocity of $\sim$3–10 km s^{-1}. The mass–loss rate of the wind, which may be material photoevaporating from the disk, implies a lifetime of the disk of $\sim$5000–10,000 yr.

The quest for rotating disks continues, and more post-AGB objects with near-IR excess are being observed in CO emissions. IW Car [10] and IRAS 08544-4133 [11] are the most recent examples of post-AGB stars in which rotating disks and winds emerging from the disk have been spatially resolved with ALMA. Rotating disks around post-AGB stars had been mapped earlier, also with the IRAM interferometer, e.g., 89 Her [12] and AC Her [13].

The properties of the disks and the X-winds (when detected) are similar in all post-AGB targets (Table 1), although a very low-number statistics still prevails. We emphasize that, given the large dimensions of the disks mapped to date (with typical outer radii of the order of 1000 AU), these structures do not represent the compact accretion disks (around the compact companion of the mass-lossing star) postulated by most wind-collimation theories. The rotating disks reported as of today are relatively large structures around the central binary system, typically with a total mass of $M_\star \sim$ 1–2 $M_\odot$. The role of these circumbinary disks in the wind collimation and/or, more generally, in the PN-shaping process, remains unclear. Recently, the first rotating disk in an AGB star, L$_2$ Pup, has been detected and spatially resolved with ALMA [14,15]. In this case, the disk is smaller and less massive than the post-AGB stars shown before, but it is also tentatively circumbinary.

Rotating disks in post-AGB objects have typical masses of $M_{disk} \approx 10^{-3}$–$10^{-2}\,M_\odot$, the Red Rectangle being the most massive detected so far, and excitation temperatures of $\approx$100 K. For objects where a rotating disk and an X-wind component are simultaneously detected, the mass of the latter is only a small fraction (10%) of the mass of the disk. These X-winds are very different from the massive ($\sim$0.1–1 $M_\odot$) and high-momentum bipolar outflows common to most pPNe [16]. It is paradoxical that none of the pPNe with massive, high-momentum bipolar outflows have shown clear evidence for rotation at their cores. Therefore, ironically, rotating disk are only found in objects where a powerful jet-launching engine does not need to be invoked. This is a mystery that ALMA will hopefully solve in the coming years, as more rotating disks and pPNe are observed.

Table 1. Keplerian circumbinary disks with CO interferometric maps.

Name	Class	Disk	X-Wind	R_{disk}	M_{disk}	$M_\star$	Reference
Red Rectangle	post-AGB	Yes	Yes	$\approx$2000 AU	$4.0 \times 10^{-2}\,M_\odot$	$\sim$1.5 $M_\odot$	[7–9]
IW Car	post-AGB	Yes	Yes	$\approx$1300 AU	$4.0 \times 10^{-3}\,M_\odot$	$\sim$1 $M_\odot$	[10]
AC Her	post-AGB	Yes	?	$\approx$1130 AU	$1.5 \times 10^{-3}\,M_\odot$	$\sim$1.5 $M_\odot$	[13]
IRAS 08544	post-AGB	Yes	Yes	$\approx$1400 AU	$6.0 \times 10^{-3}\,M_\odot$	$\sim$1.8 $M_\odot$	[11]
89 Her	post-AGB	Possibly	Yes	$\lesssim$200 AU	($\sim$10$^{-2}\,M_\odot$)	($\sim$1 $M_\odot$)	[12]
L$_2$ Pup	AGB	Yes	?	$\approx$20 AU	$2.2 \times 10^{-4}\,M_\odot$	$\sim$0.66 $M_\odot$	[14,15]

3. Nebular Architecture and Dynamics of pPNe: The ALMA View of OH 231.8+4.2

Characterization with ALMA of the nebular structure and dynamics of 'standard' pNe, i.e., with dominant bipolar outflows carrying a significant amount of mass and linear momentum, is essential to understand how they were formed. We show ALMA observations of OH 231.8+4.2, an object in a brief but key evolutionary stage of the AGB-to-PN transition.

OH 231.8+4.2 is an outstanding pPN-like nebula (Figure 1) around a Mira-type pulsating star (QX Pup). This uncommon situation makes OH 231.8+4.2 of special interest, since it demonstrates that asymmetries can develop (and become very prominent) while the central star is still on the AGB. OH 231.8+4.2 has been extensively studied at many wavelengths by several authors e.g., [17–24] (and references therein). The light from the central source inside OH 231.8's core is highly obscured by dust along the line-of-sight and is only seen indirectly scattered by the dust in the lobe walls, indicating the presence of an A 0-type main-sequence companion to the mass-losing AGB star, of spectral type M8-10 III [21]. It is believed that the bulk of the nebular mass, which is in the form of molecular gas ($\sim$1 $M_\odot$), has been ejected by the AGB star at a very high rate, $\dot{M} \approx 10^{-4}\,M_\odot\,\mathrm{yr}^{-1}$ [17]. Before ALMA came into play, two major large-scale components had been identified: (i) An equatorial waist expanding at low velocity, $<$30 km s^{-1} and (ii) a highly collimated bipolar outflow with expansion velocities that increase linearly with the distance from the centre up to $\sim$400 km s^{-1} (deprojected). The linear velocity gradient in the outflow could have resulted from the sudden interaction between collimated fast winds (CFWs) on the ambient AGB material $\sim$800 yr ago.

We present a summary of our recent observations of OH 231.8+4.2 with ALMA. These data have provided us with a high-definition image of the dominant nebular components described above, but also, more importantly, have enabled us (1) to identify the position of the mass-losing star relative to the nebula and (2) to discover a number of outflow components previously unknown. We also report the first detection of Na^{37}Cl and CH$_3$OH in OH 231.8+4.2, with CH$_3$OH also being a first detection around an AGB star. A comprehensive study of this dataset is reported by [25].

3.1. Observations

OH 231.8+4.2 was observed with the ALMA 12 m array as part of project 2015.1.00256.S on July 2016. We mapped the continuum emission and a series of spectral lines in band seven ($\sim$294–345 GHz). Our project has two major science goals: SG1 is a five-point mosaic of ^{12}CO/^{13}CO (3-2) and other molecules covering a $\sim$19$''\times$54$''$ area along the bipolar outflow, and SG2 is a single-pointing observation towards the centre to map a selection of lines, e.g., SiO and CS, amongst others. The resolution of these observations is $\sim$0$''$.2–0$''$.3. Full observational details are given in [25].

3.2. Continuum Maps

Maps of the continuum emission at four different frequencies (294, 304, 330 and 344 GHz) have been created using line-free channels. Our $\sim$294 GHz-continuum map is shown in the left panel of Figure 2 as an example. The surface brightness distribution, which is very similar at all frequencies, appears as an extended, incomplete hourglass-like structure, roughly oriented along PA $\sim$ 21° (i.e., the symmetry axis of the large-scale nebula). The continuum emission distribution is clumpy. The peak of the continuum maps is attained at one of these clumps, referred to as "*clump S*", which has coordinates R.A. = $07^h42^m16^s.915$ and Dec. = $-14°42'50''06$ (J2000). As we will show below, clump S enshrouds the central star QX Pup. We deduce a deconvolved radius of $\sim$40–70 AU for clump S, which is barely resolved. Note that clump S does not lie on the equatorial plane of the hourglass, but is clearly displaced along the axis towards the south by $\sim$0''6. The total continuum flux with ALMA is consistent with that measured with single-dish telescopes (see right panel of Figure 2). Except for clump S, the spectral index of the continuum shows no significant deviations from a $S_\nu \propto \nu^{3.3}$ power-law across the different regions. This implies a dominant contribution by optically thin $\sim$75 K dust with an emissivity index of $\alpha \sim$1.3. At clump S, the continuum obeys a $S_\nu \propto \nu^{2.1}$ dependence consistent with a population of large ($>$100 μm-sized) grains in the vicinity of QX Pup.

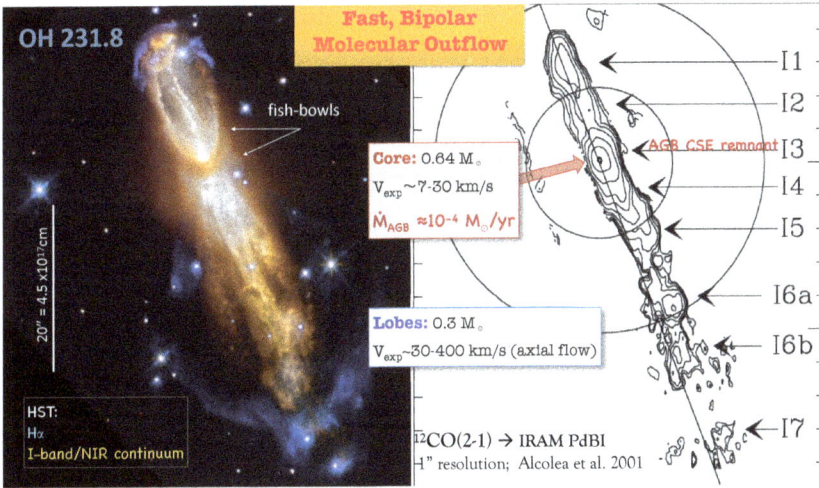

Figure 1. (**Left**) False color *HST* image of the reflection nebulosity (red-to-yellow) and Hα-emitting lobes (blue) of OH 231.8+4.2 (Credit: ESA/Hubble & NASA. Acknowledgement: Judy Schmidt). North is up. Two faint, bubble-like features (*fish-bowls*) with molecular counterparts in our ALMA data are indicated—see Section 3.3. (**Right**) ALMA order-zero moment maps of ^{12}CO (2-1) integrated over the full width of the line profile adapted from [17]. The mass and expansion velocity of the low-velocity core and the fast, bipolar lobes are indicated. The distance to OH 231.8+4.2 is $d\sim$1500 pc. The lobes are tilted with respect to the plane of the sky by $i\sim$35°, with the North lobe being the closest.

3.3. Molecular Line Maps: Nebular Components

We have mapped a large number of molecular transitions, covering a range of upper-level energies from $E_u \sim$ 20 to 1800 K and including many species (CO, CS, SO, SO$_2$, SiO, SiS, H$_3$O$^+$, CH$_3$OH, etc.). Different lines are found to selectively (or exclusively) trace different components of the outflow, which facilitates disentangling the complex nebular architecture of OH 231.8+4.2. In the following, we describe different structures traced by ALMA, starting from the most compact ones (within $\approx 10^{15}$ cm from the mass-losing star) and ending with the most extended zones (reaching out to $\approx 10^{17}$ cm).

We have detected several high-excitation lines, namely, SiS (v = 1, J = 17 − 16), ^{30}SiO (v = 1, J = 7 − 6), and Na^{37}Cl (v = 0, J = 26 − 25), with compact emission arising *entirely* from clump S. These lines are known to be produced in the warm inner ($\lesssim 10^{15}$ cm) layers of the winds of evolved mass-losing stars, where the stellar wind may have not reached its terminal expansion velocity. The emission is indeed confined to a small emitting volume of radius $\sim$60 AU ($\sim 24R_\star$), comparable to that of the continuum-emitting core. The narrow profiles imply low expansion velocities of $V_{exp} \sim 5$–7 km s^{-1} in these layers, for which we deduce a kinematic age of about 50 yr.

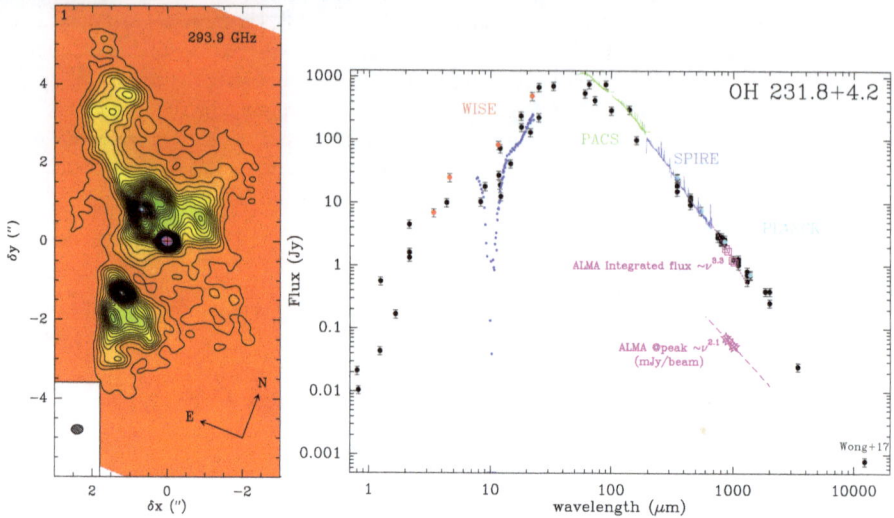

Figure 2. (**Left**) ALMA 294 GHz-continuum emission map of OH 231.8+4.2. Contour level spacing is 0.8 mJy/beam. The clean beam (HPBW = 0″31 × 0″25, PA = −84.5°) is plotted at the bottom-left corner. The compact region at the centre where the continuum emission peaks is referred to as clump *S* and it is a dust (and gas) region enshrouding the AGB star QX Pup. (**Right**) Spectral energy distribution. Pink symbols are our ALMA continuum measurements (squares = integrated flux; stars = peak-flux surface density at clump *S*) and the dotted and dashed lines are $S_\nu \propto \nu^{3.3}$ and $S_\nu \propto \nu^{2.1}$ fitted to those.

In addition to disclosing the locus of QX Pup inside clump *S*, we have discovered a compact ($\sim 1'' \times 4''$) bipolar outflow that emerges from the stellar vicinity. This outflow is exclusively sampled in our data by the SiO molecule, which is a well-known shock tracer (top panels of Figure 3). Our maps show a pair of flame-shaped lobes expanding at moderate velocities and oriented similarly to the large-scale nebula. The radial (or line-of-sight) velocity increases abruptly from the centre to two bright and compact regions located at offsets $\delta y \sim \pm 0''11$ about clump *S*, where the maximum radial velocities (~ 35 km s^{-1}) and full line widths (~ 23 km s^{-1}) are observed. These SiO-bright regions may represent bow-shocks, denoting recent ($\lesssim 50$–80 yr old) bipolar mass ejections. The lower velocities at the intermediate-to-outer regions of the flame-shaped SiO-lobes imply larger kinematic ages of $t_{kin} \sim 400$–500 yr.

Adjacent to the SiO-outflow, we identify a small-scale hourglass-shaped structure (*mini-hourglass*) that is probably made of compressed material formed as the SiO-outflow penetrates the dense, central regions of the nebula (bottom panels of Figure 3). The lobes and the equatorial waist of the mini-hourglass *both* are radially expanding with a constant velocity gradient ($V_{exp} \propto r$). The dimensions and kinematics of the mini-waist are consistent with a torus-like structure with an outer radius of about 500 AU orthogonal to, and coeval with, the SiO-outflow. The expansion velocity at the inner edge of the mini-waist (~ 150 AU) is extremely low, at $V_{exp} \sim 1$ km s^{-1}.

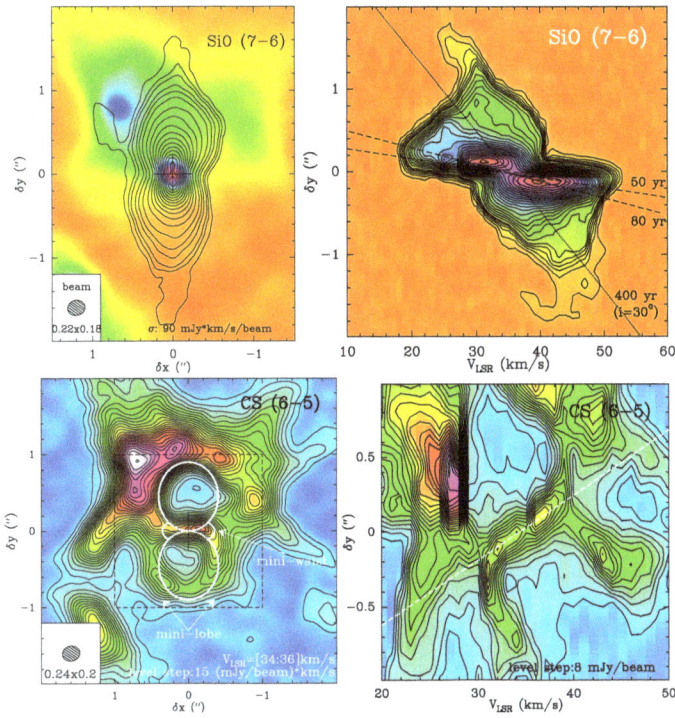

Figure 3. SiO (7−6) and CS (6−5) emission in OH 231.8+4.2. (**Left**) Zero-order moment maps (integrated over V_{LSR} = [18:53] and [34:36] km s^{-1}, for SiO and CS, respectively). The mini-hourglass centred at clump S is schematically depicted in the bottom-left panel. (**Right**) Position-velocity (PV) diagrams along PA = 21°. The lines represent different velocity gradients implying different kinematic ages at various regions of the SiO-outflow (top) and along the mini-waist (bottom).

The $\sim$8$''$ × 4$''$-sized hourglass nebula traced by the dust thermal continuum emission maps is the major emission component of most of the species mapped by us with ALMA. In Figure 4, we offer a condensed view of CS (7−6), one of the dense gas tracers observed by us, where this component is clearly seen. The large-hourglass is characterized by a dominant expansive kinematics described by a radial velocity gradient, $\nabla v \sim$ 6.0–6.5 km s^{-1} arcsec^{-1}, that is sustained from the base to the tips of the lobes, where the highest velocities are observed. The sharp outer boundary of the large-waist is now accurately determined (at a radius of $\sim$2700 AU) as well as its internal velocity field, indicating that the waist was shaped nearly simultaneously with the high-velocity (HV) lobes, about 800–900 yr ago.

We report the discovery of two large ($\sim$8$''$× 6$''$) faint, bubble-like structures surrounding the central parts of the nebula (*fish-bowls*, see Figure 5). These are the molecular counterparts of the two faint, rounded structures seen in the *HST*/NIR images (Figure 1). The north and the south fish-bowls have similar dimensions, but are oriented differently. The observed spatio-kinematic distribution is consistent with the fish-bowls being a pair of hollow, thin-walled ellipsoids (probably made of swept up ambient material) radially expanding and moving away from each other. They are relatively old structures although apparently slightly ($\sim$100–200 yr) younger than the large-scale waist and the HV-lobes. The origin of these enigmatic structures is unknown, but it may be linked to a 'critical' time period or instant of the evolution/development of the HV-lobes.

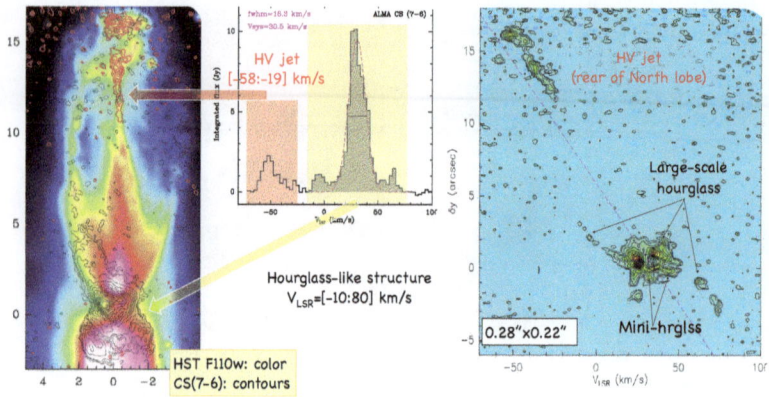

Figure 4. Overview of the CS (7−6) emission maps of OH 231.8+4.2 showing the integrated intensity map, atop the *HST*/F110W image (**left**), the one-dimensional spectral profile (**middle**), and the axial PV diagram (**right**). Different nebular components are indicated (namely, the large-hourglass, the HV-lobes, and the mini-hourglass). See text for details.

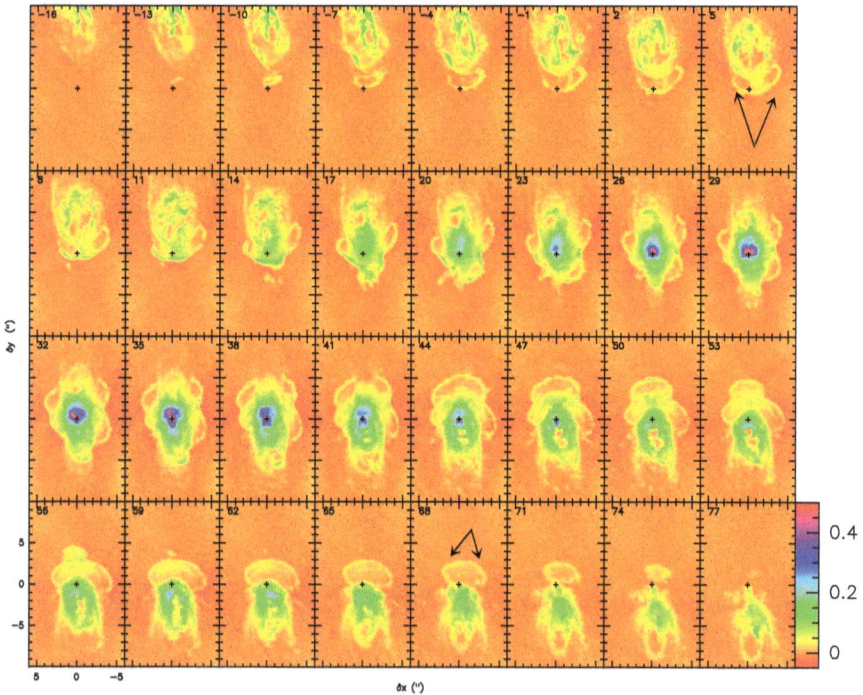

Figure 5. Velocity-channel maps of ^{12}CO (3−2) in the velocity range where two faint, elongated bubble-like features (dubbed *the fish-bowls*) are discovered in the central regions of OH 231.8.

3.4. Conclusions

Our ALMA observations of OH 231.8+4.2 unveil a series of new substructures that point to a nebular formation/shaping history significantly more complex than previously thought, and in particular, indicative of multiple non-spherical mass ejections. The origin of the bipolar ejections that led to the formation of the outstanding nebular architecture of OH 231.8+4.2 remains unknown, but the presence of present-day bipolar ejections indicate that the CFW engine is still active at its core.

There are notable differences in the velocities and overall symmetry between the large-scale ~800 yr old CO-outflow and the most current (≲80 yr old) bipolar ejections traced by SiO. These differences could indicate that the jet-launching mechanism itself has changed, or that the initial conditions under which the jets formed (presumably, involving disk-mediated mass transfer from the AGB to the companion star that blows the jets) have been modified in such a way that they now produce slower and more symmetric bipolar ejections than in the past. These changes may be intimately linked to major adjustments in the binary system configuration, which may have occurred after the powerful ejections that resulted in the formation of the massive and fast CO-outflow.

The position of QX Pup off-centre from the waist of the large-hourglass is one of the most puzzling discoveries from our ALMA data. Perhaps the combination of orbital motion and recoil of the binary system after strong asymmetrical mass ejections could explain, at least partially, its mysterious location. The orbital parameters of the central binary system are very poorly known. A loose upper limit to the orbital separation of $a < 150$ AU is deduced from the relative separation between clump *S* and the centroid of the SiO-outflow (presumably collimated by the companion).

New ~0″.05-resolution observations with ALMA by our team (2017.1.00706.S) will enable us to determine the precise location of the SiO outflow engine relative to QX Pup and the scale at which the wind collimation begins, bringing us to the closest we have ever been to watching live the launch of a bipolar outflow from a mass-losing star and to understanding how OH 231.8+4.2 assembled its complex nebular architecture.

Author Contributions: C.S.C. has written this paper. All listed coauthors participated in the design of the observations, analysis and iterpretation of the data, and have contributed with comments to this manuscript.

Funding: Supported by the Spanish MINECO through grants AYA2012-32032, AYA2016-75066-C2-1-P, and AYA2016-78994-P and by the European Research Council through grant ERC-610256: NANOCOSMOS.

Conflicts of Interest: The authors declare no conflict of interest.

References

1. Maercker, M.; Mohamed, S.; Vlemmings, W.H.; Ramstedt, S.; Groenewegen, M.A.; Humphreys, E.; Kerschbaum, F.; Lindqvist, M.; Olofsson, H.; Paladini, C.; et al. Unexpectedly large mass loss during the thermal pulse cycle of the red giant star R Sculptoris. *Nature* **2012**, *490*, 232–234. [CrossRef] [PubMed]
2. Balick, B.; Frank, A. Shapes and Shaping of Planetary Nebulae. *Annu. Rev. Astron. Astrophys.* **2002**, *40*, 439–486. [CrossRef]
3. Soker, N. Formation of Bipolar Lobes by Jets. *Astrophys. J.* **2002**, *568*, 726–732. [CrossRef]
4. Frank, A.; Blackman, E.G. Application of Magnetohydrodynamic Disk Wind Solutions to Planetary and Protoplanetary Nebulae. *Astrophys. J.* **2004**, *614*, 737–744. [CrossRef]
5. Van Winckel, H. Post-AGB Stars. *Annu. Rev. Astron. Astrophys.* **2003**, *41*, 391–427. [CrossRef]
6. Bujarrabal, V.; Alcolea, J.; Van Winckel, H.; Santander-García, M.; Castro-Carrizo, A. Extended rotating disks around post-AGB stars. *Astron. Astrophys.* **2013**, *557*, A104. [CrossRef]
7. Bujarrabal, V.; Neri, R.; Alcolea, J.; Kahane, C. Detection of an orbiting gas disk in the Red Rectangle. *Astron. Astrophys.* **2003**, *409*, 573–580. [CrossRef]
8. Bujarrabal, V.; Castro-Carrizo, A.; Alcolea, J.; Van Winckel, H.; Sánchez Contreras, C.; Santander-García, M.; Neri, R.; Lucas, R. ALMA observations of the Red Rectangle, a preliminary analysis. *Astron. Astrophys.* **2013**, *557*, L11. [CrossRef]
9. Bujarrabal, V.; Castro-Carrizo, A.; Alcolea, J.; Santander-García, M.; van Winckel, H.; Sánchez Contreras, C. Further ALMA observations and detailed modeling of the Red Rectangle. *Astron. Astrophys.* **2016**, *593*, A92. [CrossRef] [PubMed]
10. Bujarrabal, V.; Castro-Carrizo, A.; Alcolea, J.; Van Winckel, H.; Sánchez Contreras, C.; Santander-García, M. A second post-AGB nebula that contains gas in rotation and in expansion: ALMA maps of IW Carinae. *Astron. Astrophys.* **2017**, *597*, L5. [CrossRef]

11. Bujarrabal, V.; Castro-Carrizo, A.; Van Winckel, H.; Alcolea, J.; Sánchez Contreras, C.; Santander-García, M.; Hillen, M. High-resolution observations of IRAS 08544-4431. Detection of a disk orbiting a post-AGB star and of a slow disk wind. *Astron. Astrophys.* **2018**, *614*, A58. [CrossRef] [PubMed]

12. Bujarrabal, V.; Van Winckel, H.; Neri, R.; Alcolea, J.; Castro-Carrizo, A.; Deroo, P. The nebula around the post-AGB star 89 Herculis. *Astron. Astrophys.* **2007**, *468*, L45–L48. [CrossRef]

13. Bujarrabal, V.; Castro-Carrizo, A.; Alcolea, J.; Van Winckel, H. Detection of Keplerian dynamics in a disk around the post-AGB star AC Herculis. *Astron. Astrophys.* **2015**, *575*, L7. [CrossRef]

14. Kervella, P.; Homan, W.; Richards, A.M.S.; Decin, L.; McDonald, I.; Montargès, M.; Ohnaka, K. ALMA observations of the nearby AGB star L2 Puppis. I. Mass of the central star and detection of a candidate planet. *Astron. Astrophys.* **2016**, *596*, A92. [CrossRef]

15. Homan, W.; Richards, A.; Decin, L.; Kervella, P.; de Koter, A.; McDonald, I.; Ohnaka, K. ALMA observations of the nearby AGB star L2 Puppis. II. Gas disk properties derived from ^{12}CO and ^{13}CO $J = 3 - 2$ emission. *Astron. Astrophys.* **2017**, *601*, A5. [CrossRef]

16. Bujarrabal, V.; Castro-Carrizo, A.; Alcolea, J.; Sánchez Contreras, C. Mass, linear momentum and kinetic energy of bipolar flows in protoplanetary nebulae. *Astron. Astrophys.* **2001**, *377*, 868–897. [CrossRef]

17. Alcolea, J.; Bujarrabal, V.; Sánchez Contreras, C.; Neri, R.; Zweigle, J. The highly collimated bipolar outflow of OH 231.8+4.2. *Astron. Astrophys.* **2001**, *373*, 932–949. [CrossRef]

18. Bujarrabal, V.; Alcolea, J.; Sánchez Contreras, C.; Sahai, R. HST observations of the protoplanetary nebula OH 231.8+4.2: The structure of the jets and shocks. *Astron. Astrophys.* **2002**, *389*, 271–285. [CrossRef]

19. Sánchez Contreras, C.; Bujarrabal, V.; Miranda, L.F.; Fernández-Figueroa, M.J. Optical long-slit spectroscopy and imaging of OH 231.8+4.2. *Astron. Astrophys.* **2000**, *355*, 1103–1114.

20. Sánchez Contreras, C.; Desmurs, J.F.; Bujarrabal, V.; Alcolea, J.; Colomer, F. Submilliarcsecond-resolution mapping of the 43 GHz SiO maser emission in the bipolar post-AGB nebula OH231.8+4.2. *Astron. Astrophys.* **2002**, *385*, L1–L4. [CrossRef]

21. Sánchez Contreras, C.; Gil de Paz, A.; Sahai, R. The Companion to the Central Mira Star of the Protoplanetary Nebula OH 231.8+4.2. *Astrophys. J.* **2004**, *616*, 519–524. [CrossRef]

22. Matsuura, M.; Chesneau, O.; Zijlstra, A.A.; Jaffe, W.; Waters, L.B.F.M.; Yates, J.A.; Lagadec, E.; Gledhill, T.; Etoka, S.; Richards, A.M.S. The Compact Circumstellar Material around OH 231.8+4.2. *Astrophys. J.* **2006**, *646*, L123–L126. [CrossRef]

23. Forde, K.P.; Gledhill, T.M. Discovery of shocked H2 around OH 231.8+4.2. *Mon. Not. R. Astron. Soc.* **2012**, *421*, L49–L53. [CrossRef]

24. Balick, B.; Frank, A.; Liu, B.; Huarte-Espinosa, M. Models of the Hydrodynamic Histories of Post-AGB Stars. I. Multiflow Shaping of OH 231.8+04.2. *Astrophys. J.* **2017**, *843*, 108. [CrossRef]

25. Sánchez Contreras, C.; Alcolea, J.; Bujarrabal, V.; Castro-Carrizo, A.; Velilla Prieto, L.; Santander García, M.; Quintana-Lacaci, G.; Cernicharo, J. Through the magnifying glass: ALMA acute viewing of the intricate nebular architecture of OH231.8+4.2. *Astron. Astrophys.* **2018**, forthcoming. [CrossRef]

galaxies

MDPI

Article

Understanding the Spatial Distributions of the Ionic/Atomic/Molecular/Dust Components in PNe

Toshiya Ueta [1,*], Masaaki Otsuka [2,†] and the HerPlaNS consortium

[1]   Department of Physics & Astronomy, University of Denver, 2112 E. Wesley Ave., Denver, CO 80208, USA
[2]   Institute of Astronomy and Astrophysics, Academia Sinica, P.O. Box 23-141, Taipei 10617, Taiwan;
       otsuka@kusastro.kyoto-u.ac.jp
*   Correspondence: toshiya.ueta@du.edu
†   Current address: Okayama Observatory, Kyoto University, Honjo, Kamogata, Asakuchi,
       Okayama 719-0232, Japan.

Received: 4 December 2018; Accepted: 29 December 2018; Published: 4 January 2019

Abstract: Planetary nebulae (PNe) are often recognized as the hallmark of compact H II regions in the Universe. However, there exist dusty neutral regions extending beyond the central ionized region. We demonstrate that such dusty neutral regions (also known as photo-dissociation regions, or PDRs) around the central ionized region are significant parts of PNe in terms of energetics and mass. We do so by using our latest dusty photoionization model of NGC 6781 (of 13 parameters) based on one of the most comprehensive panchromatic data sets ever assembled for a PN encompassing from X-ray to radio (of 136 constraining data, including 19 flux densities, 78 line fluxes, and 37 band fluxes). We find that NGC 6781, evolved out of a 2.25–3.0 $M_\odot$ star located 460 pc away from us, possesses a massive concentration of neutral gas (molecular hydrogen) just beyond the central ionized region and that the amount of ionized gas in NGC 6781 is only 22% of the observationally accounted amount of matter in the circumstellar environment, which itself does not even account for the amount of mass presumably ejected by the central star during the last thermal pulse event according to the latest evolutionary models. This means that the observed nebula in this PN is only the tip of the iceberg.

Keywords: dust; extinction; ISM: abundances; planetary nebulae: individual (NGC 6781)

---

1. Herschel Planetary Nebula Survey

The life cycle of matter in the Universe is synonymous with the stellar evolution, because the chemical evolution of galaxies is made possible by stellar mass loss that would expel nucleosynthesized matter into the interstellar environments. Planetary nebulae (PNe) are low to intermediate initial mass stars that have completed mass loss during the preceding asymptotic giant branch (AGB) phase. PNe, consisting of a hot central star (>30,000 K; evolving to become a white dwarf) and an extensive circumstellar shell, are in the evolutionary stage during which the object becomes the most luminous and the circumstellar shell reaches its largest extent before the shell begins to dissipate into the ISM.

While PNe are famous for their spectacular circumstellar structures seen often via bright optical emission lines arising from the ionized gas component of the nebula, the ionized part of PNe is surrounded by the neutral gas and dust components of lower temperature (i.e., the photo-dissociated region, or the PDR). Hence, each of these ionic/atomic/molecular/dust components in the circumstellar shell of a PN contains variable clues about the history of mass loss from the central star. By investigating spatially extended emission from each of the ionized, atomic, and molecular gas and dust components of PNe, one can infer ionic, elemental, and molecular/dust abundances and the mass loss and evolutionary histories of the central star. Therefore, PNe provide unique laboratories to further our understanding of the stellar evolution and the chemical evolution of galaxies.

Historically, there is a wealth of archival PN data in the UV and optical. The bright ionized gas in PNe is also bright in the radio continuum. With the advent of new technologies, PN observations in the X-ray and IR follow suit. A new window of opportunity in the far-IR is opened lately by a suite of space telescopes. We conduct the Herschel Planetary Nebula Survey (HerPlaNS) [1] and its follow-up archival study, HerPlaNS+, using PN data collected with the Herschel Space Observatory [2].

2. NGC 6781

NGC 6781 is a PN exhibiting its almost pole-on bipolar nebula structure via its optical, near-IR, and radio images [3–5] as well as our own far-IR images [1], as shown in Figure 1. Because the observed distributions of emission at different wavelengths representing distinct nebular components appear very similar to one another, a highly steep temperature gradient (and hence, density gradient) is expected in the observed parts of the circumstellar nebula. This probable presence of a steep density gradient is actually consistent with expectations from two-wind interactions, in which fast wind emanating from the central star during the PN phase catches up with the slower-moving AGB wind, interacting and piling up at the interface between the two winds.

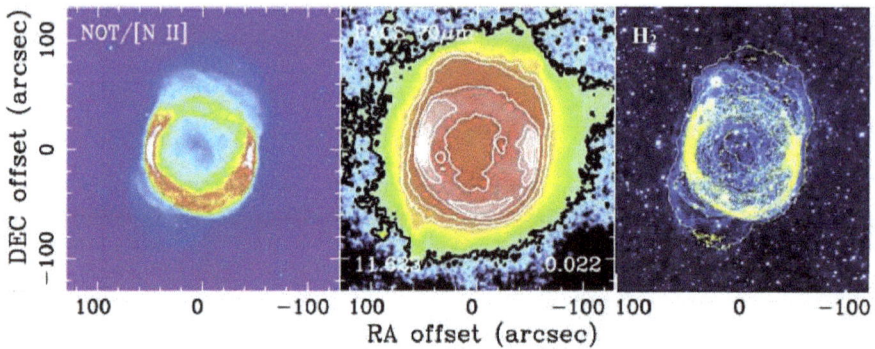

Figure 1. NGC 6781 in the optical (**left**, in [N II]; [3]), far-IR (**middle**, in dust continuum at 70 μm; [1]), and near-IR (**right**, in H_2; [4] in blue hue; yellow contours are [N II] emission) showing almost co-spatial distributions of ionized gas, dust grains, and molecular hydrogen, from left to right, respectively.

Our own observations of NGC 6781 in the far-IR with Herschel [1] nearly complete the full spectral coverage for this PN, providing one of the most comprehensive panchromatic data set ever assembled for a PN. By taking advantage of this extensive data set, we set out to generate a coherent model of NGC 6781 that satisfies the adopted panchromatic data as comprehensively as possible. To this end, we derive the empirical characteristics of the central star and its nebula with a greater amount of self-consistency, and use the derived quantities as input parameters and/or constraints to construct a dusty photoionization model consisting of ionized, atomic, and molecular gas plus dust grains ([6] to which readers are encouraged to refer, as a fair amount of details is left out from this contribution).

2.1. Plasma Diagnostics

We determine the electron density and temperature (n_e and T_e) of the ionized gas component of NGC 6781 from 9 diagnostic lines based on 15 line ratios computed from 28 line fluxes of collisionally excited and recombination lines out of 81 individual lines measured from the adopted spectra of the object. By iterating the n_e-T_e diagnostics, we compute n_e first while fixing $T_e = 10,000$ K, and re-compute T_e using the derived n_e for lines of similar excitation energies.

These diagnostics yield n_e between 100 and 10,000 cm^{-3} and T_e between 7070 and 10,800 K. Also, diagnostics of mid-IR H_2 thermal emission lines, guided by resolved images, suggest that the inner cavity of the bipolar nebula is 300 cm^{-3}, the nebula waist is 960 cm^{-3}, and there is a physically thin pile-up of molecular gas at 10,000 cm^{-3} surrounding the nebula waist.

2.2. Abundance Analyses

The measured line fluxes yield ionic abundances for 19 species, from which elemental abundances for 9 species are determined. In comparison with abundance patterns of theoretical AGB models [7,8], the derived abundances suggest that the central star of NGC 6781 was initially a 2.25–3.0 M$_\odot$ star and that Si and S depletion is consistent with the presence of dust grains.

2.3. Properties of the Central Star

The measured photometry of the central star is fit with a grid of synthesized spectra [9] to determine the luminosity (L_*) as a function of the distance (D) and the effective temperature of the star (T_{eff}). Also in comparison with the AGB nucleosynthesis models [7,8], we conclude that the best fit quantities for NGC 6781 are 0.46 kpc, 104–196 L$_\odot$, and 10–140 kK. Our luminosity fitting results turn out to be fairly robust as Gaia finds the object to be located at 0.49 kpc [10].

3. Cloudy Best-Fit Modeling of NGC 6781

We perform best-fit dusty photoionization modeling for NGC 6781 using the Cloudy code [11]. The analyses outlined in the previous section yield input model parameters, while the 3-D shape of the nebula is inferred from the apparent nebula morphology seen in resolved images at various wavelengths. To find the best-fit spectral energy distribution (SED) model, we vary 13 parameters—T_{eff}, L_*, the inner radius of the shell (R_{in}), elemental abundances (ϵ(He/N/O/Ne/Si/Cl/Ar), except for ϵ(C) which is fixed), dust and PAH mass fraction, and the minimum temperature of the PDR—within given ranges of respective uncertainties.

We terminate iterations when the predicted flux density at 250 μm reaches the corresponding observed value. Practically, this condition sets the maximum radius of the shell (R_{out}) as this quantity controls the amount of dust in the outer regions of the shell, and hence, the amount of dust continuum in the far-IR. In this sense, R_{out} is not a free parameter. The goodness of the fit is determined by χ^2 (16 for the best-fit model) calculated from the deviation of 136 constraints (37 broadband fluxes, 78 gas emission line fluxes relative to Hβ as well as I(Hβ), 19 flux densities in mid-IR, far-IR, and radio wavelengths, and the ionization boundary radius, R_{IB}) between a model and the adopted data.

4. Highlights of the Results

We briefly highlight a number of notable characteristics that are found in the best-fit model.

4.1. The Presence of the PDR in PNe

As the observed SED clearly indicates, a significant amount of energy output from NGC 6781 happens in the PDR via molecular and dust continuum emission (Figure 2). In the best-fit model, 70% of the dust component is found to exist beyond the ionization boundary, suggesting that the density distribution continues beyond the observed ionized nebula as the PDR. In addition, a model without the PDR is also constructed and is found to underestimate the far-IR fluxes significantly. Therefore, while PNe are generally known to be almost synonymous to compact H II regions in the Universe, they also contain a significant amount of cold atomic and molecular gas and dust grains.

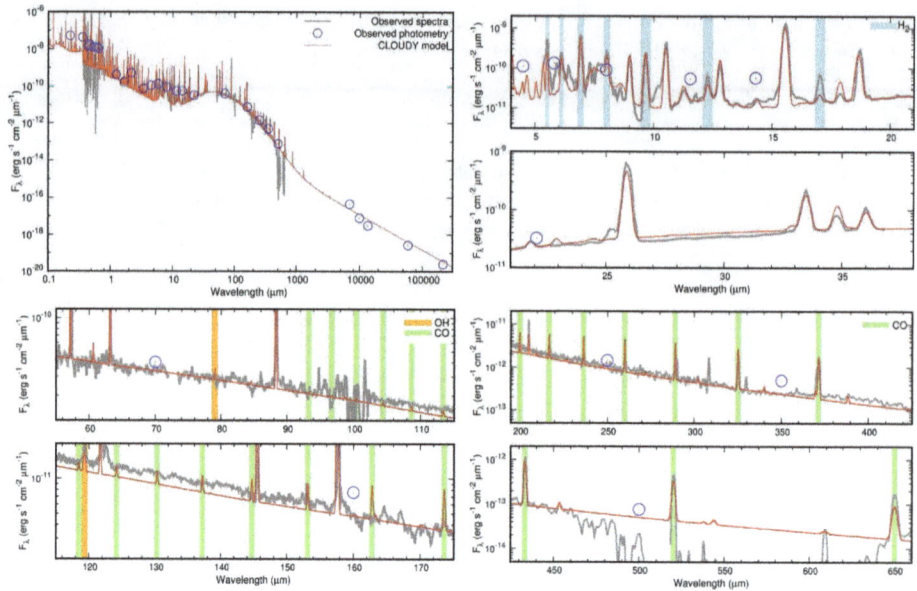

Figure 2. Comparisons between the best-fit model (red line) and the observational constraints including photometry data (blue circles) and spectroscopy data (gray line) for the entire SED ($R = 300$; **top left**), in the mid-IR region for Spitzer/IRS ($R = 100$; **top right**), in the far-IR region for Herschel/PACS ($R = 240$; **bottom left**), and in the far-IR region for Herschel/SPIRE ($R = 240$; **bottom right**). Data include those from GALEX, ING/INT, WHT/ISIS, ESO/NTT, UKIRT, WISE, Spitzer, ISO, Herschel, and various radio telescopes. Molecular lines are highlighted for rotational H_2 lines (light blue), OH (yellow), and ^{12}CO (light green).

4.2. The Amount of Matter in the Circumstellar Nebula

From the best-fit model we can account for the amount of "observed" matter in the nebula. The best-fit model yields that the nebula harbors $0.09 \, M_\odot$ of ionized gas, $0.12 \, M_\odot$ of atomic gas, $0.20 \, M_\odot$ of molecular gas, plus $0.0015 \, M_\odot$ of dust grains, totaling $0.41 \, M_\odot$ of circumstellar matter. Assuming that the central star is a $2.5 \, M_\odot$ initial mass star, it should have evolved to become a white dwarf of $0.63 \, M_\odot$ by now [7]. This implies that we expect $1.87 \, M_\odot$ of circumstellar matter, and hence, means that we observationally account for only 22% of the circumstellar matter expected to exist in the circumstellar nebula of NGC 6781. Implications of this finding are enormous.

AGB evolutionary models predict that a $2.5 \, M_\odot$ initial mass star would experience 25 AGB thermal pulse (TP) episodes while ejecting the total mass of about $1.25 \, M_\odot$. However, the predicted amount of the mass-loss ejecta would remain small ($M > 0.01 \, M_\odot$) until the 22nd TP episode. This means that the amount of the ejecta would increase precipitously over the last three TP episodes, reaching $0.70 \, M_\odot$ during the last TP episode. Therefore, our best-fit model accounts for roughly 60% of the amount of mass theoretically predicted to have been ejected during the last TP episode.

It has been naively expected that the history of mass loss can be learned from the density distribution in the circumstellar shells of PNe and AGB stars. However, the present result indicates that the instrument sensitivities is still not sufficient to probe the history of stellar mass loss, even beyond one TP ejection. Hence, there is still a long way to go to observationally account for the full mass-loss history of these stars. Alternatively, this could mean that theoretical models may need to be re-calibrated in terms of mass-loss prescriptions so that less mass gets ejected.

4.3. Gas-to-Dust Mass Ratio

The integrated gas-to-dust mass ratio (of the last TP) turns out to be 273 (=0.41/0.0015). This value is much smaller than the value of 386 ± 90 found from a group of C-rich AGB stars [12]. Note, however, that for this previous work, the amount of dust was estimated from N-band photometry of warm dust while the amount of gas was derived from CO J = 1-0 measurements of cold gas. This means that these gas and dust components may not have been sampled from the same location in the observed AGB circumstellar shells. So, the ratio may not be as high but certainly higher than 100.

5. Future Prospects

Our HerPlaNS+ archival survey will yield a fair number of PNe showing resolved structures in the far-IR, in dust continuum and/or fine-structure lines of atomic gas. Also, we are conducting an optical 2-D spectroscopy campaign for a handful of PNe for which far-IR PN data already exist. Hence, we can perform similar investigations based on comprehensive panchromatic data sets for a number of objects in near future to see if the observationally accounted amount of circumstellar mass in PNe is systematically lower than theoretical predictions or not. If this trend is found to persist among PNe, it could really mean that theoretical predictions need to be re-calibrated accordingly. Implications of this possibility is significant, as our understanding of the life cycle of matter in the Universe relies on our understanding of the stellar evolution.

Author Contributions: Writing original draft, T.U.; Investigation, M.O.

Funding: Masaaki Otsuka was supported by the research funds 104-2811-M-001-138 and 104-2112-M-001-041-MY3 from the Ministry of Science and Technology (MOST), Toshiya Ueta was partially supported by an award to the original Herschel observing program (OT1_tueta_2) under Research Support Agreement (RSA) 1428128 issued through JPL/Caltech, and by NASA under grant NNX15AF24G issued through the Science Mission Directorate.

Acknowledgments: This work is partly based on observations made with the Herschel Space Observatory, a European Space Agency (ESA) Cornerstone Mission with significant participation by National Aeronautics and Space Administration (NASA).

Conflicts of Interest: The authors declare no conflict of interest.

References

1. Ueta, T.; Ladjal, D.; Exter, K.M.; Otsuka, M.; Szczerba, R.; Siódmiak, N.; Aleman, I.; van Hoof, P.A.M.; Kastner, J.H.; Montez, R., Jr.; et al. The Herschel Planetary Nebula Survey (HerPlaNS) I. Data overview and analysis demonstration with NGC6781. *Astron. Astrophys.* **2014**, *565*, A36. [CrossRef]

2. Pilbratt, G.L.; Riedinger, J.R.; Passvogel, T.; Grone, G.; Doyle, D.; Gageur, U.; Heras, A.M.; Jewell, C.; Metcalfe, L.; Ott, S.; et al. Herschel Space Observatory. An ESA facility for far-infrared and submillimetre astronomy. *Astron. Astrophys.* **2010**, *518*, L1. [CrossRef]

3. Phillips, J.P.; Ramos-Larios, G.; Guerrero, M.A. Optical and mid-infrared observations of the planetary nebula NGC 6781. *Mon. Not. R. Astron. Soc.* **2011**, *415*, 513–524. [CrossRef]

4. Puget, P.; Stadler, E.; Doyon, R.; Gigan, P.; Thibault, S.; Luppino, G.; Barrick, G.; Benedict, T.; Forveille, T.; Rambold, W.; et al. WIRCam: The infrared wide-field camera for the Canada-France-Hawaii Telescope. *Proc. SPIE* **2004**, *5492*, 978.

5. Bachiller, R.; Huggins, P.J.; Cox, P.; Forveille, T. The spatio-kinematic structure of the CO envelopes of evolved planetary nebulae. *Astron. Astrophys.* **1993**, *267*, 177.

6. Otsuka, M.; Ueta, T.; van Hoof, P.A.M.; Sahai, R.; Aleman, I.; Zijlstra, A.A.; Chu, Y.-H.; Villaver, E.; Leal-Ferreira, M.L.; Kastner, J.; et al. The Herschel Planetary Nebula Survey (HerPlaNS): A Comprehensive Dusty Photoionization Model of NGC6781. *Astrophys. J. Suppl. Ser.* **2017**, *231*, 22. [CrossRef] [PubMed]

7. Karakas, A.I. Updated stellar yields from asymptotic giant branch models. *Mon. Not. R. Astron. Soc.* **2010**, *403*, 1413–1425. [CrossRef]

8. Vassiliadis, E.; Wood, P.R. Evolution of low- and intermediate-mass stars to the end of the asymptotic giant branch with mass loss. *Astrophys. J.* **1993**, *413*, 641–657. [CrossRef]

9. Rauch, T. A grid of synthetic ionizing spectra for very hot compact stars from NLTE model atmospheres. *Astron. Astrophys.* **2003**, *403*, 709. [CrossRef]

10. Brown, A.G.A.; Vallenari, A.; Prusti, T.; de Bruijne, J.H.J.; Babusiaux, C.; Bailer-Jones, C.A.L.; Biermann, M.; Evans, D.W.; Eyer, L.; Jansen, F.; et al. Gaia Data Release 2. Summary of the contents and survey properties. *arXiv* **2018**, arXiv:1804.09365.

11. Ferland, G.J.; Porter, R.L.; van Hoof, P.A.M.; Williams, R.J.R.; Abel, N.P.; Lykins, M.L.; Shaw, G.; Henney, W.J.; Stancil, P.C. The 2013 Release of Cloudy. *Revista Mexicana Astronomía Astrofísica* **2013**, *49*, 137–163.

12. Knapp, G.R. Mass loss from evolved stars. IV—The dust-to-gas ratio in the envelopes of Mira variables and carbon stars. *Astrophys. J.* **1985**, *293*, 273–280. [CrossRef]

galaxies

MDPI

Article

On the Origin of Morphological Structures of Planetary Nebulae

Sun Kwok [1,2]

[1]   Department of Earth, Ocean, and Atmospheric Sciences, University of British Columbia, Vancouver,
      BC V6T 1Z4, Canada; skwok@eoas.ubc.ca or sunkwok@hku.hk; Tel.: +1-778-858-5752
[2]   Laboratory for Space Research, The University of Hong Kong, Hong Kong, China

Received: 24 May 2018; Accepted: 22 June 2018; Published: 26 June 2018

Abstract: We suggest that most of the mass in planetary nebulae (PNe) resides in the equatorial region and the spherical envelope and the optically bright lobes of PNe are in fact low-density cavities cleared out by fast outflows and photoionized by UV photons leaked from the torus. The nature of multi-polar PNe is discussed under this framework.

Keywords: planetary nebulae; asymptotic giant branch stars; mass loss

1. Introduction

The diverse morphological shapes of planetary nebulae (PNe) were recognized since the beginning of PNe research [1]. Does this diversity suggest that every PN is different, or is there an intrinsic structure common to all PNe? The possibility that there is a uniform 3-D structure of PNe was explored by models such as an open-ended toroid [2] or an ellipsoidal shell [3] inclined w.r.t. the line of sight. Ionization penetrating to different depths could create the different apparent shapes of PNe [4].

A separate question is what is the physical mechanism that creates such non-spherically symmetric structures? The realization that PNe are formed by the process of interacting winds [5] led to the suggestion that nebular shaping can be achieved by asymmetry in one or more of the stellar winds [6].

Observations with CCD cameras, space-based observations, and imaging beyond the visible wavelengths, have greatly expanded our capabilities of mapping the morphological structures of PN. In this paper, we discuss whether these new observations can lead to a new paradigm in our understanding of the origin of morphological structures of PNe.

2. Problems with Morphological Classifications

Our morphological classifications of PNe are primarily based on the apparent morphological shapes of optical images of PNe. These classification schemes suffer from the following problems:

(i).   Orientation effects (PNe are 3D but images are 2D). A near-pole-on bipolar nebula may appear
        as a ring because of the prominence of the torus. Many well-known PNe (the Ring, the Helix,
        the Dumbbell, NGC 7027) turn out to have similar intrinsic bipolar structures [7–10].

(ii).   Dynamic range: if we go deep enough we see more/different structures. The well-known Messier
        PN M76 turned out to be bipolar when observed by CCD detectors, when earlier photographic
        plate images only showed the central torus. The bipolar lobes of Sh 1-89 were only found with
        narrow-band CCD imaging [11].

(iii). Limited field of view (FoV). We may miss large outer structures because the limited FoV of
        observations (examples: IPHAS PN-1 [12], M1-41 [13]).

(iv).   limited wavelength coverage. Bipolar structures not obvious in visible images may reveal
        themselves in the infrared [14].

(v). Internal dust extinction. Optical morphology of PNe may be affected by effects of circumstellar dust extinction [15].

3. Multipolar Nebulae

As the result of high dynamic-range imaging, more and more PNe are found to be multipolar [16,17]. *HST* observations have found that many compact PNe have multi-polar lobes [18]. Interestingly, many of the multi-polar lobes are roughly equal in length [19,20]. Some PNe with prominent bipolar lobes also show secondary multi-polar lobes in other directions [21]. The existence of multi-polar nebulae suggests collimated ejections formed simultaneously or episodically [18]. Precessing bipolar, rotating episodic jets have been proposed to explain the observed morphology [22].

What is the fraction of multi-polar PNe among the PNe population? Current statistics suggests a range between 10–20% [23,24], but these are almost certainly lower limits as high-dynamic-range imaging of PNe has only been done for a small fraction of PNe. If a majority of PNe are multi-polar, what are the consequences? We have performed an exercise assuming that PNe have 3 pairs of identical-length lobes oriented at random directions and simulate their apparent images when observed from different lines of sight to different degrees of sensitivity. We found that many different observed morphological shapes of PNe can be reproduced with this simulation. For example, tori and double inner–outer bipolar lobes are the result of overlapping multi-polar lobes, and ansae and knots may be the bright tips of undetected lobes [25]. Nearly aligned pairs of lobes can give the appearance of point-symmetric S-shape morphology, which does not necessarily imply precession. It is remarkable that morphology of very diverse PNe can be simulated by a single, very simple 3-D model when orientation and sensitivity effects are taken into account.

In the past, we have relied on slit spectroscopy to infer the kinematic structure of PNe. The best way to test the multipolar hypothesis is through integral field spectroscopy. The velocity maps from such observations can be compared with 3-D spatial-kinematic models to determine the true intrinsic structure of PNe.

4. Unseen "Dark" Matter in Planetary Nebulae

Many bipolar PNe (e.g., NGC 6302, NGC 2346) have very tight waists, suggesting that they are being confined by unseen external material. Other bipolar PNe (e.g., IC 4406, Hen 3-401, Hen 2-320) show sharp, well-defined lobe boundaries, suggesting that the optical lobes are dynamically confined by external media (Figure 1). This unseen matter is probably in the form of molecular gas and can be traced by molecular-line imaging. Mapping of the 2.12 μm H_2 line requires the molecule to be highly excited by shocks or UV radiation so this line can map only dynamically interacting regions but is not a good tracer of cold molecular gas. The H_2 ground-state rotational transition (para $S(0)$ $J = 2$–0) requires less (500 K) excitation but this 28 μm line cannot be observed from the ground [26].

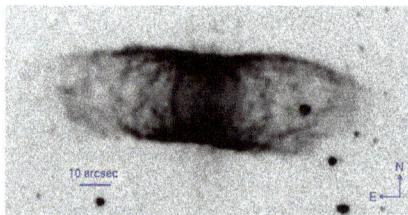

Figure 1.   Continuum-subtracted H_2 2.12 μm line image of IC 4406 obtained with the *Canada-France-Hawaii Telescope*. The bipolar lobes have sharp boundaries and are clearly confined by an external medium.

Current CO line mapping from mm/submm-wave interferometers such as *SMA* and *ALMA* mainly reveals the molecular torus around the waist of bipolar nebulae [27,28]. Mapping of the extended molecular gas will require higher surface brightness sensitivity observations.

The dust mixed with the cold molecular gas is probably too low temperature to be observed by mid-infrared (10–20 µm) imaging. Current mid-infrared imaging of PNe mainly reveals dust in the torus or in the bipolar lobes [29,30]. Although efforts have been made to map the cold dust component by *Spitzer* [31,32], *AKARI* [33], and *Herschel* [34–36], the extended cold dust component be truly revealed with high angular resolution, high sensitivity, far infrared imaging with facilities such as *SOFIA*.

5. Discussion

Many of the present theories on the origin of morphological structures of PNe are based on the assumption that the observed bright nebulosity represents ejection of physical matter. However, the visible brightness of PNe is due to recombination lines of H and He and collisionally excited lines of metals, all strong emissions originating from the ionized region of PNe that only represents a small fraction of the total mass of PNe. We suggest that the bipolar lobes of PNe are not regions of massive ejection, but low-density cavities carved out of the neutral circumstellar envelope [22]. In the AGB stage, a massive circumstellar envelope is created by mass loss via a slow wind. In the post-AGB phase, collimated fast outflows, energetic but not massive, emerge from the central star. These outflows break through less dense regions of the circumstellar torus, creating cavities in the envelope. Dust scattering of visible photons from the central star leads to bipolar and multi-polar nebulosity observed in proto-PNe [37]. In the subsequent PNe phase, UV photons from the now hot central star ionize the low-density cavity, illuminating the bipolar region. Optical morphology of PN is therefore not defined by regions of massive matter ejection, but represents holes in the matter distribution where densities are low enough for UV to ionize. The observable effects of a collimated fast wind interacting with clumpy circumstellar envelope have been explored by Steffen et al. [38].

As central stars of PNe evolve to higher temperatures, their radii contract and the escape velocities increase. These effects lead to higher power fast winds, hotter shocked bubbles, and higher thermal pressure [39]. The dynamical effects of interacting winds may wash out the multi-polar structures observed in younger PNe, leading to a lower frequency of multi-polar structures in evolved PNe.

As PNe evolve and densities drop, the ionization front will eventually breakthrough and the nebulae becomes density bounded. The morphology of the PNe will also become more spherical. The low-surface-brightness spherically-shaped nebulae found in deep Hα surveys [40,41] are probably examples of density-bounded, highly evolved PNe. The fraction of PNe that is bipolar is therefore dependent on age.

6. Conclusions

Recent high-dynamic-range optical and infrared observations have revealed that many PNe previously classified as round or elliptical are in fact bipolar. Statistical analysis of morphological classes based on apparent shapes is therefore not reliable and the true 3-D structures PNe can only be found by proper modeling of the brightness and kinematic structures. We suggest that multipolar nebulae are much more common than currently believed and more PNe will show multi-polar structure with deeper imaging. Some of the morphological features such as tori, filaments, knots, and ansae could be manifestation of multi-polar nature of the objects. The optical lobes of bipolar/multipolar PNe and proto-PNe are not volumes of concentrated mass but low-density cavities ionized or illuminated by the central star. The true mass distribution of PNe can only be revealed with future far infrared and mm/submm molecular-line imaging.

Funding: This work is supported by grants from the Research Grants Council of Hong Kong and the Natural Sciences and Engineering Research Council of Canada.

Acknowledgments: S.K. thanks Alberto López for valuable comments on an earlier version of this paper.

Galaxies **2018**, *6*, 66

Conflicts of Interest: The author declares no conflict of interest.

References

1. Curtis, H.D. The Planetary Nebulae. *Publ. Lick Obs.* **1918**, *13*, 55–74.
2. Khromov, G.S.; Kohoutek, L. Morphological Study of Planetary Nebulae. In *Symposium-International Astronomical Union*; Cambridge University Press: Cambridge, UK, 1968; Volume 34, p. 227.
3. Masson, C.R. On the Structure of Ionization-Bounded Planetary-Nebulae. *Astrophys. J.* **1990**, *348*, 580–587. [CrossRef]
4. Zhang, C.Y.; Kwok, S. A Morphological Study of Planetary Nebulae. *Astrophys. J. Suppl. Ser.* **1998**, *117*, 341. [CrossRef]
5. Kwok, S. From red giants to planetary nebulae. *Astrophys. J.* **1982**, *258*, 280–288. [CrossRef]
6. Balick, B. The evolution of planetary nebulae. I—Structures, ionizations, and morphological sequences. *Astron. J.* **1987**, *94*, 671–678. [CrossRef]
7. Bryce, M.; Balick, B.; Meaburn, J. Investigating the Haloes of Planetary Nebulae—Part Four—NGC6720 the Ring Nebula. *Mon. Not. R. Astron. Soc.* **1994**, *266*, 721–732. [CrossRef]
8. Meaburn, J.; Boumis, P.; López, J.A.; Harman, D.J.; Bryce, M.; Redman, M.P.; Mavromatakis, F. The creation of the Helix planetary nebula (NGC 7293) by multiple events. *Mon. Not. R. Astron. Soc.* **2005**, *360*, 963–973. [CrossRef]
9. Kwok, S.; Chong, S.-N.; Koning, N.; Hua, T.; Yan, C.-H. The True Shapes of the Dumbbell and the Ring. *Astrophys. J.* **2008**, *689*, 219–224. [CrossRef]
10. Latter, W.B.; Dayal, A.; Bieging, J.H.; Meakin, C.; Hora, J.L.; Kelly, D.M.; Tielens, A.G.G.M. Revealing the Photodissociation Region: HST/NICMOS Imaging of NGC 7027. *Astrophys. J.* **2000**, *539*, 783–797. [CrossRef]
11. Hua, C.T. Deep morphologies of type I planetary nebulae. *Astron. Astrophys. Suppl. Ser.* **1997**, *125*, 355–366. [CrossRef]
12. Mampaso, A.; Corradi, R.L.; Viironen, K.; Leisy, P.; Greimel, R.; Drew, J.E.; Barlow, M.J.; Frew, D.J.; Irwin, J.; Morris, R.A.H.; et al. The "Príncipes de Asturias" nebula: A new quadrupolar planetary nebula from the IPHAS survey. *Astron. Astrophys.* **2006**, *458*, 203–212. [CrossRef]
13. Zhang, Y.; Hsia, C.-H.; Kwok, S. Planetary Nebulae Detected in the Spitzer Space Telescope GLIMPSE 3D Legacy Survey. *Astrophys. J.* **2012**, *745*, 59. [CrossRef]
14. Zhang, Y.; Kwok, S. Planetary Nebulae Detected in the Spitzer Space Telescope GLIMPSE II Legacy Survey. *Astrophys. J.* **2009**, *706*, 252–305. [CrossRef]
15. Lee, T.-H.; Kwok, S. Dust Extinction in Compact Planetary Nebulae. *Astrophys. J.* **2005**, *632*, 340–354. [CrossRef]
16. Manchado, A.; Stanghellini, L.; Guerrero, M.A. Quadrupolar Planetary Nebulae: A New Morphological Class. *Astrophys. J.* **1996**, *466*, L95. [CrossRef]
17. Lopez, J.A.; Meaburn, J.; Bryce, M.; Holloway, A.J. The Morphology and Kinematics of the Complex Polypolar Planetary Nebula NGC 2440. *Astrophys. J.* **1998**, *493*, 803–810. [CrossRef]
18. Hsia, C.-H.; Chau, W.; Zhang, Y.; Kwok, S. Hubble Space Telescope Observations and Geometric Models of Compact Multipolar Planetary Nebulae. *Astrophys. J.* **2014**, *787*, 25. [CrossRef]
19. Sahai, R. The starfish twins: Two young planetary nebulae with extreme multipolar morphology. *Astrophys. J.* **2000**, *537*, L43–L47. [CrossRef]
20. Sahai, R.; Nyman, L.-Å.; Wootten, A. HE 2-113: A Multipolar Planetary Nebula with Rings around a Cool Wolf-Rayet Star. *Astrophys. J.* **2000**, *543*, 880–888. [CrossRef]
21. López, J.A.; García-Díaz, M.T.; Steffen, W.; Riesgo, H.; Richer, M.G. Morpho-kinematic Analysis of the Point-symmetric, Bipolar Planetary Nebulae Hb 5 and K 3-17, A Pathway to Poly-polarity. *Astrophys. J.* **2012**, *750*, 131. [CrossRef]
22. Lopez, J.A.; Vazquez, R.; Rodriguez, L.F. The Discovery of a Bipolar, Rotating, Episodic Jet (BRET) in the Planetary Nebula KjPn 8. *Astrophys. J.* **1995**, *455*, L63–L66. [CrossRef]
23. Manchado, A.; García-Hernández, D.A.; Villaver, E.; de Massas, J.G. Morphological Classification of Post-AGB Stars. In *Why Galaxies Care about AGB Stars II: Shining Examples and Common Inhabitants*; Kerschbaum, F., Ed.; Astronomical Society of the Pacific: San Francisco, CA, USA, 2011; p. 161.

24. Sahai, R.; Morris, M.R.; Villar, G.G. Young Planetary Nebulae: Hubble Space Telescope Imaging and a New Morphological Classification System. *Astron. J.* **2011**, *141*, 134. [CrossRef]
25. Chong, S.-N.; Kwok, S.; Imai, H.; Tafoya, D.; Chibueze, J. Multipolar Planetary Nebulae: Not as Geometrically Diversified as Thought. *Astrophys. J.* **2012**, *760*, 115. [CrossRef]
26. Kwok, S. *Physics and Chemistry of the Interstellar Medium*; University Science Books; Cambridge University Press: Cambridge, UK, 2006.
27. Trung, D.-V.; Bujarrabal, V.; Castro-Carrizo, A.; Lim, J.; Kwok, S. Massive Expanding Torus and Fast Outflow in Planetary Nebula NGC 6302. *Astrophys. J.* **2008**, *673*, 934–941. [CrossRef]
28. Santander-García, M.; Bujarrabal, V.; Alcolea, J.; Castro-Carrizo, A.; Contreras, C.S.; Quintana-Lacaci, G.; Corradi, R.L.M.; Neri, R. ALMA high spatial resolution observations of the dense molecular region of NGC 6302. *Astron. Astrophys.* **2017**, *597*, A27. [CrossRef] [PubMed]
29. Muthumariappan, C.; Kwok, S.; Volk, K. Subarcsecond Mid-Infrared Imaging of Dust in the Bipolar Nebula Hen 3-401. *Astrophys. J.* **2006**, *640*, 353–359. [CrossRef]
30. Volk, K.; Hrivnak, B.J.; Su, K.Y.L.; Kwok, S. An Infrared Imaging Study of the Bipolar Proto-Planetary Nebula IRAS 16594-4656. *Astrophys. J.* **2006**, *651*, 294–300. [CrossRef]
31. Su, K.Y.L.; Kelly, D.M.; Latter, W.B.; Misselt, K.A.; Frank, A.; Volk, K.; Engelbracht, C.W.; Gordon, K.D.; Hines, D.C.; Morrison, J.E.; et al. High spatial resolution mid- and far-infrared imaging study of NGC 2346. *Astrophys. J. Suppl. Ser.* **2004**, *154*, 302–308. [CrossRef]
32. Ueta, T. Spitzer MIPS Imaging of NGC 650: Probing the History of Mass Loss on the Asymptotic Giant Branch. *Astrophys. J.* **2006**, *650*, 228–236. [CrossRef]
33. Izumiura, H.; Ueta, T.; Yamamura, I.; Nakada, Y.; Matsunaga, N.; Ita, Y.; Matsuura, M.; Fukushi, H.; Mito, H.; Tanabe, T. Far-IR Imaging Survey of the Extended Dust Shells of Evolved Stars with AKARI. In *AKARI, a Light to Illuminate the Misty Universe*; Onaka, T., Ed.; Astronomical Society of the Pacific: San Francisco, CA, USA, 2009; p. 127.
34. Ueta, T.; Ladjal, D.; Exter, K.M.; Otsuka, M.; Szczerba, R.; Siódmiak, N.; Aleman, I.; van Hoof, P.A.M.; Kastner, J.H.; Montez, R.; et al. The Herschel Planetary Nebula Survey (HerPlaNS)-I. Data overview and analysis demonstration with NGC 6781. *Astron. Astrophys.* **2014**, *565*, A36. [CrossRef]
35. Van Hoof, P.A.M.; Van de Steene, G.C.; Exter, K.M.; Barlow, M.J.; Ueta, T.; Groenewegen, M.A.T.; Gear, W.K.; Gomez, H.L.; Hargrave, P.C.; Ivison, R.J.; et al. A Herschel study of NGC 650. *Astron. Astrophys.* **2013**, *560*, A7. [CrossRef]
36. Van de Steene, G.C.; van Hoof, P.A.M.; Exter, K.M.; Barlow, M.J.; Cernicharo, J.; Etxaluze, M.; Gear, W.K.; Goicoechea, J.R.; Gomez, H.L.; Groenewegen, M.A.T.; et al. Herschel imaging of the dust in the Helix nebula (NGC 7293). *Astron. Astrophys.* **2015**, *574*, A134. [CrossRef]
37. Koning, N.; Kwok, S.; Steffen, W. Post Asymptotic Giant Branch Bipolar Reflection Nebulae: Result of Dynamical Ejection or Selective Illumination? *Astrophys. J.* **2013**, *765*, 92. [CrossRef]
38. Steffen, W.; Koning, N.; Esquivel, A.; García-Segura, G.; García-Díaz, M.T.; López, J.A.; Magnor, M. A wind-shell interaction model for multipolar planetary nebulae. *Mon. Not. R. Astron. Soc.* **2013**, *436*, 470–478. [CrossRef]
39. Schönberner, D.; Jacob, R.; Lehmann, H.; Hildebrandt, G.; Steffen, M.; Zwanzig, A.; Sandin, C.; Corradi, R.L.M. A hydrodynamical study of multiple-shell planetary nebulae. III. Expansion properties and internal kinematics: Theory versus observation. *Astron. Nachr.* **2014**, *335*, 378–408. [CrossRef]
40. Parker, Q.A.; Phillipps, S.; Pierce, M.J.; Hartley, M.; Hambly, N.C.; Read, M.A.; MacGillivray, H.T.; Tritton, S.B.; Cass, C.P.; Cannon, R.D.; et al. The AAO/UKST superCOSMOS Hα survey. *Mon. Not. R. Astr. Soc.* **2005**, *362*, 689–710. [CrossRef]
41. Parker, Q.A.; Acker, A.; Frew, D.J.; Hartley, M.; Peyaud, A.E.J.; Ochsenbein, F.; Phillipps, S.; Russeil, D.; Beaulieu, S.F.; Cohen, M.; et al. The Macquarie/AAO/Strasbourg Hα planetary nebula catalogue: MASH. *Mon. Not. R. Astr. Soc.* **2006**, *373*, 79–94. [CrossRef]

galaxies

MDPI

Article

AGBs, Post-AGBs and the Shaping of Planetary Nebulae

Eric Lagadec

Observatoire de la Côte d'Azur, Laboratoire Lagrange, Université Côte d'Azur, Nice 06304, France;
elagadec@oca.eu

Received: 31 July 2018; Accepted: 29 August 2018; Published: 17 September 2018

Abstract: During the last decades, observations, mostly with the Hubble Space Telescope, have revealed that round Planetary Nebulae were the exception rather than rule. A huge variety of features are observed, such as jets, discs, tori, showing that the ejection of material is not due to isotropic radiation pressure on a spherical shell and that more physics is involved. This shaping process certainly occur early in the evolution of these low and intermediate mass stars and must leave imprints in the evolutionary stages prior the PN phase. Thanks to news instruments on the most advanced telescopes (e.g., the VLTI, SPHERE/VLT and ALMA), high angular resolution observations are revolutionising our view of the ejection of gas and dust during the AGB and post-AGB phases. In this review I will present the newest results concerning the mass loss from AGB stars, post-AGB stars and related objects.

Keywords: AGB stars; post-AGB stars; planetary nebulae

1. Introduction

The aim of this asymmetrical planetary conference series, started in 1994, has been to understand the cause of the spectacular morphologies displayed by planetary nebulae (PNe). It appears fairly clearly now that the departure from spherical symmetry observed for PNe is due to an extra momentum brought by a binary companion [1–4]. As it appears clear that a large fraction of PNe is being shaped by binaries, sign of interactions with companions on previous stages of stellar evolution (AGB and post-AGB) should be observable. This is made difficult by the fact that these objects are more compact and often embedded in dust. Fortunately, we now have instruments able to probe the very close environments of such stars, and even to map their surfaces. In the course of this conference series, we indeed switched from subarcsec resolution observations to the milliarcsec era. Figure 1 (courtesy of Pierre Kervella) displays the resolution achieved by some of the main current instruments at different wavelengths. We can now probe regions between 1 and 20 milliarcsec in size from the optical to the submillimetre domain. For a star at 100 parsec, this means we can map material as close 0.1 AU of the central star, i.e., that we can map the surfaces of nearby giant stars.

In this review I will show how these high angular resolution instruments are revolutionising our view of AGB and post-AGB stars and present some high-angular resolution observations of related (massive) objects that can help us understand the shaping of PNe.

Angular resolution

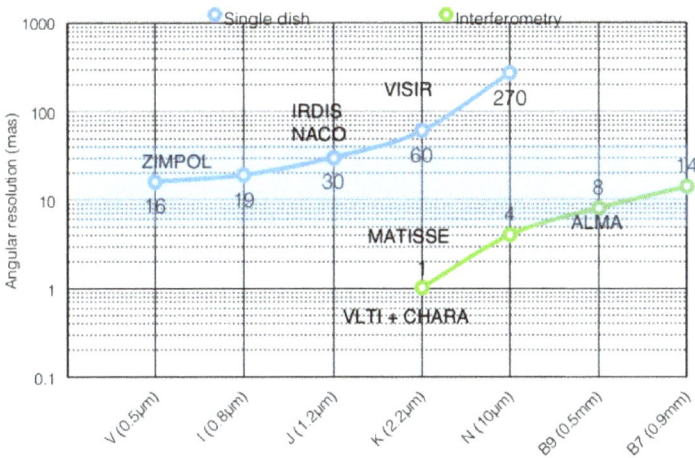

Figure 1. Angular resolution (in miliarcsec) as a function of wavelength for the instruments of the Very Large Telescope (VLT), the Very Large Telescope Interferometer (VLTI), CHARA and ALMA.

2. AGB Stars Morphologies

Before becoming a PN, stars with masses between 0.8 and 8 M$_\odot$ evolve along the Asymptotic Giant Branch (AGB). Convection and pulsation of AGB stars extend their atmosphere and can trigger shocks that can lead to dust formation. Radiation pressure on these dust grains (via absorption or scattering) can trigger the mass loss from these stars, gas being then carried along via friction [5] (and references therein). Understanding these mechanisms is the key to understand dust and gas ejection along the AGB and the formation of circumstellar envelopes that will become PNe once the central star will leave the AGB and get hotter and ionise the gas.

Optical/IR nterferometers such as the VLTI are now able to reach resolutions down to 1 milliarcsec. Combining images with more than two telescopes and using different setups (separation, position angle on the sky), it is now possible to reconstruct images in the near-infrared and to resolve the surface of the closest AGB stars. With such observation, using PIONIER/VLTI, Paladini et al. [6] mapped convective cells (with typical sizes of ~27% of the surface of the star) at the surface of the AGB star π_1Gru. When one add spectral resolution to such observations (as could be done with AMBER/VLTI), one can map molecules close to the surface and the photosphere of giant stars. This was done e.g., for the carbon stars R Scl [7], for which different gas layers (C_2H_2, CO and HCN) where shown to be more extended than the stellar surface, forming a so-called MOLSPHERE. Dust also appears to be non uniformly formed close to the star. Similar observations of W Hya also enabled to map a MOLSPHERE with CO extending up to ~3 R$_*$ [8].

Extreme adaptive optics instruments such as SPHERE/VLT can also produce images with resolution down to 15 milliarcsec and very high contrast in the optical and near-infrared. In the optical, this enables to map the light scattered by dust, and using polarimetric measurements, to study the dust properties. The closest AGB star R Dor was thus the first AGB star for which a direct image of its surface was obtained [9]. W Hya was also observed with SPHERE [8]. Three dust clumps were mapped, showing that the dust ejection was clumpy and not uniform, and that dust formation is induced by pulsation and convection. The dust properties are observed to evolve with time, with variations within 8 months. Such a behaviour, with non uniform and clumpy dust formation was also observed for the AGB star IRC+10216 [10]. Modelling of the polarization map of oxygen-rich AGB stars revealed the presence of micron-sized dust grains, indicating that the mass-loss process could be due to scattering (emission from AGB stars peak around 1 micron and scattering by dust is more efficient when grains have similar size to the peak emission of the star), as predicted by S. Hoefner [5].

Optical/NIR high angular resolution observations of AGB stars have thus revealed that dust formation was not uniform, leading to non-spherical inner envelopes of AGB stars. One could then wonder how the presence of a binary companion could affect these envelopes. Direct detection of binaries around AGB stars is made difficult by the pulsation of the star and the fact it is embedded in dust [4]. The submillimeter interferometer ALMA, thanks to its spectral and spatial resolution (see e.g., [11]) has then been a great tool to indirectly detect binaries around AGB stars. The interaction with a wide binary can indeed lead to the formation of spirals around AGB stars [12]. Such spirals are being commonly detected around AGB stars with ALMA and other millimeter interferometers such as SMA and SMT [13–18]. But these AGB stars in binary will most likely not form bipolar PNe, as they will not get sufficient angular momentum from their distant companions. Closer companions should form equatorial over densities that will favour a polar ejection of material.

The presence of discs and jets has been inferred by millimetre observations for the AGB stars V Hya and Π_1 Gru [19,20], but no AGB star was directly imaged showing the AGB star and its companion, a disc, and material outflowing perpendicular to its disc. Such a system was discovered around one of the nearest AGB star, L_2 Pup, using optical and near-infrared high angular resolution imaging [21,22]. Polarimetric observations enabled to obtain high contrast images of the disc (the star is not polarised will disc scatter and thus polarise light), with an inner rim of 6 AU. Interaction between the disc and the binary system, resolved by SPHERE (with a separation of 2 AU), form structure s that propagate in a direction perpendicular to the disc. This object is very probably a bipolar nebula in a very early stage. Combining these SPHERE observation with ALMA ones [23,24] enabled to show that the disc was in keplerian rotation. The central star had an initial mass of ∼1 $M_\odot$ and the companion could be a massive planet. The angular momentum of the disk surpasses the one expected from the star, supporting the scenario of disc formation through angular momentum transfer via a binary.

3. Shaping of Post-AGB Stars

After the AGB phase, the envelope get detached but is not ionised yet, this is the post-AGB phase. This phase is generally characterized by a double-peaked spectral energy distribution, with a peak in the optical due to emission from the star, and a peak at longer wavelength (in the infrared) due to emission from dust. However, if a post-AGB is in a binary system, a disc or a torus can be formed, leading to the presence of warm dust (at the dust condensation temperature at ∼1000–1500 K), filling the gap between the optical peak of the star and infrared dusty one [25].

I would actually like to make use of this review to raise a common problem in evolved stars nomenclature. Post-AGB and proto-Planetary Nebulae are not the same ensemble. Indeed, while all proto-PNe will give raise to a PN (by definition, as proto-PNe are PNe in the making), it is not the case for all the post-AGBs. The formation of a PN requires the presence of gas close enough to the star to be ionised once the star gets hot enough. Thus, if the envelope is removed before the star gets hot, the star will go through the post-AGB phase, but will never make a PN. Thus, the dusty RV Tau stars (a.k.a. the van Winckel's objects) are binary post-AGB stars surrounded by a disc and mostly with no circumstellar material outside the disc and most of them will most likely not form a PN.

Another issue I would like to raise is the wavelength dependence of post-AGB objects' morphologies. A bipolar post-AGB star with an equatorial disk and material outflowing in a direction perpendicular to its disc will look different in the optical and in the infrared. Indeed at short wavelengths we are more sensitive to light scattered by the disc and we thus obtain images elongated perpendicular to the disc. But if we observe in the infrared, what we see is the emission from dust and elongation along the disc direction. Similarly a bipolar nebula with holes will appear bipolar in the infrared, while searchlight beams of light scattered through the holes will be observed in the optical.

Morphological classification needs to be clear about that, as the same object observed at different wavelengths will have different morphologies.

Observationally speaking, most of the post-AGB objects seem to be aspherical [25]. They harbour two kinds of equatorial over densities. This can be either a torus, which is usually massive ($\sim$1 M$_\odot$), in slow expansion (a few km/s) and with a limited angular momentum. They are short-lived, as, if the gas supply cease, they quickly vanish. Such a torus was recently mapped with ALMA [26], around IRAS 16342, the water fountain. It harbours a dense (3×10^6 cm^{-3}), slowly expanding (20 km/s) torus, which appeared to have been formed before the observed jets (the torus has a dynamical age of 160 years vs. 110 for the jets).

The other kind of equatorial structures observed is discs in keplerian rotation, with a larger angular momentum and thus lifetime. Such discs where revealed by ALMA for a few post-AGB objects, such as the Red Rectangle [27] and AC Her [28]. The ALMA observations of IW Car [29] are a great example of what the angular resolution combined with spectral resolution of ALMA can teach us about such objects. It reveals a binary post-AGB star with a disc, with outflow perpendicular to the disc (which has a dynamical age of 10,000 years) and that there is 8 times more material in the disc than in the outflows.

Finally, optical/near-infrared interferometry is now able to produce images, making it less scary and easier to understand for non-specialists than closure phases and visibilities. With an instrument like PIONIER/VLTI, et al. [30] were thus able to obtain a very impressive result: the first image of a post-AGB system with a circumbinary disc. They detected a compact circumpanion accretion disc, where the outflow very likely originates. It would now be interesting to follow-up this object to see the time evolution of the accretion and the outflows.

4. Shaping of Related Objects

In the previous paragraphs, I have shown how high angular resolution observations with instruments such as the VLTI, SPHERE/VLT and ALMA were revolutionising our view of the shaping of evolved stars from the AGB to the PN phase. To conclude, I would like to also emphasize that similar study of related object can teach us a lot about the shaping of PNe (this will by the way be one of the topics of the next Asymmetrical Planetary Nebulae conference).

Combining SPHERE and ALMA observations of the Red Supergiant Betelgeuse, Pierre Kervella and his team completely changed the way we are now seeing this emblematic star. Plumes of gas are seen extending out the photosphere up to 3 stellar radii, and an incomplete dust shell is resolved at $\sim$3 R$_*$ [23]. The observed asymmetries are also consistent with what is predicted from 3D convection models. ALMA observations enable them to clearly "see" the rotation of the star in $\sim$36 years and to show evidence for a polar ejection certainly due to convection [31].

Similarly, the post-RSG spectroscopic binary system AFGL 4106 was observed with SPHERE (Figure 2). The binary system is clearly resolved for the first time and the amazingly complex morphology of the envelope is revealed, with signs non-continuous ejections in many directions (Lagadec et al., in prep).

Symbiotic stars are another kind of interesting related binary systems. They contain a mass-losing AGB star whose material is accreted by a compact object, usually a white dwarf. R Aqr is a prototypical object of this class and was one of the fist object to be mapped with SPHERE [32]. It countains a Mira variable, a hot companion and a spectacular jet outflow. These data reveal the inner part of the jet, showing it is emerging from the companion and precessing. They were able to measure the density of the jets, which will be of great interest for modellers to discriminate different jets scenarios. The binary system is clearly resolved (8 AU of separation), and its orbit is being monitored. This system should thus soon become a benchmark for the physics of jets in accretion systems.

Another kind of related objects are the binary Wolf-Rayet systems, were dusty spirals are formed due to the interaction of the winds of a WR star and its O companion, forming huge amounts of dust (up to 10^{-6} M$_\odot$year^{-1}). SPHERE enabled to directly image such a spiral for the first time [33] and to map 5 revolutions of the spiral and revealed that the system was not a binary but a triple system.

Finally, SPHERE also enables to map the inner part of the circumstellar material around one of the most iconic object in the sky: Eta Carinae. It is a massive binary system containing a Luminous Blue Variable star and a O-type star, with a total mass between 100 and 200 M$_\odot$, which erupted in the mid 19th century, leading to the formation of a giant bipolar nebula, nicknamed the Homunculus, with a morphology very similar to typical PNe. Another eruption, known as the "lesser eruption" occurred in 1890, creating blobs known as the Weigelt blobs. The SPHERE observations revealed many new blobs and the motion on the sky of the known blobs since their discoveries in 1988.

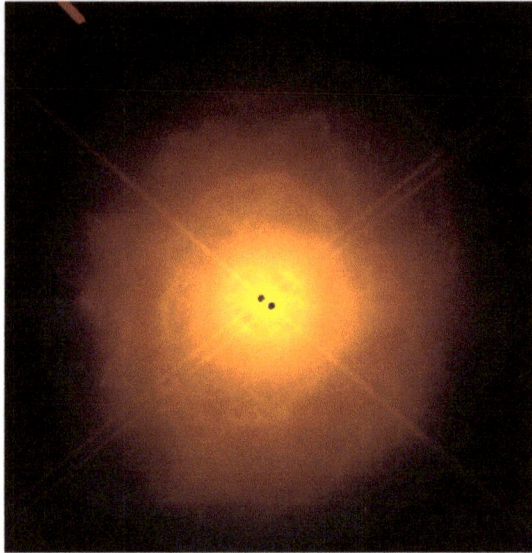

Figure 2. SPHERE/VLT observations of the post-RSG system AFGL 4106 (Lagadec et al., in prep). This system was a known spectroscopic binary. The binary system is directly resolved for the first time here (separation 0.3 arcsec). The interaction fo the post-RSG star with its massive companion lead to a very complex, non isotropic, ejection of material.

5. Conclusions

To conclude this review, in the last years, a lot of advances have been made on the study of the surface and nearby environments of AGB stars. The first maps (via interferometry and direct imaging techniques) of the surfaces have revealed convection cells covering a significant fraction of the stars' surfaces. The stars also pulsate, and this convection and pulsation lead to dust formation. The dust shells observed appear clumpy next to the surface of the stars and evolving in time, so that an onset of asymmetry is observed near the surface of AGB stars.

Spiral patterns appear to be common around AGB stars and due to companions with rather large separations: these stars more likely will not form bipolar nebulae, as the angular momentum transfer between the AGB star and it companion is not efficient enough. It is thus interesting to notice that most of the AGB stars observed most likely will not form bipolar planetary nebulae, while most of the PPNe (if not all) are likely to form bipolar PNe: does the sample of known AGB and PPNe know match? More work seems to be needed to find the AGB stars that will lead to bipolar PNe and PPNe that will not form bipolar PPNe.

Funding: This research received no external funding.

Conflicts of Interest: The authors declare no conflict of interest.

References

1. Boffin, H.M.J.; Miszalski, B.; Rauch, T.; Jones, D.; Corradi, R.L.M.; Napiwotzki, R.; Day-Jones, A.C.; Köppen, J. An Interacting Binary System Powers Precessing Outflows of an Evolved Star. *Science* **2012**, *338*, 773. [CrossRef] [PubMed]

2. Jones, D.; Boffin, H.M.J. Binary stars as the key to understanding planetary nebulae. *Nat. Astron.* **2017**, *1*, 0117. [CrossRef]

3. Soker, N. Shaping Planetary Nebulae and Related Objects. In *Asymmetrical Planetary Nebulae III: Winds, Structure and the Thunderbird*; Meixner, M., Kastner, J.H., Balick, B., Soker, N., Eds.; Astronomical Society of the Pacific Conference Series; Astronomical Society of the Pacific: San Francisco, CA, USA, 2004; Volume 313, p. 562.

4. Lagadec, E.; Chesneau, O. Observations of Binaries in AGB and Post-AGB Stars, and Planetary Nebulae. In *Why Galaxies Care about AGB Stars III: A Closer Look in Space and Time*; Kerschbaum, F., Wing, R.F., Hron, J., Eds.; Astronomical Society of the Pacific Conference Series; Astronomical Society of the Pacific: San Francisco, CA, USA, 2015; Volume 497, p. 145.

5. Höfner, S. Winds of M-type AGB stars driven by micron-sized grains. *Astron. Astrophys.* **2008**, *491*, L1–L4. [CrossRef]

6. Paladini, C.; Baron, F.; Jorissen, A.; Le Bouquin, J.B.; Freytag, B.; van Eck, S.; Wittkowski, M.; Hron, J.; Chiavassa, A.; Berger, J.P.; et al. Large granulation cells on the surface of the giant star π^1 Gruis. *Nature* **2018**, *553*, 310–312. [CrossRef] [PubMed]

7. Wittkowski, M.; Hofmann, K.H.; Höfner, S.; Le Bouquin, J.B.; Nowotny, W.; Paladini, C.; Young, J.; Berger, J.P.; Brunner, M.; de Gregorio-Monsalvo, I.; et al. Aperture synthesis imaging of the carbon AGB star R Sculptoris. Detection of a complex structure and a dominating spot on the stellar disk. *Astron. Astrophys.* **2017**, *601*, A3. [CrossRef]

8. Ohnaka, K.; Weigelt, G.; Hofmann, K.H. Clumpy dust clouds and extended atmosphere of the AGB star W Hydrae revealed with VLT/SPHERE-ZIMPOL and VLTI/AMBER. *Astron. Astrophys.* **2016**, *589*, A91. [CrossRef]

9. Khouri, T.; Maercker, M.; Waters, L.B.F.M.; Vlemmings, W.H.T.; Kervella, P.; de Koter, A.; Ginski, C.; De Beck, E.; Decin, L.; Min, M.; et al. Study of the inner dust envelope and stellar photosphere of the AGB star R Doradus using SPHERE/ZIMPOL. *Astron. Astrophys.* **2016**, *591*, A70. [CrossRef]

10. Stewart, P.N.; Tuthill, P.G.; Monnier, J.D.; Ireland, M.J.; Hedman, M.M.; Nicholson, P.D.; Lacour, S. The weather report from IRC+10216: evolving irregular clouds envelop carbon star. *MNRAS* **2016**, *455*, 3102–3109. [CrossRef]

11. Kerschbaum, F.; Maercker, M.; Brunner, M.; Lindqvist, M.; Olofsson, H.; Mecina, M.; De Beck, E.; Groenewegen, M.A.T.; Lagadec, E.; Mohamed, S.; et al. Rings and filaments: The remarkable detached CO shell of U Antliae. *Astron. Astrophys.* **2017**, *605*, A116. [CrossRef]

12. Mohamed, S.; Podsiadlowski, P. Wind Roche-Lobe Overflow: a New Mass-Transfer Mode for Wide Binaries. In *15th European Workshop on White Dwarfs*; Napiwotzki, R., Burleigh, M.R., Eds.; Astronomical Society of the Pacific Conference Series; Astronomical Society of the Pacific: San Francisco, CA, USA, 2007; Volume 372, p. 397.

13. Maercker, M.; Mohamed, S.; Vlemmings, W.H.T.; Ramstedt, S.; Groenewegen, M.A.T.; Humphreys, E.; Kerschbaum, F.; Lindqvist, M.; Olofsson, H.; Paladini, C.; et al. Unexpectedly large mass loss during the thermal pulse cycle of the red giant star R Sculptoris. *Nature* **2012**, *490*, 232–234. [CrossRef] [PubMed]

14. Decin, L.; Richards, A.M.S.; Neufeld, D.; Steffen, W.; Melnick, G.; Lombaert, R. ALMA data suggest the presence of spiral structure in the inner wind of CW Leonis. *Astron. Astrophys.* **2015**, *574*, A5. [CrossRef]

15. Kim, H.; Liu, S.Y.; Hirano, N.; Zhao-Geisler, R.; Trejo, A.; Yen, H.W.; Taam, R.E.; Kemper, F.; Kim, J.; Byun, D.Y.; et al. High-resolution CO Observation of the Carbon Star CIT 6 Revealing the Spiral Structure and a Nascent Bipolar Outflow. *Astrophys. J.* **2015**, *814*, 61. [CrossRef]

16. Guélin, M.; Patel, N.A.; Bremer, M.; Cernicharo, J.; Castro-Carrizo, A.; Pety, J.; Fonfría, J.P.; Agúndez, M.; Santander-García, M.; Quintana-Lacaci, G.; et al. IRC +10 216 in 3D: morphology of a TP-AGB star envelope. *Astron. Astrophys.* **2018**, *610*, A4. [CrossRef] [PubMed]

17. Ramstedt, S.; Mohamed, S.; Vlemmings, W.H.T.; Maercker, M.; Montez, R.; Baudry, A.; De Beck, E.; Lindqvist, M.; Olofsson, H.; Humphreys, E.M.L.; et al. The wonderful complexity of the Mira AB system. *Astron. Astrophys.* **2014**, *570*, L14. [CrossRef]

18. Ramstedt, S.; Mohamed, S.; Vlemmings, W.H.T.; Danilovich, T.; Brunner, M.; De Beck, E.; Humphreys, E.M.L.; Lindqvist, M.; Maercker, M.; Olofsson, H.; et al. The circumstellar envelope around the S-type AGB star W Aql. Effects of an eccentric binary orbit. *Astron. Astrophys.* **2017**, *605*, A126. [CrossRef] [PubMed]

19. Sahai, R.; Morris, M.; Knapp, G.R.; Young, K.; Barnbaum, C. A collimated, high-speed outflow from the dying star V Hydrae. *Nature* **2003**, *426*, 261–264. [CrossRef] [PubMed]

20. Chiu, P.J.; Hoang, C.T.; Lim, J.; Kwok, S.; Hirano, N.; Muthu, C. A Slowly Expanding Disk and Fast Bipolar Outflow from the S Star π^1 Gruis. *Astrophys. J.* **2006**, *645*, 605–612. [CrossRef]

21. Lykou, F.; Klotz, D.; Paladini, C.; Hron, J.; Zijlstra, A.A.; Kluska, J.; Norris, B.R.M.; Tuthill, P.G.; Ramstedt, S.; Lagadec, E.; et al. Dissecting the AGB star L_2 Puppis: A torus in the making. *Astron. Astrophys.* **2015**, *576*, A46. [CrossRef]

22. Kervella, P.; Montargès, M.; Lagadec, E.; Ridgway, S.T.; Haubois, X.; Girard, J.H.; Ohnaka, K.; Perrin, G.; Gallenne, A. The dust disk and companion of the nearby AGB star L_2 Puppis. SPHERE/ZIMPOL polarimetric imaging at visible wavelengths. *Astron. Astrophys.* **2015**, *578*, A77. [CrossRef]

23. Kervella, P.; Homan, W.; Richards, A.M.S.; Decin, L.; McDonald, I.; Montargès, M.; Ohnaka, K. ALMA observations of the nearby AGB star L_2 Puppis. I. Mass of the central star and detection of a candidate planet. *Astron. Astrophys.* **2016**, *596*, A92. [CrossRef]

24. Homan, W.; Richards, A.; Decin, L.; Kervella, P.; de Koter, A.; McDonald, I.; Ohnaka, K. ALMA observations of the nearby AGB star L_2 Puppis. II. Gas disk properties derived from ^{12}CO and ^{13}CO J = 3-2 emission. *Astron. Astrophys.* **2017**, *601*, A5. [CrossRef]

25. Lagadec, E.; Verhoelst, T.; Mékarnia, D.; Suáeez, O.; Zijlstra, A.A.; Bendjoya, P.; Szczerba, R.; Chesneau, O.; van Winckel, H.; Barlow, M.J.; et al. A mid-infrared imaging catalogue of post-asymptotic giant branch stars. *Mon. Not. Roy. Astron. Soc.* **2011**, *417*, 32–92. [CrossRef]

26. Sahai, R.; Vlemmings, W.H.T.; Gledhill, T.; Sánchez Contreras, C.; Lagadec, E.; Nyman, L.Å.; Quintana-Lacaci, G. ALMA Observations of the Water Fountain Pre-planetary Nebula IRAS 16342-3814: High-velocity Bipolar Jets and an Expanding Torus. *Astrophys. J. Lett.* **2017**, *835*, L13. [CrossRef] [PubMed]

27. Bujarrabal, V.; Castro-Carrizo, A.; Alcolea, J.; Santander-García, M.; van Winckel, H.; Sánchez Contreras, C. Further ALMA observations and detailed modeling of the Red Rectangle. *Astron. Astrophys.* **2016**, *593*, A92. [CrossRef] [PubMed]

28. Bujarrabal, V.; Castro-Carrizo, A.; Alcolea, J.; Van Winckel, H. Detection of Keplerian dynamics in a disk around the post-AGB star AC Herculis. *Astron. Astrophys.* **2015**, *575*, L7. [CrossRef]

29. Bujarrabal, V.; Castro-Carrizo, A.; Alcolea, J.; Van Winckel, H.; Sánchez Contreras, C.; Santander-García, M. A second post-AGB nebula that contains gas in rotation and in expansion: ALMA maps of IW Carinae. *Astron. Astrophys.* **2017**, *597*, L5. [CrossRef]

30. Hillen, M.; Kluska, J.; Le Bouquin, J.B.; Van Winckel, H.; Berger, J.P.; Kamath, D.; Bujarrabal, V. Imaging the dust sublimation front of a circumbinary disk. *Astron. Astrophys.* **2016**, *588*, L1. [CrossRef]

31. Kervella, P.; Decin, L.; Richards, A.M.S.; Harper, G.M.; McDonald, I.; O'Gorman, E.; Montargès, M.; Homan, W.; Ohnaka, K. The close circumstellar environment of Betelgeuse. V. Rotation velocity and molecular envelope properties from ALMA. *Astron. Astrophys.* **2018**, *609*, A67. [CrossRef]

32.   Schmid, H.M.; Bazzon, A.; Milli, J.; Roelfsema, R.; Engler, N.; Mouillet, D.; Lagadec, E.; Sissa, E.; Sauvage, J.F.; Ginski, C.; et al. SPHERE/ZIMPOL observations of the symbiotic system R Aquarii. I. Imaging of the stellar binary and the innermost jet clouds. *Astron. Astrophys.* **2017**, *602*, A53. [CrossRef]

33.   Soulain, A.; Millour, F.; Lopez, B.; Matter, A.; Lagadec, E.; Carbillet, M.; Camera, A.; Lamberts, A.; Langlois, M.; Milli, J.; et al. The SPHERE view of Wolf-Rayet 104. *arXiv* **2018**, arXiv:1806.08525.

galaxies   MDPI

Article

UV Monochromatic Imaging of the Protoplanetary Nebula Hen 3-1475 Using *HST* STIS

Xuan Fang [1,2,*], Martín A. Guerrero [3], Ana I. Gómez de Castro [4], Jesús A. Toalá [5], Bruce Balick [6] and Angels Riera [7]

1   Laboratory for Space Research, Faculty of Science, University of Hong Kong, Pokfulam Road, Hong Kong, China
2   Department of Physics, Faculty of Science, University of Hong Kong, Pokfulam Road, Hong Kong, China
3   Instituto de Astrofísica de Andalucía (IAA-CSIC), Glorieta de la Astronomía s/n, E-18008 Granada, Spain; mar@iaa.es
4   AEGORA Research Group, Facultad de Ciencias, Universidad Complutense, E-28040 Madrid, Spain; anai_gomez@mat.ucm.es
5   Instituto de Radioastronomía y Astrofísica, UNAM Campus Morelia, Apartado Postal 3-72, Morelia 58090, Michoacán, Mexico; j.toala@irya.unam.mx
6   Department of Astronomy, University of Washingtgon, Seattle, WA 98 195-1580, USA; balick@uw.edu
7   Departament de Física i Enginyeria Nuclear, EUETIB, Universitat Politécnica de Catalunya, E-08036 Barcelona, Spain; angels.riera@upc.edu
*   Correspondence: fangx@hku.hk; Tel.: +852-3962-1439

Received: 8 August 2018 ; Accepted: 11 December 2018; Published: 14 December 2018

Abstract: Collimated outflows and jets play a critical role in shaping planetary nebulae (PNe), especially in the brief transition from a spherical AGB envelope to an aspherical PN, which is called the protoplanetary nebula (pPN) phase. We present UV observations of Hen 3-1475, a bipolar pPN with fast, highly collimated jets, obtained with STIS on board the *Hubble Space Telescope* (*HST*). The deep, low-dispersion spectroscopy enabled monochromatic imaging of Hen 3-1475 in different UV nebular emission lines; this is the first of such attempt ever conducted for a pPN. The northwest inner knot (NW1) is resolved into four components in Mg II $\lambda2800$. Through comparison analysis with the *HST* optical narrowband images obtained 6 yr earlier, we found that these components of NW1 hardly move, despite of a negative gradient of high radial velocities, from -1550 km s^{-1} on the innermost component to ~-300 km s^{-1} on the outermost. These NW1 knot components might thus be quasi-stationary shocks near the tip of the conical outflow of Hen 3-1475.

Keywords: late stage stellar evolution; planetary nebulae; theory and observation

---

1. Introduction

Hen 3-1475 (a.k.a., the "Garden Sprinkler Nebula"; also IRAS 17423-1755) is known to harbour very fast bipolar outflows [1]. Since its nature as a protoplanetary nebula (pPN) was settled, Hen 3-1475 has been imaged several times [2–5]. It has a highly collimated bipolar structure with an *S*-shaped string of point-symmetric, [N II]-bright knots extending over $\sim17''$ along the main axis (Figure 1). High-dispersion spectroscopy revealed high-velocity jets (1200 km s^{-1}, [4]); ultra-fast (up to 2300 km s^{-1}) winds was detected very close to the central star [6]. Extended (or diffuse) X-ray emission a signature of interactions of the fast and slow stellar winds. Hen 3-1475 is so far the only pPN where diffuse X-ray emission is detected [7]; this X-ray emission mostly comes from its brightest NW1[1] knot,

---

1   Following (Borkowski et al. [3], Figure 2 therein), we hereafter refer the inner, middle, and outer pairs of knots in Hen 3-1475 as NW1/SE1, NW2/SE2, and NW3/SE3, respectively. Here NW means northwest, and SE means southeast.

whose emission is shock excited [4,8,9]. Its fast outflows are being collimated into jets far away from the central star, through two inner cones (i.e., the limb-brightened edges of conical shocks; [3]), but the collimation mechanism is not well understood.

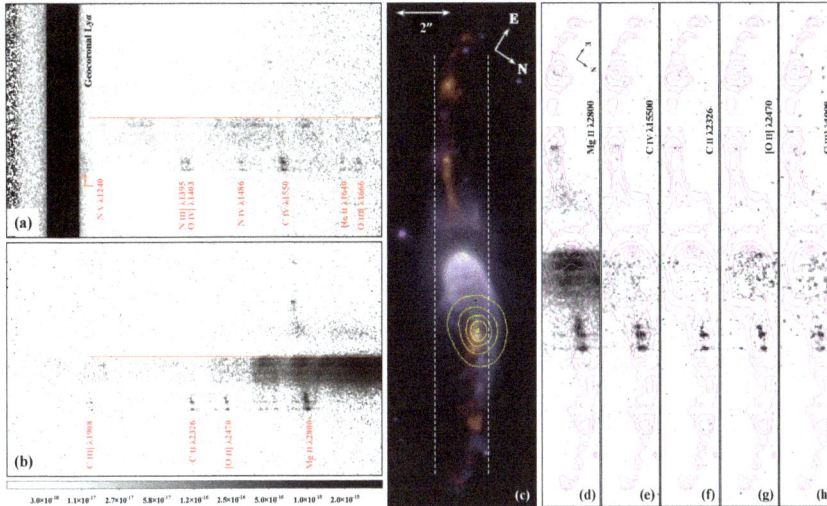

Figure 1. (**Left**) Negative grayscale images of the STIS FUV-MAMA (**a**) and NUV-MAMA (**b**) spectra of Hen 3-1475; emission features are identified. Note the bright geocoronal Lyα. Red dotted lines mark the position of the central star. The color bar below shows the grey scale coded in flux values (in units of erg s^{-1} cm^{-2} Å^{-1}). Panel (**c**): *HST* 2009 WFC3 color-composite image of Hen 3-1475 created with F658N (red), F656N (green), and F555W (blue) and overlaid with *Chandra* X-ray emission contours (yellow; Obs. ID 2580, PI: R. Sahai); white dashed lines indicate the STIS 52″ × 2″ long slit (PA = 124°.65). Panels (**d**–**h**): STIS UV spectral-line images (see legend) overlaid by WFC3 F658N emission contours (magenta); the panels are displayed in negative greyscale and slightly smoothed to reduce noise.

The shock effects of stellar wind interactions can also be traced by UV emission. *IUE* detection of UV emission lines in Hen 3-1475 was only marginal due to dust obscuration [10]. This proceeding paper reports the first UV spectroscopy of Hen 3-1475 with high spatial resolution.

2. STIS Observations and the UV Spectra

Long-slit UV spectra of Hen 3-1475 were obtained with the Space Telescope Imaging Spectrograph (STIS) on board the *Hubble Space Telescope* (*HST*) on 11 June 2015 (GO prop. #13838, PI: X. Fang, Cycle 22). The STIS 52″ × 2″ long slit was centered on the core (R.A.=17^{h}45^{m}14^{s}19, Decl.=−17°56′46″.90) with a position angle (PA, defined as the angle measured east from north) of 124°.65, along the nebular axis. All knots lie within the slit (Figure 1c). The FUV spectrum was obtained with the first-order grating G140L (∼1150–1730 Å) and STIS/FUV-MAMA (0″.0246 pixel^{-1}) at an exposure of 2800 s, and has a spectral resolution of R ∼1000–1440 with a dispersion of 0.6 Å pixel^{-1}. The NUV spectrum was obtained with the first-order grating G230L (∼1570–3180 Å) and STIS/NUV-MAMA (0″.0248 pixel^{-1}) at an exposure of 2100 s, and has a spectral resolution of R ∼500–1010 with a 1.55 Å pixel^{-1} dispersion. The STIS observations were made in the TIME-TAG mode. Data were reduced and calibrated using the *HST* STIS pipeline, and are demonstrated in Figure 1a,b. The FUV spectrum covers N v λ1240 (a blend of $\lambda\lambda$1239,1243; all wavelengths in Å), C IV λ1550 (a blend of $\lambda\lambda$1548,1551), He II λ1640, O III] λ1666, etc. In NUV we detected C III] λ1908 (a blend of $\lambda\lambda$1907,1909), C II λ2326, [O II] λ2470, and Mg II λ2800 (a blend of $\lambda\lambda$2795,2803). The deep, low-resolution spectroscopy enables *monochromatic imaging of*

Hen 3-1475 in UV nebular emission lines (Figure 1d–h). No fine-structure transition lines within any doublet are spectroscopically resolved in our UV spectra.

3. UV Morphology of Hen 3-1475

In the STIS low-dispersion UV spectra, the overall morphology of Hen 3-1475 as seen in the *HST* optical images is generally maintained in UV emission lines (Figure 1a,b), especially in Mg II λ2800, the strongest emission line in our UV spectra, where the brightest region between the NW1 and SE1 knots are shown. We created UV monochromatic images by first carefully selecting and chopping from the STIS 2D spectrograms rectangular regions (centered on UV emission lines), and then registering them with the archival WFC3 F656N image (GO prop. 11580, PI: B. Balick); NW1 and SE1 are both seen in the UV spectral-line images and thus used as reference when aligning the chopped UV images with the WFC3 F656N image. The UV morphology of Hen 3-1475 was studied in Mg II λ2800, the brightest nebular emission line in the STIS UV spectra. The WFC3 narrowband filters F656N and F658N show the Hα and [N II] λ6583 emission, respectively, from the jets and knots in Hen 3-1475 that are presumably shock excited [4].

Our *HST* STIS UV spectroscopy of Hen 3-1475, utilizing a broad (2″ width) long slit that covers the whole object, follows essentially the same technique employed in the previous STIS optical slitless spectroscopy of the Magellanic Cloud PNe [11,12]. In an optical slitless (or broad-slit) spectrum, the overall morphology of the target can be seen at different nebular emission lines; this is equivalent to monochromatic imaging of the target at individual emission lines. Our STIS UV low-dispersion spectra have the similar format, as the tilted nebular axis of Hen 3-1475 (with respect to the slit direction) that follows the optical image is also seen in the UV emission lines (see Figure 1a,b), except that, as discussed in Section 4, the radial velocities of the jet in Hen 3-1475 are so high that some features along the jet are blueshifted with respect to their counterparts in the *HST* narrow-band images (see the description below and the discussion in Section 4).

In the Mg II λ2800 spectral-line image (Figure 1d), the NW1 knot is resolved into four components, hereafter named A, B, C, and D from inside out (Figure 2, top), with angular sizes of ∼0″.12–0″.14 along jet axis; in the other UV lines, generally fainter than Mg II, only B, C, and D are seen (Figure 1e–h). The SE1 knot seen in the UV could be the counterpart of the outermost component D. The optical counterparts (in Hα) of the four UV components A, B, C, and D are located 2″.49, 2″.67, 2″.91, and 3″.30 from the central star, respectively. B, C, and D are "elongated" (in the dispersion direction) in the Mg II spectral-line image, which may be due to internal velocity gradients [4]; this effect is obvious in B and C, which are compact in the Hα image. D comprises two subcomponents in Hα that are ∼0″.15 apart and stretch by 0″.32, similar to the size in the UV. The NW conical shock "converges" on A (Figure 2), the most compact (radius∼0″.06) among the four NW1 components in the UV.

Along the dispersion direction of the STIS NUV spectrum, most of the components of the NW1 knot are displaced with respect to their optical counterparts, especially in Hα; we briefly discuss this in Section 4. Along the slit direction, positions of A, B, and C in the UV are unchanged compared to their counterparts in Hα. Component D and SE1 in the UV are slightly outside their Hα counterparts (Figure 2, top) due to jet propagation since 2009. The position of D in Mg II is shifted outward (i.e., towards NW) by 0″.069 from its position in Hα, which corresponds to a proper motion of 11.8 mas yr^{-1}, or ∼280 km s^{-1}. Similarly, SE1 in Mg II is displaced from its position in Hα by 0″.091, corresponding to ∼15.5 mas yr^{-1}, or ∼360 km s^{-1}.

In the NUV spectrum, <2″ from the central star, there are four "stripes" (or striations) that are more prominent near the red end of STIS NUV-MAMA (Figure 1b). These striations are only detected in the NW lobe, and seem to spatially align with the optical features at corresponding locations: in Hα and [N II], the hollow *U*-shaped "bowls" near the nucleus of Hen 3-1475 also seem striated (Figure 1c). The central region of Hen 3-1475 is dominated by the scattered starlight as shown in the *HST* STIS G750M spectrum ([4], see Figure 2 therein). Along the jet axis, the four UV striations in the NW lobe are ∼0″.13, 0″.5, 0″.95, and 1″.43 from the central star (Figure 2, bottom).

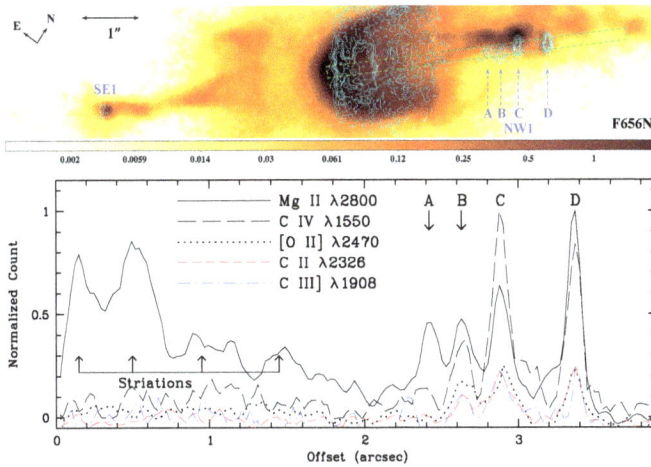

Figure 2. (**Top**) Emission contours (in cyan) of the STIS Mg II λ2800 line overlaid on the 2009 WFC3 F656N image showing the central 2″ × 10″ region of Hen 3-1475; images are displayed in logarithm and data counts are indicated on the colorbar. The NW1 knot is resolved into four components (A, B, C, and D labeled in blue) in Mg II emission. Nebular core is saturated in F656N to enhance lobe structures, including NW1/SE1 and the conical shocks. (**Bottom**) Emission profiles of different UV nebular lines along a cut (green-dashed rectangle in the top panel) through the central star and the NW1 knot components, at PA = 311°; positions of the four components of NW1 and the striations seen in Mg II emission (Figure 1d) are labeled. The other UV emission lines are shown in Figure 1e–h.

4. Jet Kinematics and Preliminary Interpretation

In the slit direction (i.e., along the nebular axis of Hen 3-1475), positions of A, B, and C in Mg II λ2800 emission generally coincide with those of their counterparts in Hα (the WFC3 F656N image obtained in 2009). Along the dispersion direction (i.e., perpendicular to the long slit) of the STIS NUV spectrum, these three inner components of NW1 show noticeable blueward displacements with respect to their counterparts in Hα (Figure 2, top); these displacements are probably due to Doppler shift, given that the NW jet of Hen 3-1475 has high approaching radial velocities [4,8].

Compared to the Hα image, A, B, and C in Mg II emission are shifted blueward by -1550 ± 160, -1200 ± 330, and -1100 ± 400 km s^{-1}, respectively. The systemic radial velocity of Hen 3-1475 ($+40$ km s^{-1}, [4]) is well within the velocity errors. At 2800 Å the kinematic resolution of the G230L grating is $\sim$310 km s^{-1} (as given by the STIS Instrument Handbook), which can be set as an upper-limit approaching speed of the outermost component D. Adopting this radial speed and the sky-projected velocity of D ($\sim$280 km s^{-1}, see Section 3), we deduce a jet inclination (i, with respect to the line of sight) $\geq$42°, which is consistent with the previous estimate of 40° [4]. The radial velocities of A, B, C, and D of the NW1 knot generally follow a trend of negative gradient, as found in previous *HST* STIS G750M optical spectroscopy of the [N II] and Hα emission lines (in 1999; [4]).

The unchanged (or very little changes in) radial positions and fast Doppler speeds of the NW1 knot components seem to resemble quasi-stationary shocks: A, B, C, and D may actually be shock interfaces where the outflowing gas moves through with high speeds, and have no (or very small) measurable proper motions; each of these clumpy shocks may brighten or fade in time owing to variations in density or speed of the gas flow. The negative radial velocity gradient in NW1 may then be explained as a series of speed losses of the (possibly unstable) flow as it propagates outward along the jet. This interpretation, although quite reasonable, is speculative and needs careful assessment with comprehensive hydrodynamic simulations.

A precessing jet model with a time-dependent ejection velocity was proposed to interpret the observed morphology and kinematics of Hen 3-1475 [13]; but this model might still be ad hoc. The actual jet dynamics within the lobes could be more complex than previously anticipated. An investigation of the multi-epoch *HST* F658N images suggests that the outer knots (NW2/SE2 and NW3/SE3) of Hen 3-1475 may not be the radially fast-moving "bullets" as previously thought, but locally excited due to Kelvin–Helmholtz instabilities [14]. If this is true, then NW2/SE2 and NW3/SE3 are probably different from NW1/SE1 in terms of formation and excitation. Better understanding of the jets in Hen 3-1475 requires better modeling.

The simultaneous detection of X-ray, UV and optical emission suggests that the NW1 knots are cooling regions with strong temperature gradient. We also expect ionization stratification in NW1, i.e., Mg II and Hα may come from different regions. However, the UV components (at least, the innermost component A) of NW1 are spatially unresolved in our STIS UV monochromatic imaging, given that their angular sizes are comparable to the actual angular resolution of the STIS/NUV-MAMA detector ($0.''06$, as provided by the STIS Instrument Handbook); thus our comparison of the NW1 knot positions in Mg II and Hα is reasonable.

5. Summary and Conclusions

We report on monochromatic imaging of Hen 3-1475 in UV nebular emission lines obtained through the *HST* STIS UV long-slit spectroscopy; this is the first of such attempt ever made for a pPN in the UV. We analyzed the UV morphology of Hen 3-1475, in conjunction with the archival *HST* optical narrowband images. The high spatial resolution of our STIS imaging enables a very sharp view of the inner region of Hen 3-1475, especially the NW1 knot near the tip of the inner conical shock. The NW1 knot is clearly resolved into four components in the Mg II $\lambda 2800$ line emission.

The four UV components of the NW1 knot are distributed roughly along the jet axis. Compared to their optical counterparts in Hα, the four components of NW1 are mostly blueshifted in the Mg II spectral-line image, probably due to their high (approaching) radial velocities, from -1550 km s^{-1} on the innermost component to ~ -300 km s^{-1} on the outermost one. Despite of their high Doppler speeds, these components of NW1 show no obvious proper motions compared to previous *HST* imaging, indicating that they might be quasi-stationary shocks where fast gas flows through.

Author Contributions: The original *HST* proposal (conceptualization) was prepared together by X.F., M.A.G., A.I.G.d.C., J.A.T., and A.R.; analysis of the *HST* STIS data was carried out by X.F.; data visualization, X.F., M.A.G., B.B., and J.A.T.; intensive discussion was made together by B.B., M.A.G., X.F., A.I.G.d.C, and J.A.T.; writing–original draft preparation, X.F.; writing–review and editing, M.A.G., B.B., A.I.G.d.C., and J.A.T.

Funding: M.A.G. acknowledges support of the grant AYA 2014-57280-P, cofunded with FEDER funds. M.A.G. and J.A.T. are funded by UNAM DGAPA PAPITT project IA100318.

Acknowledgments: This research is based on observations made with the NASA/ESA *Hubble Space Telescope*, and obtained from the *Hubble* Legacy Archive (HLA; http://hla.stsci.edu), which is a collaboration between the Space Telescope Science Institute (STScI/NASA), the Space Telescope European Coordinating Facility (ST-ECF/ESA) and the Canadian Astronomy Data Centre (CADC/NRC/CSA). This research also made use of the *Chandra* Data Archive (CDA; http://cxc.cfa.harvard.edu/cda) and NASA's Astrophysics Data System (http://adsabs.harvard.edu). X.F. acknowledges the support and hospitality of the IAA-CSIC for an academic visit in September 2017, during which part of the contents in this paper was discussed. We thank Quentin A. Parker for discussion and helpful suggestions. The coauthor Angels Riera passed away in Barcelona on 27 September 2017, when the *HST* STIS UV spectra reported in this article were analyzed; she was the first to recognize the peculiar nature of the bipolar pPN Hen 3-1475. This article is dedicated to her.

Conflicts of Interest: The authors declare no conflict of interest.

References

1. Riera, A.; García-Lario, P.; Manchado, A.; Pottasch, S.R.; Raga, A.C. IRAS 17423-1755: A massive post-AGB star evolving into the planetary nebula stage? *Astron. Astrophys.* **1995**, *302*, 137–153.

2. Bobrowsky, M.; Zijlstra, A.A.; Grebel, E.K.; Tinney, C.G.; Hekkert, P.; Van de Steene, G.C.; Likkel, L.; Bedding, T.R. He 3-1475 and its jets. *Astrophys. J. Lett.* **1995**, *446*, L89–L92. [CrossRef]

3. Borkowski, K.J.; Blondin, J.M.; Harrington, J.P. Collimation of astrophysical jets: The protoplanetary nebula He 3-1475. *Astrophys. J. Lett.* **1997**, *482*, L97–L100. [CrossRef]

4. Borkowski, K.J.; Harrington, J.P. Kinematics of 1200 kilometer per second jets in He 3-1475. *Astrophys. J.* **2001**, *550*, 778–784. [CrossRef]

5. Ueta, T.; Meixner, M.; Bobrowsky, M. A *Hubble Space Telescope* snapshot survey of proto-planetary nebula candidates: Two types of axisymmetric reflection nebulosities. *Astrophys. J.* **2000**, *528*, 861–884. [CrossRef]

6. SáchezContreras, C.; Sahai, R. A 2000 kilometer per second "pristine" post-asymptotic giant branch wind in the proto-planetary nebula He 3-1475. *Astrophys. J. Lett.* **2001**, *553*, L173–L176. [CrossRef]

7. Sahai, R.; Kastner, J.H.; Frank, A.; Morris, M.; Blackman, E.G. X-ray emission from the pre-planetary nebula Henize 3-1475. *Astrophys. J. Lett.* **2003**, *599*, L87–L90. [CrossRef]

8. Riera, A.; García-Lario, P.; Manchado, A.; Bobrowsky, M.; Estalella, R. The high-velocity outflow in the proto-planetary nebula Hen 3-1475. *Astron. Astrophys.* **2003**, *401*, 1039–1056. [CrossRef]

9. Riera, A.; Binette, L.; Raga, A.C. Shock excitation of the knots of Hen 3-1475. *Astron. Astrophys.* **2006**, *455*, 203–213. [CrossRef]

10. Gauba, G.; Parthasarathy, M. UV (IUE) spectra of hot post-AGB candidates. *Astron. Astrophys.* **2003**, *407*, 1007–1020. [CrossRef]

11. Stanghellini, L.; Shaw, R.A.; Mutchler, M.; Palen, S.; Balick, B.; Blades, J.C. Optical slitless spectroscopy of Large Magellanic Cloud planetary nebulae: A study of the emission lines and morphology. *Astrophys. J.* **2002**, *575*, 178. [CrossRef]

12. Stanghellini, L.; Shaw, R.A.; Balick, B.; Mutchler, M.; Blades, J.C.; Villaver, E. Space Telescope Imaging Spectrograph slitless observations of Small Magellanic Cloud planetary nebulae: A study on morphology, emission-line intensity, andevolution. *Astrophys. J.* **2003**, *596*, 997. [CrossRef]

13. Velázquez, P.F.; Riera, A.; Raga, C. Time-dependent ejection velocity model for the outflow of Hen 3-1475. *Astron. Astrophys.* **2004**, *419*, 991–998. [CrossRef]

14. Fang, X.; Gómez de Castro, A.I.; Toalá, J.A.; Riera, A. *HST* STIS UV spectroscopic observations of the protoplanetary nebula Hen 3-1475. *Astrophys. J. Lett.* **2018**, *865*, L23. [CrossRef]

galaxies

MDPI

Article

Sliding along the Eddington Limit—Heavy-Weight Central Stars of Planetary Nebulae

Lisa Löbling [1,2]

[1] Institute for Astronomy and Astrophysics, Kepler Center for Astro- and Particle Physics, Eberhard Karls University, Astronomy and Astrophysics, Sand 1, D-72076 Tübingen, Germany; loebling@astro.uni-tuebingen.de

[2] European Southern Observatory (ESO), Karl-Schwarzschild-Straße 2, D-85748 Garching bei München, Germany

Received: 23 April 2018; Accepted: 15 June 2018; Published: 19 June 2018

Abstract: Due to thermal pulses, asymptotic giant branch (AGB) stars experience periods of convective mixing that provide ideal conditions for slow neutron-capture nucleosynthesis. These processes are affected by large uncertainties and are still not fully understood. By the lucky coincidence that about a quarter of all post-AGB stars turn hydrogen-deficient in a final flash of the helium-burning shell, they display nuclear processed material at the surface providing an unique insight to nucleosynthesis and mixing. We present results of non-local thermodynamic equilibrium spectral analyses of the extremely hot, hydrogen-deficient, PG 1159-type central stars of the Skull Nebula NGC 246 and the "Galactic Soccerballs" Abell 43 and NGC 7094.

Keywords: stars: abundances; stars: AGB and post-AGB; stars: atmospheres; stars: individual: WD 0044−121; stars: individual: WD 2134+25; stars: individual: WD 1751+106

1. Introduction

There are different evolutionary channels a star may go through after leaving the asymptotic giant branch (AGB). About a quarter become hydrogen (H) deficient as a result of a late flash of the helium (He)-burning shell (late thermal pulse, LTP, c.f., [1]). For stars still located on the AGB at this event (AGB final thermal pulse, AFTP, Figure 1), the H-rich envelope (with a mass of $10^{-2} M_\odot$) is mixed with the helium (He) rich intershell material ($10^{-2} M_\odot$). If the final flash happens after the departure from the AGB (envelope mass $\leq 10^{-4} M_\odot$), H is either diluted by mixing in the LTP, if the nuclear fusion is still "on", or it is mixed down and totally consumed by the He-shell, if the star is already on the white dwarf cooling track, fusion is "off" and, thus, no entropy border exists any more. Predicted H mass fractions are 0.20 for an AFTP, 0.02 for an LTP and 0.00 for a VLTP ([1,2], and references therein).

The three objects presented in this work belong to the spectral type of PG 1159 stars (effective temperatures of $75,000\,\mathrm{K} \leq T_{\mathrm{eff}} \leq 250,000\,\mathrm{K}$ and surface gravities of $5.5 \leq \log(g \,/\, \mathrm{cm/s^2}) \leq 8.0$, [1]) resulting from the H-deficient evolutionary channel. The central stars of the planetary nebulae (CSPNe) NGC 7094 (PN G066.7−28.2 [3]; CS: WD 2134−125 [4]) and Abell 43 (PN G036.0+17.6 [3]; WD 1751−106 [4]), known as spectroscopic twins, belong to the sub-type of hybrid PG 1159-stars [5] exhibiting H lines in their spectra and, thus, resulting from an AFTP. In previous analyses, $T_{\mathrm{eff}} = 100 \pm 15\,\mathrm{kK}$, $\log g = 5.5 \pm 0.2$, and a deficiency in Fe and Ni of <1 dex were determined [6]. These stars are known to be fast rotators with rotational velocities of $v_{\mathrm{rot}} = 46 \pm 16\,\mathrm{km/s}$ and $42 \pm 13\,\mathrm{km/s}$, respectively [7].

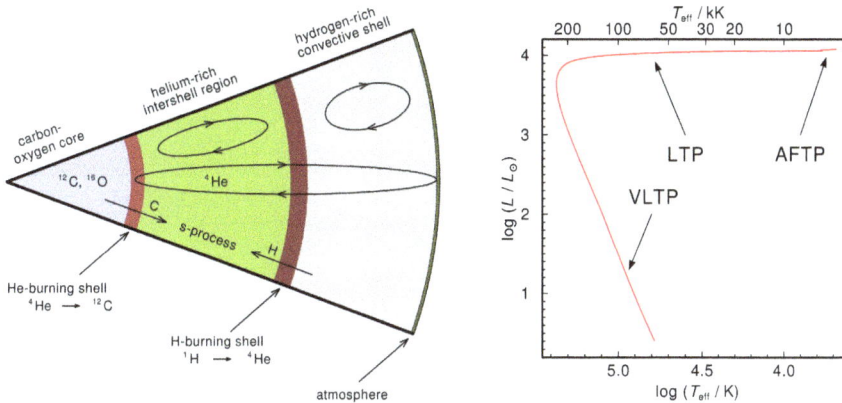

Figure 1. (**Left**): Internal structure of an AGB star (from [8]); (**Right**): Occurrence of AFTP, LTP, and VLTP in the Hertzsprung-Russell diagram.

WD 0044−121 [4] is the PG 1159-type central star of the Skull Nebula (NGC 246, PN G118.8−74.7 [3]) and with a mass of 0.7 $M_\odot$, it is among the most massive of this type. Previous analyses revealed $T_{\rm eff} = 150\,{\rm kK}$, $\log g = 5.7$, and a Fe and Ni deficiency of about 0.25 dex. It rotates with $77^{+23}_{-17}\,{\rm km/s}$ [7]. These stars are unique probes for AGB nucleosynthesis since they show the intershell material at the surface and exhibit a strong enough wind to prevent processes like gravitational settling and radiative levitaion to tamper the original composition. The aim of this analysis is to consider atomic data of trans-iron elements ($Z \geq 30$) that became available recently ([9], and references therein) for a NLTE stellar-atmosphere calculation to determine abundances for a large set of elements to draw conclusions on the evolutionary history of these stars and on AGB nucleosynthesis.

2. Observations and Model Atmospheres

Our analysis is based on high-resolution observations from the far ultraviolet (FUV) to the optical wavelength range. The log for the observations retrieved from the Barbara A. Mikulski Archive for Space Telescopes (MAST) and the European Southern Observatory (ESO) Science Archive is shown in Table 1. For the calculation of synthetic spectra, we employ the Tübingen NLTE Model Atmosphere Package (TMAP[1], [10–12]) working under the assumption of a plane-parallel geometry and hydrostatic and radiative equilibrium. For WD 2134+125 and WD 1751+106, we consider opacities of 31 elements from H to barium Ba. The models for WD 0044−121 include the elements H, He, C, N, O, F, Ne, Mg, Ar, Ca, Fe, and Ni. To analyse transiron-element abundances, we added them individually in line-formation calculations in which the temperature structure is kept fixed and the occupation numbers for the levels are calculated. We performed test calculations to confirm that these elements do not affect the atmospheric structure (see also ([9], and references therein)). Atomic data for H, He, and the light metals (atomic weight $Z < 20$) is retrieved from the Tübingen Model Atom Database (TMAD[2], [10]), for the iron group (Ca-Ni) elements, we used Kurucz's line lists[3] [13,14], and the transiron-element data is available via the Tübingen Oscillator Strength Service (TOSS[4]). For all elements with $Z \geq 20$, a statistical approach is used based on super lines and super levels [10].

---

[1] http://astro.uni-tuebingen.de/~TMAP
[2] http://astro.uni-tuebingen.de/~TMAD
[3] http://kurucz.harvard.edu/atoms.html
[4] http://dc.g-vo.org/TOSS

Table 1. Observation log for WD 2134+125, WD 1751+106, and WD 0044−121.

Object	Instrument	Dataset/ Prog. ID	Start Time (UT)	Exp. Time (s)
WD 2134+125	FUSE [a]	P1043701000	2000-11-13 08:53:28	22,754
	STIS [b]	O8MU02010	2004-06-24 20:43:30	650
	STIS	O8MU02020	2004-06-24 22:19:29	656
	STIS	O8MU02030	2004-06-24 23:55:29	655
	UVES [d]	167.D−0407(A)	2001-08-21 02:00:03	300
	UVES	167.D−0407(A)	2001-09-20 01:40:00	300
WD 1751+106	FUSE	B0520201000	2001-07-29 20:41:47	11,438
	FUSE	B0520202000	2001-08-03 22:18:20	9528
	GHRS [c]	Z3GW0304T	1996-09-08 07:00:34	4243
	UVES	167.D−0407(A)	2001-06-18 05:03:38	300
	UVES	167.D−0407(A)	2001-07-26 01:27:48	300
WD 0044−121	FUSE	E1180201000	2004-07-12 17:01:47	6505
	STIS	O8O701010	2004-05-28 10:11:51	1967
	STIS	O8O701020	2004-05-28 11:31:56	2736
	UVES	167.D−0407(A)	2002-09-06 09:32:22	300
	UVES	165.H−0588(A)	2000-12-07 02:26:29	300

a: Far Ultraviolet Spectroscopic Explorer; b: Space Telescope Imaging Spectrograph; c: Goddard High Resolution Spectrograph; d: UV-Visual Echelle Spectrograph.

3. Preliminary Results

For accurate abundance determinations, the precise knowledge of the stellar parameters is an inevitable requirement. We redetermined the surface gravity using our best fit for the observed line wings and depth increments of He II $\lambda\lambda$ 4100.1, 4338.7, 4859.3, 5411.5 Å and H I $\lambda\lambda$ 4101.7, 4340.5, 4861.3 Å. The temperature determination is based on the ionization equilibrium of O v/O vi using O v λ 1371.3 Å and O vi $\lambda\lambda$ 1080.6, 1081.2, 1122.3, 1122.6, 1124.7, 1124.9, 1126.3, 1290.1, 1290.2, 1291.8, 1291.9. The redetermination of v_{rot} results in lower values for both hybrid PG 1159 stars. The new values are summarized in Table 2.

Table 2. Parameters of WD 2134−125, WD 1751+106, and WD 0044−121.

	WD 2134−125	WD 1751+106	WD 0044−121
T_{eff} / kK	115 ± 5	115 ± 5	150 ± 10
$\log(g / cm/s^2)$	5.4 ± 0.1	5.5 ± 0.1	5.7 ± 0.1
v_{rot} / km/s	28 ± 5	18 ± 5	75 ± 15
d / kpc	$2.38^{+0.59}_{-0.70}$	$2.94^{+0.82}_{-0.91}$	$1.08^{+0.22}_{-0.26}$
$M / M_\odot$	$0.64^{+0.06}_{-0.07}$	$0.60^{+0.09}_{-0.06}$	$0.74^{+0.19}_{-0.23}$
$M_{ini} / M_\odot$	$3.30^{+0.65}_{-1.43}$	$2.87^{+0.71}_{-2.14}$	$3.91^{+2.55}_{-0.88}$
$\log(L / L_\odot)$	$4.04^{+0.32}_{-0.33}$	$3.91^{+0.34}_{-0.32}$	$4.27^{+0.41}_{-0.59}$

We found that the Fe abundance for both hybrid PG 1159 stars needed to be corrected. The lines of Fe VII and Fe VIII appear weaker in the synthetic spectra due to the higher T_{eff} and rotational broadening that was not taken into account in previous works with the result that we find Fe to be less deficient ([Fe] = −0.79 for WD 2134−125 and [Fe] = −0.43 for WD 1751+106, [X] = log (mass fraction/solar mass fraction)). Due to the challenge of its fast rotation, no unambiguous identification of Fe lines was possible for WD 0044−121. Based on Fe VII λ 1141.4 Å and Fe VIII $\lambda\lambda$ 1006.1, 1148.2 Å an upper limit of the super-solar value [Fe] = 0.33 is reasonable. No clear identification of trans-iron element lines was possible for any of the three stars. For the two hybrid PG 1159 stars, the strongest computed lines were used to determine upper abundance limits for Zn, Ga, Ge, Kr, Zr, Te, I, and Xe. This was not possible for WD 0044−121, since the atomic data for the most prominent ionization stages of these elements in this high temperature range is still lacking. The abundance determinations are summarized in Figure 2. Mass and luminosity were determined using He-burning post-AGB tracks [15] (Figure 3). Using the flux calibration of [16], we find spectroscopic distances for the stars. The values are summarized in Table 2.

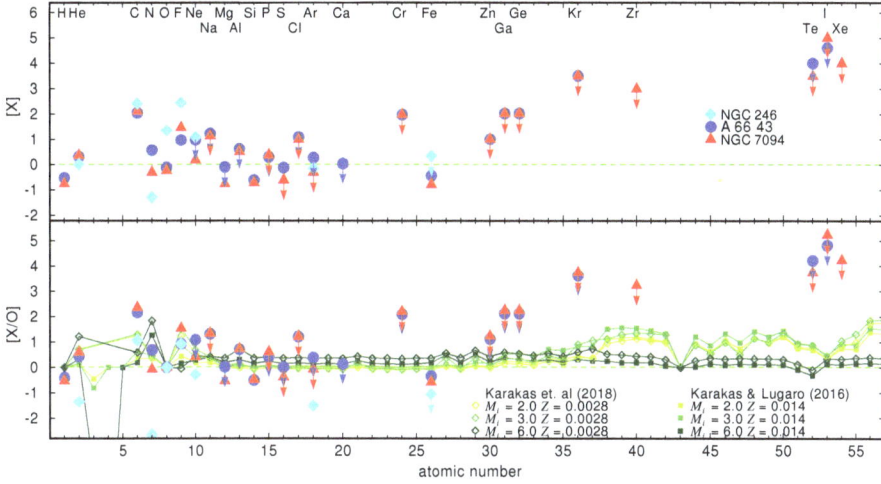

Figure 2. Photospheric abundances (estimated errors $+/-0.1$ dex) of WD 2134$-$125, WD 1751+106, and WD 0044$-$121. Arrows indicate upper limits. In the lower panel, the abundances in $[X/O] = \log(X/O)_{surf} - \log(X/O)_{solar}$ with the number fractions of element X and O are compared to the yields of the models from [17,18] for different initial masses.

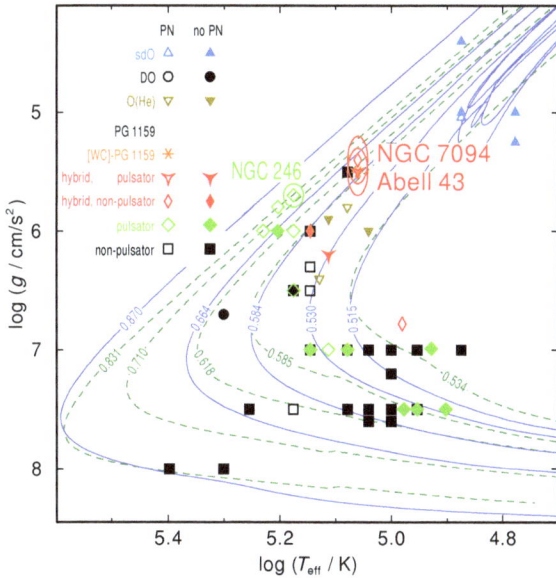

Figure 3. Positions of the CSPNe of NGC 7094, A66 43, and NGC 246 (within their error ellipses) and related objects in the $\log T_{eff} - \log g$ plane compared with evolutionary tracks (labeled with the respective masses in $M_\odot$) of VLTP stars ([15], full lines) and of hydrogen-burning post-AGB stars ([19], calculated with initial solar metallicity; dashed lines).

4. Discussion

In our detailed spectroscopic analysis, we could confirm the high T_{eff} and low $\log g$ of WD 0044−121 which places this star close to the Eddington limit for PG 1159 stars (Figure 3) and approves its classification as one of the heaviest stars of this type. In return, this causes difficulties due to instabilities in calculating model atmospheres for this star. As improvement, models including wind effects will be employed in a further analysis of all these stars. The Fe deficiency is not explained by nucleosynthesis models [17,18] that predict solar Fe abundances. A speculative reason for low Fe abundances is the conversion into Ni and heavier elements due to neutron capture [20]. An enhanced Ni/Fe ratio or a clear enhancement of trans-iron elements would indicate this. Unfortunately, no lines of these elements were identified. Given the high distances above and below the galactic plane, the progenitors of these stars could also belong to a low metallicity halo population. We included the low metallicity models for the Small Magellanic Cloud [18] in Figure 2 for comparison but they do not reproduce the negative [Fe/O] values for these stars and therefore cannot be consulted to explain the Fe deficiency. Parallaxes of (0.431 ± 0.061) ″ and (0.615 ± 0.059) ″ for WD 1751+106 and WD 2134−125, respectively, were published in the second data release of the Gaia mission from which we can derive distances of (2.32 ± 0.33) kpc and (1.63 ± 0.16) kpc. The values lie below our spectroscopically measured distances but for WD 1751+106, the values still agree within the error limits. This might lead to the speculation of potentially too high surface gravity values.

5. Conclusions

In our NLTE spectral analysis of the CSPNe WD 2134−125, WD 1751+106, and WD 0044−121, we improved the determination of the stellar parameters and found, for the first two objects, upper abundance limits for s-process elements that have never been analysed in any of these stars before. WD 0044−121 shows no clue for s-process element enhancement which leads to the speculation that the s-process becomes less effective for higher mass and metallicity. Unfortunately, the abundance limits cannot be used to constrain the nucleosynthesis models. In a forthcoming analysis, we plan to search for the radioactive element technetium and for s-process signatures in the ejected nebula material.

Funding: L.L. is supported by the German Research Foundation (DFG, grant WE 1312/49-1) and by the ESO studentship programme. The GAVO project had been supported by the Federal Ministry of Education and Research (BMBF) at Tübingen (05 AC 6 VTB, 05 AC 11 VTB). **Acknowledgments:** The author thanks the reviewers

for there fruitful comments and remarks to improve this proceedings paper. Sincere thanks are given to the SOC for awarding the follow-up price for the Lyra award for the best oral contribution of an early career scientist to this conference contribution. This research has made use of NASA's Astrophysics Data System and the SIMBAD database, operated at CDS, Strasbourg, France.

Conflicts of Interest: The author declares no conflicts of interest.

References

1. Werner, K.; Herwig, F. The elemental abundances in bare planetary nebula central stars and the shell burning in AGB stars. *Publ. Astron. Soc. Pac.* **2006**, *118*, 183. [CrossRef]
2. Herwig, F. Evolution of asymptotic giant branch stars. *Ann. Rev. Astron. Astrophys.* **2005**, *43*, 435. [CrossRef]
3. Acker, A.; Marcout, J.; Ochsenbein, F.; Stenholm, B.; Tylenda, R.; Schohn, C. *The Strasbourg-ESO Catalogue of Galactic Planetary Nebulae*; Parts I, II; European Southern Observatory: Garching, Germany, 1992; p. 1047
4. McCook, G.P.; Sion, E.M. A catalog of spectroscopically identified white dwarfs. *Astrophys. J. Suppl. Ser.* **1999**, *121*, 1. [CrossRef]
5. Napiwotzki, R. Spectroscopic investigation of old planetaries IV. Model atmosphere analysis. *Astron. Astrophys.* **1999**, *350*, 101.
6. Ziegler, M.; Rauch, T.; Werner, K.; Koesterke, L.; Kruk, J.W. (F) UV spectroscopy of the hybrid PG1159-type central stars of the planetary nebulae NGC 7094 and Abell43. *J. Phys. Conf. Ser.* **2009**, *172*, 012032. [CrossRef]

7. Rauch, T.; Köper, S.; Dreizler, S.; Werner, K.; Heber, U.; Reid, I.N. The rotational velocity of helium-rich pre-white dwarfs. *Stellar Rotat.* **2004**, *215*, 573. [CrossRef]
8. Rauch, T.; Werner, K.; Ziegler, M.; Koesterke, L.; Kruk, J.W. Non-LTE Spectral Analysis of Extremely Hot Post-AGB Stars: Constraints for Evolutionary Theory. *Art Model. Stars 21st Century* **2008**, *252*, 223. [CrossRef]
9. Rauch, T.; Quinet, P.; Knörzer, M.; Hoyer, D.; Werner, K.; Kruk, J.W.; Demleitner, M. Stellar laboratories-IX. New Se v, Sr iv–vii, Te vi, and I vi oscillator strengths and the Se, Sr, Te, and I abundances in the hot white dwarfs G191–B2B and RE 0503–289. *Astron. Astrophys.* **2017**, *606*, A105. [CrossRef]
10. Rauch, T.; Deetjen, J.L. Handling of atomic data. *Stellar Atmos. Model.* **2003**, *288*, 103.
11. Werner, K.; Deetjen, J.L.; Dreizler, S.; Nagel, T.; Rauch, T.; Schuh, S.L. Model photospheres with accelerated lambda iteration. *Stellar Atmos. Model.* **2003**, *288*, 31.
12. Werner, K.; Dreizler, S.; Rauch, T. *TMAP: Tübingen NLTE Model-Atmosphere Package.* ascl:1212.015. Astrophysics Source Code Library. 2012. Available online: http://ascl.net/1212.015 (accessed on 18 June 2018).
13. Kurucz, R.L. Including All the Lines. In *Recent Directions in Astrophysical Quantitative Spectroscopy and Radiation Hydrodynamics: Proceedings of the International Conference in Honor of Dimitri Mihalas for His Lifetime Scientific Contributions on the Occasion of His 70th Birthday, Boulder, CO, USA, 30 March–3 April 2009;* Hubeny, I., Stone, J.M., MacGregor, K., Werner, K., Eds.; Canadian Science Publishing: Ottawa, Canada, 2009; Volume 1171.
14. Kurucz, R.L. Including all the lines. *Can. J. Phys.* **2011**, *89*, 417. [CrossRef]
15. Miller Bertolami, M.M.; Althaus, L.G. Full evolutionary models for PG 1159 stars. Implications for the helium-rich O (He) stars. *Astron. Astrophys.* **2006**, *454*, 845. [CrossRef]
16. Heber, U.; Hunger, K.; Jonas, G.; Kudritzki, R.P. The atmosphere of subluminous B stars. *Astron. Astrophys.* **1984**, *130*, 119.
17. Karakas, A.I.; Lugaro, M. Stellar Yields from Metal-rich Asymptotic Giant Branch Models. *Astron. J.* **2016**, *825*, 26. [CrossRef]
18. Karakas, A.I.; Lugaro, M.; Carlos, M.; Cseh, B.; Kamath, D.; García-Hernández, D.A. Heavy-element yields and abundances of asymptotic giant branch models with a Small Magellanic Cloud metallicity. *Mon. Not. RAS* **2018**, *477*, 421. [CrossRef]
19. Miller Bertolami, M.M. New models for the evolution of post-asymptotic giant branch stars and central stars of planetary nebulae. *Astron. Astrophys.* **2016**, *588*, A25. [CrossRef]
20. Herwig, F.; Lugaro, M.; Werner, K. Planetary Nebulae: Their Evolution and Role in the Universe. In Proceedings of the IAU Symposium, Sydney, Australia, 22–25 July 2003; Kwok, S., Dopita, M., Sutherland, R., Eds.; Volume 209.

Article

Planets, Planetary Nebulae, and Intermediate Luminosity Optical Transients (ILOTs)

Noam Soker [1,2]

[1]   Department of Physics, Technion, Israel Institute of Technology, Haifa 32000, Israel;
      soker@physics.technion.ac.il
[2]   Guangdong Technion Israel Institute of Technology, Shantou 515069, China

Received: 16 April 2018; Accepted: 24 May 2018; Published: 28 May 2018

Abstract: I review some aspects related to the influence of planets on the evolution of stars before and beyond the main sequence. Some processes include the tidal destruction of a planet on to a very young main sequence star, on to a low-mass main sequence star, and on to a brown dwarf. This process releases gravitational energy that might be observed as a faint intermediate luminosity optical transient (ILOT) event. I then summarize the view that some elliptical planetary nebulae are shaped by planets. When the planet interacts with a low-mass, upper asymptotic giant branch (AGB) star, it both enhances the mass-loss rate and shapes the wind to form an elliptical planetary nebula, mainly by spinning up the envelope and by exciting waves in the envelope. If no interaction with a companion, stellar or substellar, takes place beyond the main sequence, the star is termed a *Isolated star*, and its mass-loss rates on the giant branches are likely to be much lower than what is traditionally assumed.

Keywords: planetary systems; planetary nebulae; stars: binaries; stars: AGB and post-AGB; stars: variables: general

1. Introduction

Planetary nebulae (PNe) can be shaped by stellar and substellar companions (see Jones & Boffin 2017 for a recent review [1]). One open question that has been with us for more than two decades (e.g., Soker 1996 [2]) is to what extent substellar objects, and in particular, planets, also shape PNe (see De Marco & Izzard 2017 for a recent review [3]). De Marco & Soker [4] took that about one-quarter of all stars in the initial mass range $1 - 8M_\odot$ do form PNe, and estimate that about 20% of all PNe were shaped via planets and brown dwarfs. This amounts to about 5% of all $1 - 8M_\odot$ stars. In light of the general interest in the manner by which planets can influence stellar evolution (e.g., [5–10]), I discuss some issues related to star–planet interaction. The paper is based on a talk I gave at the Asymmetrical Planetary Nebulae (APN) VII meeting (Hong Kong, December 2017), and all figures are from my presentation at the meeting.

2. Engulfment of Planets by Asymptotic Giant Branch (AGB) Stars

For a planet to influence the envelope of an asymptotic giant branch (AGB) star, the envelope cannot be too massive. This implies a low-mass star. However, for traditionally used mass-loss rates, low-mass stars reach very large radii already on their red giant branch (RGB; Figure 1 [11]) and are likely to swallow close planets before they even reach the AGB. The way to have planets interact with their host star on the AGB is if the mass-loss rate on the giant branches is lower than what traditional values are. In that case, the stellar core on the AGB and consequently the stellar radius are larger than in standard theoretical calculations, and the star is much more likely to swallow a planet on its upper AGB. Sabach & Soker [12,13] assume that *Isolated stars*, i.e., those that are not spun up in their post-main-sequence evolution, lose mass at a rate that is less than about one-third of the traditional one (for justification see [12]). They then show that in that case, low-mass AGB stars reach much larger radii and are much more likely to swallow planets than in standard theoretical calculations. The assumption of lower mass-loss rates of Jsolated stars needs further examination by future observations and theoretical studies.

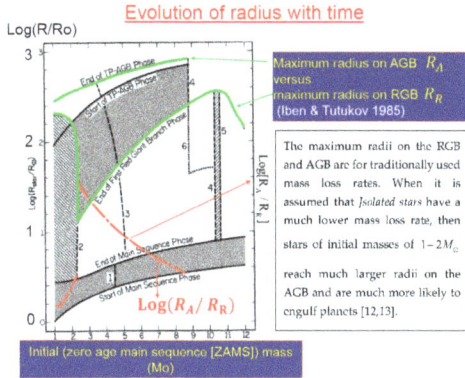

Figure 1. Maximum radii stars reach on their RGB and AGB as function of their initial mass for traditional mass-loss rates (from [11]).

3. The Fate of a Planet: Tidal Destruction versus Engulfment

The fate of the planet as it comes close to the envelope of a star depends on the density ratio. If the density of the planet is larger than that of the star, it dives in to the envelope as one entity and starts a common envelope evolution [14,15]. It will later be destroyed near the core by either tidal forces or evaporation. Light planets are evaporated before they reach the core. Planets that are more massive can reach closer to the core and then suffer tidal destruction with part of their material accreted onto the core (marked "destruction on core" in Figure 3). If the density of the planet is lower than that of the star, it is tidally destroyed and forms an accretion belt (or a disk) around the star. In Figure 2, I present a schematic evolution of the planet and the stellar radii and densities, and mark which of the two outcomes takes place. I give more details of the outcomes in Figure 3.

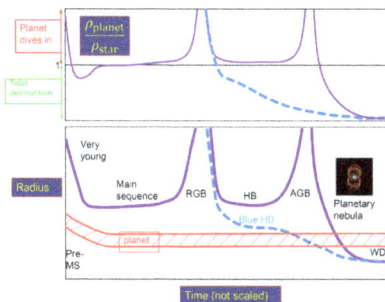

Figure 2. A schematic evolution of the radii and densities of planets and stars from the pre-main-sequence phase to the WD phase. **Upper panel**: the ratio of the planet density to the stellar density. If the ratio is above 1, the planet dives in to the envelope as one entity. If the density ratio is below 1, the planet is tidally destroyed and forms an accretion belt/disk around the star. **Lower panel**: The planet and stellar radii as function of time.

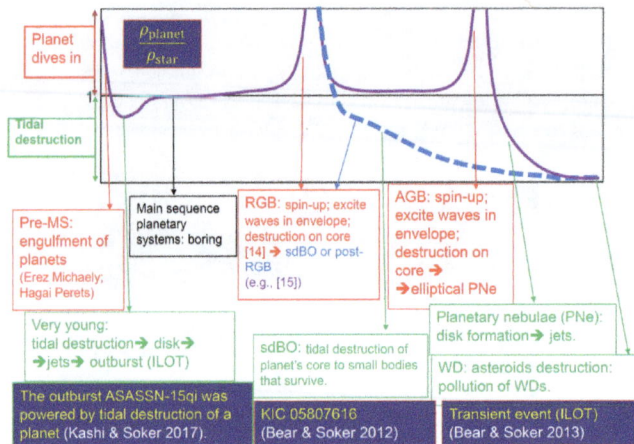

Figure 3. Outcomes of planet–star interaction.

4. Intermediate Luminosity Optical Transients (ILOTs) with Planets

ILOTs are outbursts of a star or a binary system with a peak luminosity mostly between those of novae and supernovae (other names for these events are red novae, luminous red novae, and intermediate luminous red transients). Several studies have proposed that the interaction of a planet with a star can account for a minority of ILOTs. Retter & Marom (2003 [16]) proposed that V838 Mon was a result of planets entering a common envelope with a star. Bear et al. (2011 [17]) proposed that the destruction of a planet by a brown dwarf or a low-mass main sequence star can result in an ILOT event. Kashi & Soker (2017 [18]) proposed that the outburst of the young stellar object ASASSN-15qi was an ILOT event where a sub-Jupiter young planet was tidally destroyed by a young main sequence star. Because the system was young, the density of the planet was smaller than that of the star (Figure 3) and the planet was tidally destroyed. This, they suggested, resulted in the formation of an accretion disc and a gravitationally powered ILOT. The mass of the planet was too small to inflate a giant envelope and hence the ILOT was hot, rather than red. As well, its energy was below that of classical novae.

Bear et al. (2013 [19]) discussed the possibility of observing the transient event that might result from the tidal destruction process of an asteroid near a WD. However, this event is much weaker than typical ILOTs.

5. Intermediate Luminosity Optical Transients (ILOTs)

In Figure 4, I present the energy–time diagram of ILOTs that Amit Kashi and I have been developing in the last several years (see http://phsites.technion.ac.il/soker/ilot-club/ for an updated diagram).

We suggest ([18] and references therein) that these ILOTs are powered by gravitational energy in one of several types of processes: (1) The secondary star is completely destroyed and part of its mass is accreted onto the primary star, e.g., as a planet destruction onto a brown dwarf; (2) the secondary star enters the envelope of a companion but stays intact and forms a common envelope, e.g., as Retter & Marom (2003 [16]) suggested; (3) the secondary star accretes mass while outside the envelope of the primary star, e.g., as in our model for the Great Eruption of Eta Carinae or our suggested scenario for some PNe [18].

With Amit Kashi [20], we suggest that the binary progenitors of some bipolar PNe experienced ILOT events that shaped the PN. The several-months-long to several-years-long outbursts were powered by mass transfer from an AGB star onto a main sequence companion that orbits outside the

AGB envelope. Jets launched by an accretion disk around the main sequence companion shaped the bipolar lobes. Four such bipolar PNe are marked on Figure 4. They are marked with a long horizontal line because we know more or less the kinetic energy of the nebulae, which is about the same as the ILOT energy, but we cannot tell how long the mass ejection phase lasted.

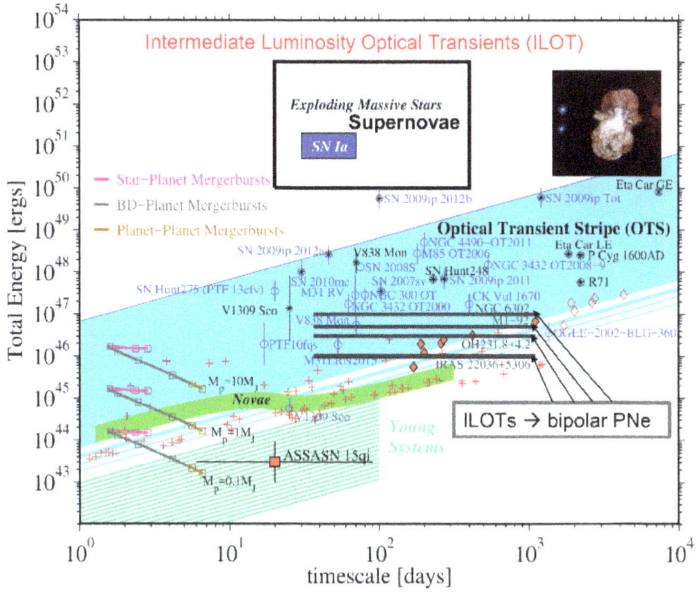

Figure 4. Observed transient events on the energy–time diagram. Blue empty circles represent the total (radiated plus kinetic) energy of the observed transients as a function of the duration of their eruptions, i.e., usually the time for the visible luminosity to decrease by 3 magnitudes. The Optical Transient Stripe is populated by ILOT events that we [18] suggest are powered by gravitational energy of complete merger events or vigorous mass-transfer events. For details of this figure, see http://phsites.technion.ac.il/soker/ilot-club/.

6. Planet-Shaped Planetary Nebulae

When a planet spirals in inside the loosely bound envelope of an upper AGB star, it can excite waves in the envelope and spin up the envelope, both of which can cause asymmetrical mass loss. Finally, when the planet is destroyed near the core, it might lead to further asymmetrical mass loss from inside the envelope, e.g., jets that might be launched by the core. For example, a Jupiter-like planet might form a disk that launches jets with about a few percent of the mass of the planet. This amounts to the jets' mass of about few $\times 10^{-5} M_\odot$. This mass is sufficient to form two opposite bullets ('ansae') along the symmetry axis of the nebula.

I started the paper with the discussion of the general interaction of planets with evolving stars, I then moved to discuss the formation of ILOTs with planet companions, and in Section 5, I mentioned some bipolar PNe that can be shaped by ILOT events with a stellar companion to the AGB progenitor of the PN. I now end the discussion of planet-shaped PNe by listing the evolutionary channels and the resulting PN types. For that I use a table from De Marco & Soker (2011 [4]), which I present here as Figure 5.

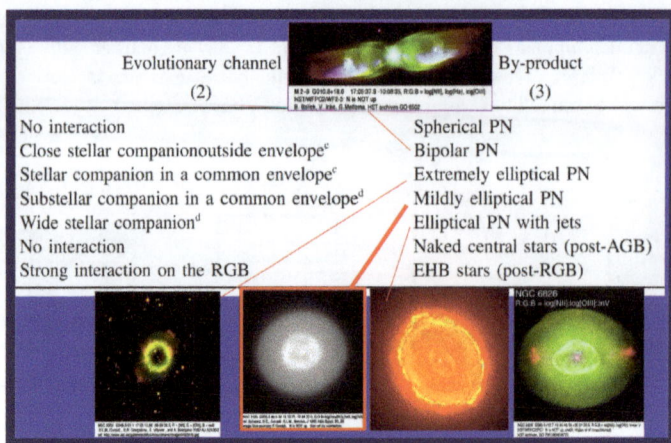

Figure 5. The evolutionary channels and the PN types that result from them (based on De Marco & Soker 2011 [4]).

7. Discussion

I discussed some aspects of the influence of planets on late stellar evolution and its relation to some aspects of stellar binary interaction. The main claim of this presentation is that planets can influence the evolution of low-mass stars in, e.g., enhancing the mass-loss rate of RGB and AGB stars (that without any companion are *Isolated stars* that have low mass-loss rates, lower than what is usually assumed), forming some ILOTs, and shaping PNe. There are other aspects I did not get into, such as the strength of the tidal interaction between the planet and the star, some aspects of which can be found in reviews from earlier related meetings (e.g., [21,22]).

Acknowledgments: I thank the referees for useful comments.

Conflicts of Interest: The authors declare no conflict of interest.

References

1. Jones, D.; Boffin, H.M.J. Binary stars as the key to understanding planetary nebulae. *Nat. Astron.* **2017**, *1*, 0117. [CrossRef]
2. Soker, N. What Planetary Nebulae Can Tell Us about Planetary Systems. *Astrophys. J. Lett.* **1996**, *460*, L53–L56. [CrossRef]
3. De Marco, O.; Izzard, R.G. Dawes Review 6: The Impact of Companions on Stellar Evolution. *Publ. Astron. Soc. Aust.* **2017**, *34*, e001. [CrossRef]
4. De Marco, O.; Soker, N. The Role of Planets in Shaping Planetary Nebulae. *Publ. Astron. Soc. Pac.* **2011**, *123*, 402–411. [CrossRef]
5. Nordhaus, J.; Spiegel, D.S. On the orbits of low-mass companions to white dwarfs and the fates of the known exoplanets. *Mon. Notices R. Astron. Soc.* **2013**, *432*, 500–505. [CrossRef]
6. Villaver, E.; Livio, M.; Mustill, A.J.; Siess, L. Hot Jupiters and Cool Stars. *Astrophys. J.* **2014**, *794*, 3. [CrossRef]
7. Aguilera-Gomez, C.; Chaname, J.; Pinsonneault, M.H.; Carlberg, J.K. On Lithium-rich Red Giants. I. Engulfment of Substellar Companions. *Astrophys. J.* **2016**, *829*, 127. [CrossRef]
8. Staff, J.E.; De Marco, O.; Wood, P.; Galaviz, P.; Passy, J.C. Hydrodynamic simulations of the interaction between giant stars and planets. *Mon. Notices R. Astron. Soc.* **2016**, *458*, 832–844. [CrossRef]
9. Schaffenroth, V.; Barlow, B.; Geier, S.; Vučković, M.; Kilkenny, D.; Schaffenroth, J. News from the Erebos Project. *Open Astron.* **2017**, *26*, 208–213. [CrossRef]

10. Mustill, A.J.; Villaver, E.; Veras, D.; Gansicke, B.T.; Bonsor, A. Unstable low-mass planetary systems as drivers of white dwarf pollution. *Mon. Notices R. Astron. Soc.* **2018**, *476*, 3939–3955. [CrossRef]

11. Iben, I., Jr.; Tutukov, A.V. On the evolution of close binaries with components of initial mass between 3 solar masses and 12 solar masses. *Astrophys. J. Suppl.* **1985**, *58*, 661–710. [CrossRef]

12. Sabach, E.; Soker, N. The Class of Jsolated Stars and Luminous Planetary Nebulae in old stellar populations. *arXiv*, **2017**, arXiv:1704.05395.

13. Sabach, E.; Soker, N. Accounting for planet-shaped planetary nebulae. *Mon. Notices R. Astron. Soc.* **2018**, *473*, 286–294. [CrossRef]

14. Soker, N. Can Planets Influence the Horizontal Branch Morphology? *Astrophys. J.* **1998**, *116*, 1308–1313. [CrossRef]

15. Geier, S.; Kupfer, T.; Schaffenroth, V.; Heber, U. Hot subdwarf stars in the Galactic halo Tracers of prominent events in late stellar evolution. *Proc. Int. Astron. Union* **2015**, *11*, 302–303. [CrossRef]

16. Retter, A.; Marom, A. A model. *Astron. Soc.* **2003**, *345*, L25–L28. [CrossRef]

17. Bear, E.; Kashi, A.; Soker, N. Mergerburst transients of brown dwarfs with exoplanets. *Mon. Notices R. Astron. Soc.* **2011**, *416*, 1965–1970. [CrossRef]

18. Kashi, A.; Soker, N. An intermediate luminosity optical transient (ILOTs) model for the young stellar object ASASSN-15qi. *Mon. Notices R. Astron. Soc.* **2017**, *468*, 4938–4943. [CrossRef]

19. Bear, E.; Soker, N. Transient outburst events from tidally disrupted asteroids near white dwarfs. *New Astron.* **2013**, *19*, 56–61. [CrossRef]

20. Soker, N.; Kashi, A. Formation of Bipolar Planetary Nebulae by Intermediate-luminosity Optical Transients. *Astrophys. J.* **2012**, *746*, 100. [CrossRef]

21. Villaver, E. Planets, evolved stars, and how they might influence each other. *Proc. Int. Astron. Union* **2012**, *283*, 219. [CrossRef]

22. Villaver, E. The Fate of Planets. *AIP Conf. Proc.* **2011**, *1331*, 21.

MDPI

St. Alban-Anlage 66

4052 Basel

Switzerland

Tel. +41 61 683 77 34

Fax +41 61 302 89 18

www.mdpi.com

Galaxies Editorial Office

E-mail: galaxies@mdpi.com

www.mdpi.com/journal/galaxies